CAMBRIDGE LIBRARY COLLECTION
Books of enduring scholarly value

Technology

The focus of this series is engineering, broadly construed. It covers technological innovation from a range of periods and cultures, but centres on the
technological achievements of the industrial era in the West, particularly in
the nineteenth century, as understood by their contemporaries. Infra-structure
is one major focus, covering the building of railways and canals, bridges and
tunnels, land drainage, the laying of submarine cables, and the construction
of docks and lighthouses. Other key topics include developments in industrial
and manufacturing fields such as mining technology, the production of iron
and steel, the use of steam power, and chemical processes such as photography
and textile dyes.

Lives of the Engineers

A political and social reformer, Samuel Smiles (1812–1904) was also a noted
biographer in the Victorian period. Following the engineer's death in 1848,
Smiles published his highly successful *Life of George Stephenson* in 1857
(also reissued in this series). His interest in engineering evolved and he began
working on biographies of Britain's most notable engineers from the Roman
to the Victorian era. Originally published in three volumes between 1861 and
1862, this work contains detailed and lively accounts of the educations, careers
and pioneering work of seven of Britain's most accomplished engineers. These
volumes stand as a remarkable undertaking, advancing not only the genre, but
also the author's belief in what hard work could achieve. Volume 2 includes
accounts of the lives of three important engineers of the eighteenth and early
nineteenth centuries: John Smeaton (1724–92), John Rennie (1761–1821) and
Thomas Telford (1757–1834).

Cambridge University Press has long been a pioneer in the reissuing of out-of-print titles from its own backlist, producing digital reprints of books that are still sought after by scholars and students but could not be reprinted economically using traditional technology. The Cambridge Library Collection extends this activity to a wider range of books which are still of importance to researchers and professionals, either for the source material they contain, or as landmarks in the history of their academic discipline.

Drawing from the world-renowned collections in the Cambridge University Library and other partner libraries, and guided by the advice of experts in each subject area, Cambridge University Press is using state-of-the-art scanning machines in its own Printing House to capture the content of each book selected for inclusion. The files are processed to give a consistently clear, crisp image, and the books finished to the high quality standard for which the Press is recognised around the world. The latest print-on-demand technology ensures that the books will remain available indefinitely, and that orders for single or multiple copies can quickly be supplied.

The Cambridge Library Collection brings back to life books of enduring scholarly value (including out-of-copyright works originally issued by other publishers) across a wide range of disciplines in the humanities and social sciences and in science and technology.

Lives of the Engineers

With an Account of Their Principal Works

VOLUME 2

SAMUEL SMILES

CAMBRIDGE UNIVERSITY PRESS

Cambridge, New York, Melbourne, Madrid, Cape Town,
Singapore, São Paolo, Delhi, Mexico City

Published in the United States of America by Cambridge University Press, New York

www.cambridge.org
Information on this title: www.cambridge.org/9781108052931

© in this compilation Cambridge University Press 2012

This edition first published 1861
This digitally printed version 2012

ISBN 978-1-108-05293-1 Paperback

LIVES

OF

THE ENGINEERS.

Volume II.

John Smeaton, F.R.S.

Engraved by W Holl, after the portrait by Mather Brown.

Published by John Murray, Albemarle Street, 1861.

LIVES

OF

THE ENGINEERS,

WITH

AN ACCOUNT OF THEIR PRINCIPAL WORKS;

COMPRISING ALSO

A HISTORY OF INLAND COMMUNICATION IN BRITAIN.

By SAMUEL SMILES.

" Bid Harbours open, Public Ways extend;
Bid Temples, worthier of God, ascend;
Bid the broad Arch the dang'rous flood contain,
The Mole projected, break the roaring main;
Back to his bounds their subject sea command,
And roll obedient rivers through the land.
These honours, Peace to happy Britain brings;
These are imperial works, and worthy kings."

POPE.

WITH PORTRAITS AND NUMEROUS ILLUSTRATIONS.

VOL. II.

LONDON:

JOHN MURRAY, ALBEMARLE STREET.

1861.

LONDON: PRINTED BY W. CLOWES AND SONS, DUKE STREET, STAMFORD STREET,
AND CHARING CROSS.

CONTENTS OF VOL. II.

———◇———

PART VI.—LIFE OF JOHN SMEATON.

CHAPTER I.

John Smeaton's birth and education — Leeds at the beginning of last century — Road communications of the neighbourhood — Austhorpe Lodge — The boy's mechanical amusements — Leeds Grammar-school — Smeaton's workshop — Hindley's account of his boyish occupations .. .. .. Page 3–9

CHAPTER II.

Placed in an attorney's office — Attends the Law Courts in Westminster Hall — Learns the trade of mathematical instrument maker — Frequents meetings of the Royal Society — His mechanical contrivances and inventions — His paper on the Natural Powers of Wind and Water to Turn Mills — An indefatigable student — Turns his attention to civil engineering — His tour in Holland 10–14

CHAPTER III.

Dangers of the Eddystone Rock — Necessity for a lighthouse — Henry Winstanley — His eccentricities — Designs and erects the first lighthouse on the Eddystone — Is washed away in a storm — John Rudyerd — Builds the second lighthouse — Is destroyed by fire — Mr. Smeaton applied to for a design 15–25

CHAPTER IV.

Lord Macclesfield's recommendation of Smeaton as engineer of the new lighthouse on the Eddystone — His investigation of the subject — Decides that it must be built of stone — The design — His journey to Plymouth — His visits to the rock — Makes a model of the proposed building — Mr. Jessop appointed resident engineer, and the excavations commenced — Dangers of the work — Smeaton narrowly escapes shipwreck — Progress of the work — Smeaton's courage — His carefulness as to details — Smeaton on the Hoe — The lighthouse finished and the light exhibited — Its uses — The lights in the English Channel 26–48

PART VI.—*Continued.*

CHAPTER V.

Smeaton appointed receiver for the Derwentwater estates — The roads and commerce of England — General want of capital — Smeaton extensively employed as an engineer — Improvement of navigations — Calder Navigation — His Report on the drainage of the Lincoln and Cambridge Fens — Various drainage works — Repairs London Bridge — Designs pumping-engines — Erects bridge at Perth — Constructs Forth and Clyde Canal — Erects bridge at Coldstream — Designs improvements for the Carron Works — Banff Bridge — Hexham Bridge — St. Ives Harbour — Ramsgate Harbour — Use of the Diving-bell — Eyemouth Harbour — Mills and machinery erected by Smeaton — His improvements in Newcomen's steam-engine

Page 49–73

CHAPTER VI.

Smeaton's home at Austhorpe — His study and workshop — His blacksmith — Papers contributed to Royal Society — His tools — His lathe — His mechanical ingenuity and skill — His visits to London — Engineers' first club — His views of money — Refuses an engagement to serve the Empress of Russia — Determines to publish an account of his works — His opinion of literary composition — His manners — Anecdote of Smeaton and the Duchess of Queensberry — His domestic character — His benevolence — Recognition of his eminent intellectual ability — His great industry — Failure of his health — Death   74–89

PART VII.—LIFE OF JOHN RENNIE.

CHAPTER I.

Rennie born at Phantassie, East Lothian — Scotland at the middle of last century — State of agriculture in the Lothians — The population — Their indolence — Their poverty — Wages of labour — County of Ayr — The Highland border — Want of roads — Communication between Edinburgh and Glasgow — Stage-coaches and carriers — Post-road between Edinburgh and London — The first Scotch Turnpike Act passed — Ancient Scotch bridges — Low state of the mechanical arts in Scotland   93–104

CHAPTER II.

Fletcher of Saltoun introduces barley-mills and fanners into Scotland — James Meikle — Popular prejudice against 'artificial wind' — Andrew Meikle, millwright — Progress of agricultural improvement in East Lothian — Mr. Cockburn of Ormiston — Meikle's mills — Clumsy methods employed in thrashing grain — Various attempts made to invent a thrashing-mill — Mr. Kinloch's models — Andrew Meikle's invention of the thrashing-machine — His improvements in windmills — Drainage of Kincardine Moss — Meikle's mechanical ingenuity — His death   105–117

PART VII.—*Continued.*

CHAPTER III.

The Rennie family — Early mechanical bias of John Rennie — Frequents Andrew Meikle's workshop — Attends the parish school of Prestonkirk — Learns carpentry and mill-work under Andrew Meikle — Attends Dunbar Grammar-school — Is offered the appointment of teacher — Begins business as a millwright — Attends the University of Edinburgh —Dr. Robison — Rennie's studies and amusements — Mills erected by him in Scotland — Tour in England — Visits James Watt at Birmingham — Rennie builds his first bridge near Edinburgh — Removes to Birmingham — Is engaged to superintend the erection of the Albion Mills, Southwark .. Page 118–133

CHAPTER IV.

London in 1785 — Coaches and turnpike roads — Trade — The shipping of the Thames — Erection of the Albion Mills — The first employment of the steam-engine in driving mill machinery — Rennie's extensive use of iron-work in their erection — The Albion Mills destroyed by fire — Rennie's employment on similar works — Earl Stanhope and steam navigation — Rennie undertakes works of civil engineering — Constructs the Kennet and Avon Canal — The Rochdale Canal — The Lancaster Canal — Various canal works — The Royal Canal, Ireland 134–151

CHAPTER V.

Recommends the employment of the steam-engine in Fen drainage — Drowned state of the Lincoln Fens — Arthur Young's account of them — The East Fen — Sir Joseph Banks resolves upon their drainage — Mr. Rennie employed to devise a plan for draining Wildmore Fen — His comprehensive view of the subject of Fen drainage — His catchwater system — His report — The works executed — Great Hobhole Drain — Effects of the drainage on agriculture — His proposed improvement of the Witham at Boston — Reports on the drainage of the Great Level — Eau Brink Cut — Characteristics of Fen scenery 152–169

CHAPTER VI.

Dr. Robison visits Rennie in London — Theory and practice in bridge-building — Early designs of bridges — Designs Kelso Bridge — Musselburgh Bridge— Projected cast-iron bridge over the Menai Straits — Boston Bridge — Rennie's various designs of bridges — Waterloo Bridge — Its distinctive features — Foundation of the piers — The centering — Mode of fixing the centres — The bridge road — Completion of the work — Southwark Bridge — The iron-work — Details of construction — Characteristics of the bridge 170–194

PART VII.—*Continued.*

CHAPTER VII.

Growth of the Trade of London — Police of the Thames — Necessity for docks — Rennie's London Docks — Construction of the lock entrances — Arrangement of the working details — The East India Docks — Improved methods of working — Report on Wick Harbour — The River Clyde — Grimsby Docks — Invention of hollow quay walls — Holyhead Harbour — Hull Harbour and Docks — Improvement of the dredging-machine — Leith Docks — Various harbour works — Rennie's principles of harbour construction — Ramsgate Harbour — Improvement of the Diving-bell   ..   Page 195–222

CHAPTER VIII.

Dangers of the Bell Rock — Scotch lighthouses — Plans of a lighthouse on the Bell Rock proposed — Rennie requested to report — His recommendations adopted — Appointed engineer — Prepares detailed plans of the lighthouse — The excavations commenced — Dangers of the work described by Mr. Stevenson, resident engineer — Rennie's visits to the rock — Recommends modifications in the plans, which are adopted — The lighthouse finished — Mr. Rennie's claims as chief engineer vindicated   223–234

CHAPTER IX.

Rennie extensively employed by Government — Defences of the coast — Defensive works on the river Lea — The Hythe Military Canal constructed after his designs — Fulton's Torpedo — Report on the Government dockyards — Recommends extensive improvements and concentration of dockyard machinery — Plymouth, Portsmouth, Deptford, Woolwich, Chatham — His plan of a naval arsenal at Northfleet — Sheerness dockyard works — Construction of the great dockyard wall — Design of Medway improvement   235–251

CHAPTER X.

Plymouth Sound — Plans for its protection from southerly winds — Mr. Rennie's report thereon — His plan of a breakwater adopted — The works commenced — Methods employed — Effects of storms — Modifications of the plan — The works completed by Sir John Rennie — Uses of the breakwater   252–263

CHAPTER XI.

Rennie's extensive and various employment as civil and mechanical engineer — Advises the introduction of steam-power into the Royal Navy — The 'Comet' built — New London Bridge, his last great design — His private life — Failure of his health — Short continental tour — His close application — Death — His portrait — Habits — Conscientiousness — Truthfulness — Anecdote of his handiness as a mechanic — Love of old books — Solidity of his structures — Conclusion   264–284

PART VIII.—LIFE OF THOMAS TELFORD.

CHAPTER I.

Eskdale — Langholm — Former lawlessness of the Border population — Johnnie Armstrong — Border energy — Westerkirk — Telford's birthplace — Glendinning — Valley of the Meggat — The " unblameable shepherd " — Telford's mother — Early years — " Laughing Tam " — Put to school — His school-fellows .. .. Page 287–297

CHAPTER II.

Telford apprenticed to a stonemason — Runs away — Re-apprenticed to a mason at Langholm — Building operations in the district — Miss Pasley lends books to young Telford — Attempts to write poetry — Becomes village letter-writer — Works as a journeyman mason — Employed on Langholm Bridge — Manse of Westerkirk — Poem of ' Eskdale ' — Hews headstones and doorheads — Works as a mason at Edinburgh — Study of architecture — Revisits Eskdale — His ride to London 298–307

CHAPTER III.

Telford a working man in London — Obtains employment as a mason at Somerset House — His impressions of Sir Robert Chambers and Mr. Robert Adam, architects — Correspondence with Eskdale friends — Observations on his fellow-workmen — Proposes to begin business, but wants money — Mr. Pulteney — Becomes foreman of builders at Portsmouth Dockyard — Continues to write poetry — Employment of his time — Prints letters to his mother 308–315

CHAPTER IV.

Superintends repairs of Shrewsbury Castle — Appointed Surveyor for county of Salop — Superintends erection of new gaol — Interview with John Howard — His studies in science and literature — Poetical exercises — Fall of St. Chad's church, Shrewsbury — Discovery of the Roman city of Uriconium — Superintends the excavations — Overseer of felons — Mrs. Jordan at Shrewsbury — Telford's indifference to music — Telford's support of his mother — Miscellaneous reading — His politics — Paine's ' Rights of Man ' — Reprints his poem of ' Eskdale ' 316–329

CHAPTER V.

Advantages of mechanical training to an engineer — Erects Montford Bridge — Erects St. Mary Magdalen church, Bridgenorth — Telford's design — Architectural tour — Bath — Studies in British Museum — Oxford — Birmingham — Study of architecture — Makes friends — Telford and Pulteney — Appointed Engineer to the Ellesmere Canal — John Wilkinson, ironmaster — Telford's salary 330–339

PART VIII.—*Continued.*

CHAPTER VI.

Course of the Ellesmere Canal — Success of the early canals — The Act obtained and working survey made — Chirk Aqueduct — Pont-Cysylltau Aqueduct — Telford's hollow walls — His cast-iron trough at Pont-Cysylltau — The canal works completed — Revisits Eskdale — Interview with Frank Beattie — Early impressions corrected — Tour in Wales — Conduct of Ellesmere Canal navigation — His literary compositions .. .. .. Page 340–354

CHAPTER VII.

Use of Iron in bridge-building — Design of a Lyons architect — First iron bridge erected at Coalbrookdale — Wear iron bridge, Sunderland — Telford's iron bridge at Buildwas — His iron lock-gates and turn-bridges — Projects a one-arched bridge of iron over the Thames — Bewdley stone bridge — Tongueland Bridge — Extension of Telford's engineering business — Literary friendships — Thomas Campbell — Miscellaneous reading 355–373

CHAPTER VIII.

Progress of Scotch agriculture — Romilly's account — State of the Highlands — Want of roads — Use of the Cas-chrom — Emigration — Telford's survey of Scotland — His report — Want of bridges — Lord Cockburn's account of the difficulties of travelling the North Circuit — State of Caithness and Sutherland — Parliamentary Commission of Highland Roads and Bridges appointed — Dunkeld Bridge built — 920 miles of new roads constructed — Craigellachie Bridge — Travelling promoted — Agriculture improved — Moral results of Telford's Highland contracts 374–389

CHAPTER IX.

Highland harbours — Wick and Pulteney Town — Columnar pier work — Peterhead Harbour — Frazerburgh Harbour — Banff Harbour — Old history of Aberdeen, its witchburning and slave-trading — Improvement of its harbour — Telford's design carried out — Dundee Harbour.. .. 390–408

CHAPTER X.

Canal projected through the Great Glen of the Highlands — Survey by James Watt — Survey by Telford — Tide-basin at Corpach — Neptune's Staircase — Dock at Clachnagarry — The chain of lochs — Construction of the works — Commercial failure of the canal — Telford's disappointment — Glasgow and Ardrossan Canal — Weaver Navigation — Gotha Canal, Sweden — Gloucester and Berkeley, and other canals — Harecastle Tunnel — Birmingham Canal — Macclesfield Canal — Birmingham and Liverpool Junction Canal — Telford's pride in his canals 409–426

PART VIII.—*Continued.*

CHAPTER XI.

Increase of road-traffic — Improvement of the main routes between the principal towns — Carlisle and Glasgow road — Telford's principles of road-construction — Macadam — Cartland Crags Bridge — Improvement of the London and Edinburgh post road — Communication with Ireland — Wretched state of the Welsh roads — Telford's survey of the Shrewsbury and Holyhead road —Its construction — Roads and railways — London and Shrewsbury post road — Roads near London — Coast road, North Wales
Page 427-443

CHAPTER XII.

Bridges projected over the Menai Straits — Telford's designs — Ingenious plan of suspended centering — Design of a suspension bridge over the Mersey at Runcorn — Design of suspension bridge at Menai — The works begun — The main piers — The suspension chains — Hoisting of the first main-chain — Progress of the works to completion — The bridge formally opened — Conway Suspension Bridge 444-460

CHAPTER XIII.

Résumé of English engineering — General increase in trade and population — The Thames — St. Katherine's Docks — Tewkesbury Bridge — Gloucester Bridge — Dean Bridge, Edinburgh — Glasgow Bridge — Telford's works of drainage in the Fens — The North Level — The Nene Outfall — Effects of Fen drainage 461-472

CHAPTER XIV.

Telford's residence in London — Leaves the Salopian — First President of Institute of Civil Engineers — Consulted by foreign Governments as to roads and bridges — His views on railways — Failure of health — Consulted as to Dover Harbour — Illness and death — His character — His friends — Integrity — Views on money-making — Benevolence — Patriotism — His Will — Libraries in Eskdale supported by his bequests 473-493

INDEX 495

LIST OF ILLUSTRATIONS.

———◆◇◆———

PORTRAIT OF JOHN SMEATON .. .. *Frontispiece.*
„ JOHN RENNIE .. .. *to face page* 91.
„ THOMAS TELFORD .. .. „ *page* 285.

	PAGE		PAGE
Banff Bridge	2	Map of the Lincolnshire Fens before	
Map of Smeaton's Native District ..	4	their Drainage by Rennie	155
View of Leeds, early in the 18th		Plan of Eau Brink Cut	168
century	6	Kelso Bridge	173
Whitkirk, near Leeds	9	Musselburgh Bridge	174
Map of coast of Devon and Cornwall	15	Boston Bridge	176
Winstanley's Lighthouse (Eddystone)	18	Section of Waterloo Bridge .. .	180
Rudyerd's Lighthouse (Eddystone)	22	Centering of Waterloo Bridge arching	182
Smeaton's Lighthouse (Eddystone)	26	Waterloo Bridge (Skelton's)	186
Section of Smeaton's Lighthouse ..	34	Southwark Bridge	190
Progress of the works to the 15th		Waterloo Bridge (Leitch's)	194
course	41	Portrait of William Jessop	198
Smeaton on the Hoe	42	Plan of London Docks	199
Plan of the 46th course	44	Plan of East and West India Docks	203
Light at the Nore	48	Plan of Holyhead Harbour	210
Old London Bridge, before the altera-		View of Holyhead Harbour	211
tion of 1758	54	Method of using the Diving Bell ..	221
Old London Bridge, after the alteration	55	Section of Bell Rock Lighthouse ..	232
Coldstream Bridge	60	View of Bell Rock Lighthouse ..	233
Plan of St. Ives Harbour	64	Plan of proposed Docks at Northfleet	244
View of St. Ives Harbour	65	Plan of Sheerness Docks	247
Map of Ramsgate and Harbour ..	66	Plan of proposed Medway improve-	
View of Ramsgate Harbour	67	ment	250
Plan of Eyemouth Harbour .. ' ..	70	Map of Plymouth Sound	253
View of Eyemouth Harbour	71	View of Plymouth Breakwater ..	262
Smeaton's House at Austhorpe ..	74	Section of the Breakwater	263
Smeaton's Lathe	77	Portrait of Captain Huddart	265
Monumental Tablet in Whitkirk ..	89	View of New London Bridge	272
Smeaton's Burial-place, Whitkirk ..	90	Section of Bridge	273
Plymouth Breakwater from Mount		View of Glasgow Bridge	286
Edgcumbe	92	Map of Telford's Native District ..	288
Map of Rennie's Native District ..	93	Telford's Birthplace	293
Houston Mill	107	Cottage at the Crooks	295
Portrait of Andrew Meikle	113	Westerkirk Church and School ..	297
Rennie's Birthplace, Phantassie ..	118	Telford's Tool-mark	302
Rennie's first Bridge	131	Valley of Eskdale	303
The Albion Mills	138	Valley of the Meggat	307
Locks on the Rochdale Canal ..	146	Shrewsbury Castle	329
Lune Aqueduct, Lancaster	148	Medallion of John Wilkinson	337

	PAGE
St. Mary Magdalen, Bridgenorth	339
Map of Ellesmere Canal	340
View of Chirk Aqueduct	343
Section of Pier of Chirk Aqueduct	344
Iron Trough at Pont-Cysylltau	346
Section of Aqueduct	347
View of Pont-Cysylltau Aqueduct	348
First Iron Bridge, Coalbrookdale	357
Wear Bridge, Sunderland	358
Buildwas Bridge	360
Telford's proposed one arch Iron Bridge over the Thames	364
Bewdley Bridge	367
Tongueland Bridge	368
The Cas-Chrom	376
Map of Highland Roads	382
Dunkeld Bridge	384
Craigellachie Bridge	387
Folkestone Harbour	392
View of Peterhead	394
Map of Peterhead Harbour	395
Map of Banff Harbour	398
Map of Aberdeen Harbour	404
View of Aberdeen Harbour	405
Section of Pier-head Work	405
Plan of Dundee Harbour	406
View of Dundee Harbour	407
Map of Caledonian Canal	412
Lock, Caledonian Canal	414
Cross Section, Harecastle Tunnel	422
Galton Bridge, Birmingham Canal	424
Portrait of J. L. Macadam	430
Cartland Crags Bridge	431
Road descent to the Lugwy (N. Wales)	439
Road, Nant Ffrancon (N. Wales)	440
Map of Menai Strait	444
Telford's proposed Cast Iron Bridge over the Menai	445
Plan of suspended Centering for ditto	445
Outline of proposed Runcorn Bridge	447
Outline of Menai Suspension Bridge	448
Section of Main Pier	451
Mode of fixing Chains in the Rock	453
State of Suspension Chain before hoisting	456
View of Menai Bridge	458
View of Conway Bridge	460
Dean Bridge, Edinburgh	466
Section of Polish Road	476
Telford's Burial-place, Westminster Abbey	481
View in Eskdale	493

LIVES OF THE ENGINEERS.

LIFE OF JOHN SMEATON.

BANFF BRIDGE. [By R. P. Leitch]

LIVES OF THE ENGINEERS.

LIFE OF JOHN SMEATON.

CHAPTER I.

THE engineer of the Eddystone Lighthouse was Brindley's
junior by only eight years. They frequently met in
consultation upon important engineering undertakings;
sometimes Smeaton advising that Brindley should be
called in, and Brindley, on his part, recommending
Smeaton. They were, in fact, during their lifetime, the
leading men in their profession; and at Brindley's
death Smeaton succeeded to much of his business as
consulting engineer in connection with the construction
of canals and of public works generally.

Smeaton had the great advantage over Brindley of a
good education and bringing up. He had not, like the
Macclesfield millwright, to force his way up through the
obstructions of poverty, toil, and parental neglect; but
was led gently by the hand from his earliest years, and
carefully trained and cultured after the best methods
then known. But Smeaton, not less than Brindley,
was impelled to the professional career on which he
entered by a like innate genius for construction which
displayed itself at a very early age; and, being permitted to follow his own bent, his force of character and
strong natural ability, diligently cultivated by study and

B 2

experience, eventually carried him to the very highest
eminence as an engineer.

John Smeaton was born at Austhorpe Lodge, near
Leeds, on the 8th of June, 1724, his father being

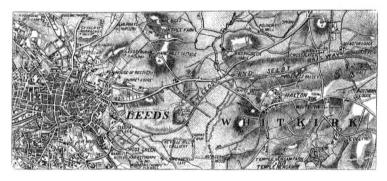

SMEATON'S NATIVE DISTRICT   [Ordnance Survey.]

a respectable attorney practising in that town.   The
house in which the future engineer was born was built
by his grandfather John Smeaton, who is described on
the tablet to his memory erected in the neighbouring
parish church of Whitkirk, as "late of York."   Leeds
was then a place of small importance, compared with
what it now is.   The principal streets were those still
known as Briggate, leading to the bridge; Kirkgate,
leading to the parish church; and Swinegate, leading to
the old castle.   Beyond those streets there lay a wide
extent of open fields.   Boar Lane, now nearly the centre
of the town, was a kind of airy suburb, in which the
principal merchants resided; and the back of the houses
in the upper part of Briggate, now the main street,
looked into the country,[1] or "the Park," on which Park
Square, Park Row, and Park Lane (now containing the
principal architectural ornament of the place, the new
Town Hall), have since been built.   There were also green
fields, with pleasant footpaths, between the parish church

---

[1] Whitaker's Thoresby, 'Loidis and Elmete,' p. 89.

and the river side, through certain gardens, then, as now, named "The Calls," but gardens no longer. The clothing trade of the town was then so small that the cloth market was held in the open air upon the bridge, where the cloth was exposed for sale on the parapets. The homely entertainment of the clothiers at that day was a "brigend shot," consisting of a noggin of porridge and a pot of ale, followed by a twopenny trencher of meat. Down to the year 1730, the bridge was so narrow that only one cart could pass over it at a time. But the number of wheeled vehicles then in use was so small that the inconvenience was scarcely felt. The whole of the cloth was brought to market on men's and horses' backs.[1] Coals were in like manner carried from the pits on horseback, the stated weight of a "horse-pack" being eighteen stone, or equal to two hundredweight and a quarter.[2] In the rural districts of Yorkshire manure was also carried a-field on horses' backs, and sometimes on women's backs, while the men sat at home knitting.[3] The cloth-packs were carried by the "bell-horses," or pack-horses; and this mode of conveyance continued until towards the end of last century. Scatcherd says the pack-horses only ceased to travel about the year 1794.

The Leeds men, it seems, were not considered so "quick" as those of Bradford, then a much smaller place and comparatively of the dimensions of a village; and it was long before they provided themselves with a

---

[1] This is clear from an allusion made by Thoresby to an Act passed in 1714, regulating the manufacture of broad-cloth, by which the length was increased from four or six-and-twenty to sixty and even seventy yards, "to the great oppression," says Thoresby, "both of man and beast in carriage."

[2] Smeaton's 'Reports,' vol. iii., p. 410. Mr. Smeaton says that before the invention of rail or waggon roads at Newcastle, "all the coals that were carried down to the ships must have been conveyed on horses' backs." What was called "a bowl of coals," was reckoned a horse-load; and in Yorkshire (where the first waggonway was laid within Smeaton's recollection) the load of coals and the "horse-pack" were readily substituted the one for the other.

[3] Brockett's 'Glossary of North Country Words.' Newcastle, 1825.

proper market for their cloth, first on Mill Hill, and
afterwards in the Calls; finally, in 1757, erecting a
large hall for the markets in the Parks, which is now
known as the Coloured Cloth Hall.   But even then the
place remained comparatively rural in point of size and
surroundings.

VIEW OF LEEDS, EARLY PART OF 18TH CENTURY. 1

[From Thoresby's ' Ducatus Leodensis.']

Smeaton was greatly favoured in his home and his
family.   He received his first education at his mother's
knees, and when not occupied with his lessons he led the
life of a healthy, happy country boy.   Austhorpe was
then quite in the country, the only houses in the neighbourhood being those of the little hamlet of Whitkirk,
with the large old mansion of Temple Newsam, surrounded by its noble park and woods, close at hand.
Young Smeaton was not much given to boyish sports,
early displaying a thoughtfulness beyond his years.
Most children are naturally fond of building up miniature fabrics, and perhaps still more so of pulling them
down.   But the little Smeaton seemed to have a more
than ordinary love of contrivance, and that mainly for

---

¹ The principal buildings shown in
the above view of Leeds, about the
time when Smeaton was born, are the
Parish Church (described by Thoresby
as " black, but comely "), St. John's
Church, and Call Lane Chapel.

its own sake. He was never so happy as when put in possession of any cutting-tool, by which he could make his little imitations of houses, pumps, and windmills. Even whilst a boy in petticoats he was continually dividing circles and squares, and the only playthings in which he seemed to take any real pleasure were his models of things that would "work." When any carpenters or masons were employed in the neighbourhood of his father's house, the inquisitive boy was sure to be amongst them, watching the men, observing how they handled their tools, and frequently asking them questions. His life-long friend, Mr. Holmes,[1] who knew him in his youth, has related that having one day observed some millwrights at work, shortly after, to the great alarm of his family, he was seen fixing something like a windmill on the top of his father's barn. On another occasion, when watching some workmen fixing a pump in the village, he was so lucky as to procure from them a piece of bored pipe, which he succeeded in fashioning into a working pump that actually raised water. His odd cleverness, however, does not seem to have been appreciated; and it is told of him that amongst the other boys he was known as "Fooely Smeaton;" for, though forward enough in putting questions to the workpeople, amongst boys of his own age he was remarkably shy, and, as they thought, stupid.

At a proper age the boy was sent to school at Leeds. That town then possessed, as it still does, the great advantage of an excellent free grammar school, founded by the benefactions of Catholics in early times, afterwards greatly augmented by the endowment of one John Harrison, a native of the town, about the period of

---

[1] An eminent clock and watchmaker in the Strand, afterwards Smeaton's partner in the Deptford Waterworks. His 'Short Narrative of the Genius, Life, and Works of the late Mr. John Smeaton, C.E., F.R.S.,' published in 1793, contains the gist of nearly all the notices of Smeaton's life which have since been published; but it is a very meagre account of only a few pages in length.

the Reformation. At this school Smeaton is supposed
to have received the best part of his school instruction,
and it is said that his progress in geometry and arithmetic was very decided; but, as before, the chief part
of his education was conducted at home, amongst his
tools and his model machines. There he was incessantly busy whenever he had a spare moment. Indeed,
his mechanical ingenuity sometimes led him to play
tricks which involved him in trouble. Thus, it happened that some mechanics came into the neighbourhood to erect a "fire-engine," as the steam-engine was
then called, for the purpose of pumping water from the
Garforth coal-mines; and Smeaton made daily visits to
them for the purpose of watching their operations.
Carefully observing their methods, he proceeded to
make a miniature engine at home, provided with
pumps and other apparatus, and he even succeeded
in getting it set to work before the colliery engine
was ready. He first tried its powers upon one of the
fish-ponds in front of the house at Austhorpe, which he
succeeded in very soon pumping completely dry, and so
killed all the fish in it, very much to the surprise as
well as annoyance of his father. But the latter seems,
on the whole, to have been very indulgent, for he
provided the boy with a workshop in an outhouse,
where he hammered, filed, and chiselled away very
much to his heart's content. Working on in this way,
by the time he had arrived at his fifteenth year, young
Smeaton had contrived to make a turning-lathe, on
which he turned wood and ivory, and he delighted in
making presents of little boxes and other articles to his
friends. He also learned to work in metals, which he
fused and forged himself, and by the age of eighteen he
could handle his tools with the expertness of any regular
smith or joiner.

"In the year 1742," says his friend, Mr. Holmes, "I
spent a month at his father's house; and being intended

myself for a mechanical employment, and a few years younger than he was, I could not but view his works with astonishment. He forged his iron and steel, and melted his metal. He had tools of every sort for working in wood, ivory, and metals. He had made a lathe, by which he cut a perpetual screw in brass,—a thing little known at that day, and which, I believe, was the invention of Mr. Henry Hindley, of York, with whom I served my apprenticeship. Mr. Hindley was a man of the most communicative disposition, a great lover of mechanics, and of the most fertile genius. Mr. Smeaton soon became acquainted with him, and spent many a night at Mr. Hindley's house till daylight, conversing on these subjects."

WHITKIRK, NEAR LEEDS.

[By E. M. Wimperis, after a Sketch by T. Sutcliffe.]

CHAPTER II.

SMEATON LEARNS THE TRADE OF MATHEMATICAL INSTRUMENT MAKER.

YOUNG SMEATON left school in his sixteenth year, and from that time he was employed in his father's office, copying legal documents, and passing through the necessary preliminary training to fit him to follow the profession of an attorney. Mr. Smeaton, having a good connection in his native town, naturally desired that his only son should succeed him. But the youth took no pleasure in the law: his heart was in his workshop amongst his tools; and though he mechanically travelled to the office daily, worked assiduously at his desk, and then travelled back again to Austhorpe, he more and more felt the irksomeness of his intended vocation. Partly to wean him from his mechanical pursuits at home, which often engrossed his attention half the night, and partly to give him the best legal education which it was in his power to bestow, Mr. Smeaton sent his son to London towards the end of the year 1742, and for a short time he occupied himself, in conformity with his parent's wishes, in attending the Courts in Westminster Hall. But at length he could not repress his strong desire to follow some mechanical occupation, and in a strong but respectful memorial to his father, he fully set forth his views as to the calling which he wished to pursue in preference to that of the law.

The father's heart was touched, and probably also his good sense was influenced, by the son's earnest appeal; and he wrote back, giving his assent, though not without his strong expression of regret as to the course which his son desired to adopt. No doubt he thought that by giving up the position of a member of a learned and lucrative profession, and descending to the level of a mechanical workman, his son was performing an act of great folly, for there was no such thing then as the profession of a civil engineer. Almost the only mechanical work of importance done at that time was executed by millwrights and others, at labourers' wages, as we have already seen in the Life of Brindley. The educated classes eschewed mechanical callings, which were neither regarded as honourable nor remunerative; and that Smeaton should have felt so strongly impelled to depart from the usual course and enter upon a line of occupation, must be attributed entirely to his innate love of construction, or, as he himself expressed it to his father, the strong " bent of his genius."

When he received his father's letter, the young man experienced the joy of a prisoner on hearing of his reprieve, and he lost no time in exercising his new-found liberty. He sought out for himself a philosophical instrument maker, who could give him instruction in the business he proposed to follow, and entered into his service, his father being at the expense of his maintenance. In due course of time, however, he was enabled to earn enough to maintain himself; but his father continued to assist him liberally on every occasion when money was required either for purposes of instruction or of business.

Young Smeaton did not live a mere workman's life, but frequented the society of educated men, and was a regular attendant of the meetings of the Royal Society. In 1750, he lodged in Great Turnstile, a passage leading from the south side of Holborn to the east side of Lincoln's Inn Fields; and shortly after, when he commenced business as a mathematical instrument maker on his own account, he lodged in Furnival's Inn Court, from which his earlier papers read before the Royal

Society were dated. The very same year in which he began business, when he was only twenty-six, he read a communication before the Royal Society, descriptive of his own and Dr. Gowin Knight's improvements in the mariner's compass. In the year following (1751) we find him engaged in a boat on the Serpentine river, performing experiments with a machine of his invention, for the purpose of measuring the way of a ship at sea. With the same object he made a voyage down the Thames, in a small sailing vessel, to several leagues beyond the Nore; and he afterwards made a short cruise in the *Fortune* sloop of war, testing his instruments by the way.

His attention as yet seems to have been confined chiefly to the improvement of mathematical instruments used for purposes of navigation or astronomical observation. In the year 1752, however, he enlarged the range of his experiments; for we find him, in April, reading a paper before the Royal Society, descriptive of some improvements which he had contrived in the air-pump. On the 11th of June following, he read a second paper, descriptive of an improvement which he had made in ship-tackle by a construction of pulleys, by means of which one man might easily raise a ton weight; and on the 9th of November following, he read a third paper, descriptive of M. De Moura's experiments on Savary's steam-engine. In the course of the same year he was busily occupied in performing a series of experiments, on which his admirable paper, read before the same Society, was founded—for which he received their Gold Medal in 1759—entitled ' An Experimental Inquiry concerning the Natural Powers of Water and Wind to turn Mills and other Machines depending on a Circular Motion.' This paper was very carefully elaborated, and is justly regarded as the most masterly report that has ever been published on the subject.

To accomplish all this, and at the same time to carry

on his business, necessarily involved great application and industry. Indeed, Smeaton was throughout life an indefatigable student, bent, above all things, on self-improvement. One of his maxims was, that " the abilities of the individual are a debt due to the common stock of public happiness ;" and the steadfastness with which he devoted himself to useful work, in which he at the same time found his own true happiness, shows that this maxim was no mere lip-utterance on his part, but formed the very mainspring of his life. From an early period he carefully laid out his time with a view to getting the most good out of it : so much for study, so much for practical experiments, so much for business, and so much for rest and relaxation.

We infer that Smeaton could never have had a large business as a philosophical instrument maker from the large portion of his time that he devoted to study and experiments. Probably he already felt that, in the course of the development of English industry, a field was opening before him of a more important character than any that was likely to present itself in the mathematical instrument line. He accordingly seems early to have turned his attention to engineering, and, amongst other branches of study, he devoted several hours in every day to the acquisition of French, in order that he might be enabled to read for himself the works on that science which were then only to be found in that and the Italian language. He had, however, a further object in studying French, which was to enable him to make a journey which he contemplated into the Low Countries, for the purpose of inspecting the great canal works of foreign engineers.

Accordingly, in 1754, he set out for Holland, and traversed that country and Belgium, travelling mostly on foot and in treckschuyts, or canal boats, both for the sake of economy, and that he might more closely inspect the engineering works of the districts through which he passed.   He found himself in a country which had been, as it were, raked out of the very sea,—for which Nature had done so little, and skill and industry so much.   From Rotterdam he went by Delft and the Hague to Amsterdam, and as far north as Helder, narrowly inspecting the vast dykes raised around the land to secure it against the hungry clutches of the sea from which it was originally won.   At Amsterdam he was astonished at the amount of harbour and dock accommodation, existing at a time when London as yet possessed no conveniences of the sort, though indeed it always had its magnificent Thames.   Passing round the country by Utrecht, he proceeded to the great sea-sluices at Brill and Helvoetsluys, by means of which the inland waters discharged themselves, while the sea-waters were securely dammed out.   Seventeen years later, he made use of the experience which he had acquired in the course of his careful inspection of these great works, in illustrating and enforcing the recommendations contained in his elaborate report on the best means of improving Dover Harbour. He made careful memoranda during his journey, to which he was often accustomed to refer, and they proved of great practical value to him in the course of his subsequent extensive employment as a canal and harbour engineer.

Shortly after his return to England in 1755, an opportunity occurred for the exercise of that genius in construction which Smeaton had thus so carefully disciplined and cultivated; and it proved the turning-point in his fortunes, as well as the great event of his professional life.

CHAPTER III.

The Eddystone Rock—Winstanley's and Rudyerd's Lighthouses.

The Eddystone forms the crest of an extensive reef of rocks which rise up in deep water about fourteen miles S.S.W. of Plymouth Harbour. Being well out at sea, they are nearly in a line with Lizard Head and Start Point, and besides being in the way of ships bound for Plymouth Sound, they lie in the very direction of vessels coasting up and down the English Channel. At low

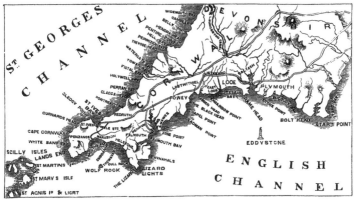

COAST OF DEVON AND CORNWALL.

water, several long low reefs of gneiss are visible, jagged and black; but at high water they are almost completely submerged. Lying in a sloping manner towards the south-west quarter, from which the heaviest seas come, the waves in stormy weather come tumbling up the slope and break over their crest with tremendous violence. The water boils and eddies amongst the reefs, and hence the name which they have borne from the earliest times of the Eddystone Rocks.

It may readily be imagined that this reef, whilst unprotected by any beacon, was a source of much danger to the mariner. Many a ship coming in from the Atlantic was dashed to pieces there, almost within sight of land, and all that came ashore was only dead bodies and floating wreck. To avoid this terrible rock, the navigator was accustomed to give it as wide a berth as possible, and homeward-bound ships accordingly entered the Channel on a much more southern parallel of latitude than they now do. In his solicitude to avoid the one danger, the sailor too often ran foul of another; and hence the numerous wrecks which formerly occurred along the French coast, more particularly upon the dangerous rocks which surround the Islands of Jersey, Guernsey, and Alderney.

We have already described the rude expedients adopted in early times to light up certain of the more dangerous parts of the coast, and referred to the privilege granted to private persons who erected lighthouses, of levying tolls on passing shipping.[1] But it was long before any private adventurer was found ready to undertake so daring an enterprise as the erection of a lighthouse on the Eddystone, where only a little crest of rock was visible at high water, scarcely capable of affording foothold for a structure of the very narrowest basis. At length, however, one Mr. Henry Winstanley (a mercer and country gentleman), of Littlebury, in the county of Essex, obtained the necessary powers, in the year 1696, to erect a lighthouse on the Eddystone. That gentleman seems to have possessed a curious mechanical genius, which first displayed itself in devising sundry practical jokes for the entertainment of his guests.

---

[1] Nearly all the private lights first erected—amongst which were those on Dungeness, the Skerries (off the Isle of Anglesey), the Eddystone, Harwich, Wintertonness and Orfordness, Hunstanton Cliff, &c.—have been purchased by the Trinity House, some of them at very large sums. The revenue of the Skerries Light alone, previous to its purchase by the Trinity House, amounted to about 20,000l. a year.

Smeaton tells us[1] that in one room there lay an old slipper, which, if a kick was given it, immediately raised a ghost from the floor; in another, the visitor sat down upon a chair, which suddenly threw out two arms and held him a fast prisoner; whilst, in the garden, if he sought the shelter of an arbour and sat down upon a particular seat, he was straightway set afloat into the middle of the adjoining canal. These tricks must have rendered the house at Littlebury a somewhat exciting residence for the uninitiated guest. The amateur inventor exercised the same genius to a certain extent for the entertainment of the inhabitants of the metropolis, and at Hyde Park Corner he erected a variety of jets d'eau, known by the name of Winstanley's Waterworks, which he exhibited at stated times at a shilling a-head.[2]

This whimsical character of the man in some measure accounts for the oddity of the wooden building afterwards erected by him for the purpose of a lighthouse on the Eddystone rock; and it is a matter of some surprise that it should have stood the severe weather of the English Channel for several seasons. The building was begun in the year 1696, and finished in four years. It must necessarily have been a work attended with great difficulty as well as danger, as operations could only be carried on during fine weather, when the sea was comparatively smooth. The first summer was wholly spent in making twelve holes in the rock, and fastening twelve irons in them by which to hold the superstructure. "Even in summer," Winstanley says, "the weather would at times prove so bad, that for ten or fourteen days together the sea would be so raging about these rocks, caused by outwinds and the running of the ground seas coming from the main ocean, that although the weather should seem and be

---

[1] 'Narrative of the Building and a Description of the Construction of the Eddystone Lighthouse with Stone.' By John Smeaton, Civil Engineer, F.R.S. Second Edition. London, 1813.

[2] They continued to be exhibited for some time after Mr. Winstanley's death. See 'Tatler,' for September, 1709.

most calm in other places, yet here it would mount and fly more than two hundred feet, as has been so found since there was lodgment on the place, and therefore all our works were constantly buried at those times, and exposed to the mercy of the seas."

The second summer was spent in making a solid pillar, twelve feet high and fourteen feet in diameter, on which to set the lighthouse. In the third year, all the upper work was erected to the vane, which was eighty feet above the foundation. In the midsummer of that year Winstanley ventured to take up his lodging with the workmen in the lighthouse; but a storm arose, and eleven days passed before any boats could come near them. During that period the sea washed in upon Winstanley and his companions, wetting all their clothing and provisions, and carrying off many of their materials. By the time the boats could land, the party

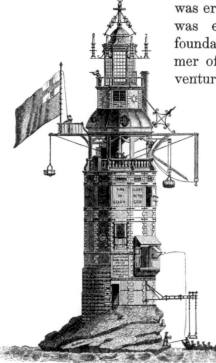

WINSTANLEY'S LIGHTHOUSE.

were reduced almost to their last crust; but happily the building stood, apparently firm. Finally, the light was exhibited on the 14th of November, 1698.

The fourth year was occupied in strengthening the building round the foundations, making all solid nearly to a height of twenty feet, and also in raising the upper part of the lighthouse forty feet, to keep it well out of the wash of the sea. This timber erection, when finished, somewhat resembled a Chinese pagoda, with open galleries and numerous fantastic projections. The main gallery under the light was so wide and open, that an old gentleman who remembered both Mr. Winstanley and his lighthouse, afterwards told Mr. Smeaton, that it was "possible for a six-oared boat to be lifted up on a wave, and driven clear through the open gallery into the sea on the other side." In the perspective print of the lighthouse, published by the architect after its erection, he complacently represented himself as fishing out of the kitchen-window!

When Winstanley had brought his work to completion, he is said to have expressed himself so satisfied as to its strength, that he only wished he might be there in the fiercest storm that ever blew. In this wish he was not disappointed, though the result was directly the reverse of its builder's anticipations. In November, 1703, Winstanley went off to the lighthouse to superintend some repairs which had become necessary, and he was still in the place with the lightkeepers, when, on the night of the 26th, a storm of unparalleled fury burst along the coast. As day broke on the morning of the 27th, people on shore anxiously looked in the direction of the rock to see if Winstanley's structure had withstood the fury of the gale ; but not a vestige of it remained. The lighthouse and its builder had been swept completely away.

The building had, in fact, been deficient in every element of stability, and its form was such as to render it peculiarly liable to damage from the violence both of wind and water. "Nevertheless," as Smeaton generously observes, " it was no small degree of heroic merit in Winstanley to undertake a piece of work which had before been deemed impracticable, and, by the success which attended his endeavours, to show mankind that the erection of such a building was not in itself a thing of that kind." He may indeed be said to have paved the way for the more successful enterprise of Smeaton himself; and his failure was not without its influence in inducing that great builder to exercise the care which he did in devising a structure that should withstand the most violent force of the sea on that coast. Shortly after Winstanley's lighthouse had been swept away, the *Winchelsea*, a richly-laden homeward-bound Virginiaman, was wrecked on the Eddystone rocks, and almost every soul on board perished; so that the erection of a lighthouse upon the dangerous reef remained as much a necessity as ever.

A new architect was not long in making his appearance. He did not, however, come from the class of architects, or builders, or even of mechanics: as for the class of engineers, it had not yet sprung into existence. Again the bold projector of a lighthouse for the Eddystone was a London mercer, who kept a silk-shop on Ludgate Hill. John Rudyerd—for such was his name—was, however, a man of unquestionable genius, and possessed of much force of character. He was originally the son of a Cornish labourer, whom nobody would employ, his character was so bad; and the rest of the family were no better, being looked upon in their neighbourhood as "a worthless set of ragged beggars." John seems to have been the one sound chick in the whole brood. He had a naturally clear head and honest heart, and succeeded in withstanding the bad example of his family. When his brothers went out a-pilfering, he refused to accompany them, and hence they regarded him as sullen and obstinate. They ill-used him, and he ran away. Fortunately he succeeded in getting into the service of a gentleman at Plymouth, who saw something promising in his appearance. The boy conducted himself so well in the capacity of a servant, that he was permitted the opportunity of learning reading, writing, and accounts; and he proved so quick and intelligent, that his kind master eventually placed him in a situation where his talents could have better scope for exercise than in his service, and he thus succeeded in laying the foundations of John Rudyerd's success in life. We are not informed of the steps by which he worked his way upward, until we find him called from his silk-mercer's shop to undertake the rebuilding of the Eddystone Lighthouse. But it is probable that by this time he had become known for his mechanical skill in design, if not in construction, as well as for his thoroughly practical and reliable character as a man of business; and that for these reasons, amongst others, he was selected to conduct this difficult and responsible undertaking.

After the lapse of about three years from the destruction of Winstanley's fabric, the Brethren of the Trinity, in 1706, obtained an Act of Parliament enabling them to rebuild the lighthouse, with power to grant a lease to the undertaker. It was taken by one Captain Lovet for a period of ninety-nine years, and he it was that found out and employed Mr. Rudyerd. His design of the new structure was simple but masterly. He selected the form that offered the least possible resistance to the force of the winds and the waves, avoiding the open galleries and projections of his predecessor. Instead of a polygon he chose a cone for the outline of his building, and he carried up the elevation in that form. In the practical execution of the work he was assisted by two shipwrights from the King's yard at Woolwich, who worked with him during the whole time that he was occupied in the erection.

The main defect of the lighthouse consisted in the faultiness of the material of which it was built; for, like Winstanley's, it was of wood. The means employed to fix the work to its foundation proved quite efficient; dove-tailed holes were cut out of the rock, into which strong iron bolts or branches were keyed,[1] and the interstices were afterwards filled with molten pewter. To these branches were firmly fixed a crown of squared oak balks, and across these a set of shorter balks, and so on, till a basement of solid wood was raised, the whole being firmly fitted and tied together with trenails and screw-bolts. At the same time, to increase the weight and vertical pressure of the building, and thereby present a greater resistance to any disturbing external force, Rudyerd introduced numerous courses of Cornish moorstone, as well jointed as possible, and

RUDYERD'S LIGHTHOUSE.

cramped with iron. It is not necessary to follow the details of the construction further than to state, that outside the solid timber and stone courses strong upright timbers were fixed, and carried up as the work proceeded, binding the whole firmly together.

Within these upright timbers the rooms of the lighthouse were formed, the floor of the lowest, the storeroom, being situated twenty-seven feet above the highest

---

[1] Mr. Smeaton says that the instrument now called the Lewis, though an invention of old date, was for the first time made use of by Rudyerd in fixing his iron branches firmly to the rock. "Mr. Rudyerd's method," he says, "of keying and securing, must be considered as a material accession to the practical part of engineering, as it furnishes us with a secure method of fixing ring-bolts and eye-bolts, stanchions, &c., not only in rocks of any known hardness, but into piers, moles, &c., that have been already constructed, for the safe mooring of ships, or fixing additional works, whether of stone or wood."— Smeaton's 'Narrative,' p. 22.

side of the rock.   The upper part of the building comprehended four rooms, one above another, chiefly formed by the upright outside timbers, scarfed—that is, the ends overlapping, and then they were firmly fastened together.   The whole building was, indeed, an admirable piece of ship-carpentry, excepting the moorstone, which was only introduced, as it were, by way of ballast.   The outer timbers were tightly caulked with oakum, like a ship, and the whole was payed over with pitch.   Upon the roof of the main column Mr. Rudyerd fixed his lantern, which was lit by candles, seventy feet above the highest side of the foundation, which was of a sloping form.   From its lowest side to the summit of the ball fixed on the top of the building was ninety-two feet, the timber-column resting on a base of twenty-three feet four inches.   " The whole building," says Mr. Smeaton, " consisted of a simple figure, being an elegant frustum of a cone, unbroken by any projecting ornament, or anything whereon the violence of the storms could lay hold."   The structure was completely finished in 1709, though the light was exhibited in the lantern as early as the 28th of July, 1706.[1]

That the building erected by Mr. Rudyerd was on the whole exceedingly well adapted for the purpose for which it was intended, was proved by the fact that it served as a lighthouse for vessels navigating the English Channel, and withstood the fierce storms which rage along that part of the coast, for a period of nearly fifty years.[2]   The

---

[1] An anecdote is told of a circumstance which occurred during its erection, so creditable to Louis XIV., then King of France, that we repeat it here.   There being war at the time between France and England, a French privateer took the opportunity of one day seizing the men employed upon the rock, and carrying them off prisoners to France.   But the capture coming to the ears of the King, he immediately ordered that the prisoners should be released

and sent back to their work with presents, declaring that, though he was at war with England, he was not at war with mankind; and, moreover, that the Eddystone Lighthouse was so situated as to be of equal service to all nations having occasion to navigate the channel that divided France from England.—Smeaton's ' Narrative,' p. 28.

[2] Mr. Smeaton, in his quaint and interesting ' Narrative,' relates some curious anecdotes of the early light-

chief defect, as we have already observed, consisted in the material of which it was composed. It was combustible, yet it could only be made useful as a lighthouse by the constant employment of fire in some shape. Though the heat of the candles used in the lantern may not have been very great, still it was sufficient to produce great dryness and inflammability in the timbers lining the roof, and these being covered with a crust of soot, must have proved a constant source of danger. The immediate cause of the accident by which the lighthouse was destroyed was never ascertained. All that became known was, that about two o'clock in the morning of the 2nd December, 1755, the lightkeeper on duty, going into the lantern to snuff the candles, found it full of smoke. The lighthouse was on fire! In a few minutes the wooden fabric was in a blaze. Water could not be brought up the tower by the men in sufficient quantities to be thrown with any effect upon the flames raging above their heads: the molten lead fell down upon the

---

keepers. Rudyerd's house was at first attended by only two men, as the duty required no more. During the night they kept watch by turns for four hours alternately, snuffing and renewing the candles. It happened, however, that one of the keepers took ill and died, and only one man remained to do the work. He hoisted the flag as a signal to those on land to come off to his assistance; but the sea was running so wild at the time that no boat could live in the vicinity of the rock; and the same rough weather lasted for nearly a month. What was the surviving man to do with the dead body of his comrade? The thought struck him that if he threw it into the sea, he might be charged with murder. He determined, therefore, to keep the corpse in the lighthouse until the boat could come off from the shore. One may imagine the horrors endured by the surviving lightkeeper during that long, dismal month. At last the boat came off, but the weather was still so rough that a landing was only effected with the greatest difficulty. By this time the effluvia rising from the corpse was most overpowering; it completely filled the chambers of the lighthouse, and it was all that the men could do to get the body disposed of by throwing it into the sea. This circumstance induced the proprietors for the future to employ a third man to supply the place of a disabled or dead keeper, though the occupation proved exceedingly healthy on the whole. There was always a large number of candidates for any vacant office, probably of the same class to which pike-keepers belong. They must have been naturally morose, and perhaps slightly misanthropic; for Mr. Smeaton relates that, some visitors having once landed at the rock, one of them observed to the lightkeeper how comfortably they might live there in a state of retirement: "Yes," replied the man, "very comfortably, if we could have the use of our tongues; but it is now a full month since my partner and I have spoken to each other!"

lightkeepers, into their very mouths,[1] and they fled from room to room, the fire following them down towards the sea. From Cawsand and Rame Head the unusual glare of light proceeding from the Eddystone was seen in the early morning, and fishing-boats with men went off to the rock, though a fresh east wind was blowing. By the time they reached it, the lightkeepers had not only been driven from all the rooms, but, to protect themselves from the molten lead and red-hot bolts and falling timbers, they had been compelled to take shelter under a ledge of the rock on its eastern side, and, after considerable delay, the poor fellows were taken off, more dead than alive. And thus was Rudyerd's lighthouse also completely destroyed.

As the necessity for protecting the navigation of the Channel by a light on the Eddystone was greater than ever, in consequence of the increasing foreign as well as coasting trade of the kingdom, it was immediately determined by the proprietors to take the necessary steps for rebuilding it; and it was at this juncture that Mr. Smeaton was applied to. As on the two previous occasions, when a mercer and country gentleman, and next a London silk-mercer, had been called upon to undertake this difficult work, the person now applied to was not a builder, nor an architect, nor engineer, but a mathematical instrument maker. Mr. Smeaton had, however, by this time gained for himself so general an estimation amongst scientific men as a painstaking observer, an able mechanic, and one who would patiently master, and, if possible, overcome difficulties, that he was at once pointed out as the person of all others the most capable of satisfactorily rebuilding this important beacon on the south-eastern coast.

---

[1] It appears that a post-mortem examination of one of the lightkeepers, who died from injuries received during the fire, took place some thirteen days after its occurrence, and a flat oval piece of lead, some seven ounces in weight, was taken out of his stomach, having proved the cause of his death.

EDDYSTONE LIGHTHOUSE. [By Percival Skelton.]

CHAPTER IV.

Smeaton's Lighthouse on the Eddystone.

Captain Lovet, the lessee of the lighthouse, having died in 1715, his property was sold, and Mr. Robert Weston, in company with two others, became the purchasers of the lease. On the destruction of Rudyerd's timber building, Mr. Weston applied to the Earl of Macclesfield, President of the Royal Society, for his advice under the circumstances, and requested him to point out an architect capable of undertaking the reconstruction of the lighthouse in an efficient manner. Mr. Smeaton's account of the reply made by the Earl to Mr. Weston is so characteristic of him, that we quote his own words. Lord Macclesfield told him "that there was one of their body whom he could venture to recommend to the business; yet that the most material part of what he knew of him was, his having within the compass of the last seven years recommended himself to the Society by the communication of several mechanical inventions and improvements; and that though he had at first made it his business to execute things in the instrument way (without having ever been bred to the trade), yet on account of the merit of his performances he had been chosen a member of the Society; and that for about three years past, having found the business of a philosophical instrument maker not likely to afford an adequate recompense, he had wholly applied himself to such branches of mechanics as he (Mr. Weston) had appeared to want; that he was then somewhere in Scotland, or in the north of England, doing business in that line; that what he had to say of him further was, his never having known him undertake anything but what he completed to the satisfaction of those who employed him, and that Mr. Weston might rely upon it, when the business was stated to him, he would not undertake it unless he clearly saw himself capable of performing it." [1]

This description seems to have been enough for Mr. Weston, who immediately addressed Mr. Smeaton on the subject. News then travelled so slowly, and the particulars which had got abroad relating to the accident at the Eddystone were so meagre, that Smeaton did not know that the lighthouse had been totally destroyed. When he at length received Mr. Weston's letter, more than a month after the accident, he fancied that it was only the repair of some of the upper works that was required of him, and he replied that he had engagements

---

[1] Smeaton's 'Narrative,' &c., p. 38.

on hand that he could not leave upon an uncertainty.
The answer he received was, that the former building
was totally destroyed—that the lighthouse must be rebuilt—and the letter concluded with the words, "thou
art the man to do it."

Smeaton then returned to town and proceeded to take
the matter in hand. The subject was wholly new to him;
but he determined to investigate it thoroughly, and he
lost no time in doing so. One of the earliest conclusions
he arrived at was, that stone was the proper material
with which to rebuild the lighthouse, though the superiority of timber was strongly urged upon him. The
popular impression, which also prevailed amongst the
Brethren of the Trinity House, was, that "nothing but
wood could possibly stand on the Eddystone;" and many
were the predictions uttered as to the inevitable failure
of a structure composed of any other material. The
first thing which our engineer did in the matter was to
examine carefully the plans and models of the two former
lighthouses; by which he sought to ascertain their
defects, with a view to their avoidance in the intended
new erection. In the course of this inquiry, he became
more and more convinced that a great defect in the late
building had been its want of Weight, through which it
had rocked about in heavy storms, and would probably
have been washed away before long if it had not been
burnt; and he came to the conclusion, that if the lighthouse was to be contrived so as not to give way to the
sea, it must then be made so strong as that the sea
must be compelled to give way to the building. He
also had regard to durability as an important point in
its re-erection. To quote his own words: "In con-
templating the use and benefit of such a structure as
this, my ideas of what its duration and continued
existence ought to be, were not confined within the
boundary of one Age or two, but extended themselves
to look towards a possible Perpetuity."

Thus, before Smeaton had proceeded very far, he had come to the firm conviction that the new lighthouse must be built of stone. Nevertheless, he resolved to preserve the conical form of Rudyerd's building, but to enlarge considerably the diameter of the foundation, and thus increase the stability of the whole superstructure. The idea of the bole of a large spreading oak-tree presented itself to his mind as the natural model of a column, presenting probably the greatest possible strength. Another point which he long and carefully studied, was the best mode of bonding the blocks of stone to the rock and to each other, in such a way as that not only every individual piece, but the whole fabric, should be rendered proof against external force. Binding the blocks together by iron cramps was considered, but dismissed as insufficient, as well as impracticable. Then the process of dove-tailing occurred to him—a practice then generally applied to carpentry, though scarcely as yet known in masonry. Still more suitable for his purpose was the method which he had observed adopted in fixing the kerbs along the London footpaths, by which the long pieces or stretchers were retained between the two headers or bond-pieces, whose heads being cut dovetail-wise, adapted themselves to and bound in the stretchers; and the tye being as good at the bottom as at the top, this arrangement, he conceived, was the very best that could be devised for his purpose.

From these beginnings he was readily led to think that if the blocks themselves, both inside and out, were all formed into large dovetails, they might be managed so as mutually to lock one another together, being primarily engrafted into the rock; and in the round and entire courses, along the top of the rock, they might all proceed from and be locked to one large centre stone. By thus rooting the foundations into the rock, and also binding every stone by a similar dovetailing process to every other stone in each course, upon which the sea could only act edgeways, he conceived that he would be enabled to erect a building of a strength sufficient to resist the strongest force of winds and waves that was likely to be brought against it.

Having thus thought out the subject, and deliberately matured his views—carefully studying, amongst other works, 'Wren's Parentalia' and Price's account of the building of Salisbury Cathedral—he proceeded to design a lighthouse on the principles we have thus summarily indicated; and, with a few modifications rendered necessary by the situation and the various circumstances which presented themselves in the course of the work, he proceeded to carry his design into effect in the building of the third Eddystone Lighthouse.

All this had been done before Mr. Smeaton had even paid a visit to the site on which the lighthouse was to be built. The difficulty of reaching the place was great, and his time was precious. Besides, he thought it best to prepare himself for his first visit by completing his thorough preliminary investigation of the whole case. It was not until the end of March, 1756, that he set out from London to Plymouth for the purpose of making his first inspection of the rock. He was no less than six days in performing the journey, of which he says, " I had nothing to regret but the loss of time that I suffered, which was occasioned chiefly by the badness of the roads." At Plymouth he met Mr. Josias Jessop, to whom he had been referred for information as to the previous lighthouse. Mr. Jessop was then a foreman of shipwrights, called a quarterman, in Plymouth Dock—a man of much modesty, integrity, and ingenuity in mechanical matters.[1] Mr. Smeaton also found him to be a competent draughtsman and an excellent modeller, and he cheerfully acknowledged the great assistance which he obtained from him during the progress of the work. Smeaton showed

---

[1] His son, William Jessop, the engineer, became a pupil of Smeaton's, and afterwards rose to great eminence in the profession.

Jessop the plan of the stone building which he had already made. The foreman expressed his great surprise on first looking at it, having made up his mind that the lighthouse could only be reconstructed of wood. But he readily admitted the superiority of a stone structure, if it could be made to stand in so very exposed and dangerous a situation.

Mr. Smeaton was anxious to go off to the rock at once; but the wind had been blowing fresh for several days, and there was so heavy a sea in the Channel, that it was not until the 2nd of April that he could set sail. On reaching the Eddystone, the sea was breaking upon the landing-place with such violence, that it was impossible to land. All that Smeaton could do was to view the cone of bare rock—the mere crest of the mountain whose base was laid so far down in the sea-deeps beneath —over which the waves were lashing, and to form a more adequate idea of the very narrow as well as turbulent site on which he was expected to erect his building. Three days later he made a second voyage, and he rejoiced on this occasion to be able to set his foot for the first time upon the Eddystone. He stayed there for more than two hours, and thoroughly examined the rock; being at length compelled to leave it by the sea, which began to break over it as the tide rose. The only traces that he could find of the two former lighthouses were the iron branches fixed by Rudyerd, and numerous traces of those fixed by Winstanley. On a third attempt to make the rock, Mr. Smeaton was foiled by the wind, which compelled him to re-land without even having got within sight of it. After five more days—during which the engineer was occupied in looking out a proper site for a work-yard,[1] and examining the granite in the neighbourhood for the purposes of the building—he made a

---

[1] The work-yard eventually fixed upon was in a field adjacent to Mill Bay, situated about midway between Plymouth and Devonport, behind Drake's Island.

fourth voyage, and although the vessel reached the rock, the wind was blowing so fresh and the breakers were so wild, that it was again found impossible to land. He could only direct the boat to lie off and on, for the purpose of watching the breaking of the sea and its action upon the reef. A fifth trial, made after the lapse of a week, proved no more successful. After rowing about all day with the wind ahead, the party found themselves at night about four miles from the Eddystone, near which they anchored until morning; but wind and rain coming on, they were forced to return to harbour without accomplishing their object.

The sixth attempt was successful, and on the 22nd of April, after the lapse of seventeen days, Mr. Smeaton was able to effect his second landing at low water. After a further inspection, the party retreated to their sloop which lay off until the tide had fallen, when Smeaton again landed, and the night being perfectly still, he says, "I went on with my business till nine in the evening, having worked an hour by candle-light." On the 23rd he landed again and pursued his operations; but this time he was interrupted by the ground swell, which sent the waves upon the reef, and, the wind rising, the sloop was forced to put back to Plymouth. Mr. Smeaton had, however, during this visit, secured some fifteen hours' occupation on the rock, and taken dimensions of all its parts, to enable him to construct an accurate model of the foundation of the proposed building. He succeeded in obtaining such measurements as he thought would enable him to carry out his intention; but to correct the drawing, which he made to a scale, he determined upon attempting a seventh and final voyage of inspection on the 28th of April. But again the sea was found so turbulent, that a landing was impossible. Another fortnight passed, the weather still continuing unfavourable; but meanwhile the engineer had been maturing his design, and making all requisite preliminary arrangements to proceed with the work. Among the other facilities required for carrying on the operations, was the provision of an improved landing-place, which he regarded as of essential importance. He also drew up a careful code of regulations for the guidance and government of the artificers and others who were to be employed in constructing the lighthouse. Having done all this, he arranged to proceed to London, but not until he had paid three more visits to the rock for the purpose of correcting his measurements, in one of which he got thoroughly drenched by the spray.

On his return to town, Mr. Smeaton made his report to the proprietors, and was fully authorised by them to carry out the design which he had now matured. He accordingly proceeded to make a complete model[1] of the lighthouse, as he intended it to be built. His expertness in handling tools now proved of the greatest use to him. As every course of stones in it involved fresh adaptations and the invention of new forms to give the requisite firmness and stability to the whole, it is obvious that he secured greater accuracy by executing the work with his own hands, than if he had entrusted it to any model-maker to carry out after given dimensions and drawings, however accurately they might

---

[1] He thus states the reasons which prevailed with him in undertaking the construction of this model with his own hands: " Those who are not in the practice of handling mechanical tools themselves, but are under the necessity of applying to the manual operations of others, will undoubtedly conclude that I would have saved much time by employing the hands of others in this matter; and on the idea of the design being already fixed, and fully and accurately as well as distinctly made out—that is, supposing the thing done that was wanted to be done—it certainly would have been so; and had I wanted a duplicate of any part, or of the whole, when done, I should certainly have had recourse to the hands of others. But such as are in the use of handling tools for the purpose of contrivance and invention, will clearly see that, provided I could work with as much facility and despatch as those I might happen to meet with and employ, I should save all the time and difficulty, and often the vexation, mistakes, and disappointments that arise from a communication of one's own ideas to others; and that when steps of invention are to follow one another in succession and dependence on what preceded, under such circumstances it is not eligible to make use of the hands of others."

have been laid down upon paper. After more than two months' close work, the model was ready, when it

SECTION OF SMEATON'S LIGHTHOUSE.

was submitted to a meeting of the proprietors and unanimously approved, as also by the Lords of the Admiralty, before whom it was afterwards laid. The engineer then set out for Plymouth to enter upon the necessary arrangements for preparing the foundations,—arranging with Mr. Roper, at Dorchester, on his way, for a supply of Portland stone, of which it was finally determined that the lighthouse should principally be built.

Artificers and foremen were appointed, working companies arranged, vessels provided for the transport of men and materials, work-yards hired and prepared, and Mr. Jessop was appointed the general assistant, or, as it is now termed, the Resident Engineer, of the building. Mr. Smeaton himself fixed the centre and laid down the lines on the afternoon of the 3rd of August, 1756, and from that time forward the work proceeded, though with many interruptions, caused by bad weather and heavy seas. At most, only about six hours' labour could be done at a time; and when the weather was favourable, in order that no opportunity should be lost, the men proceeded by torchlight. The principal object of the first season was to get the dovetail recesses cut out of the rock for the reception of the foundation-stones. To facilitate this process, and avoid the delay and loss of time involved by frequent voyages between the Eddystone and the shore, the *Neptune* buss was employed as a store-vessel, and rode at anchor, at a convenient distance from the rock, in about twenty fathoms water.   But, as the season advanced, it became more and more difficult to carry on the operations.   For many days together the men could not land, and, even if they had been able to do so, they must have been washed off the rock unless lashed to it.   At such times the provisions in the *Neptune* occasionally ran short, no boat being able to come off from Plymouth in consequence of the roughness of the weather.   Towards the end of October, the yawl riding at the stern of the buss broke loose by stress of weather, and was thus lost.   Mr. Smeaton was most anxious, however, to finish the boring of the foundation-holes during that season, so as to commence getting in the lower courses at the beginning of the next. The men, therefore, still persevered when the weather permitted, though sometimes they were only able to labour for two hours out of the twenty-four.   About the end of November, the whole of the requisite cutting in the rock had been accomplished without accident, and the party prepared to return to the yard on shore, and proceed with the dressing of the stones for the work of the ensuing year.

The voyage of the buss to port, however, proved a very dangerous one, and the engineer and his men narrowly escaped shipwreck.   Not being able, in consequence of the gale that was blowing, to make Plymouth Harbour, the *Neptune* was steered for Fowey, on the coast of Cornwall.   The wind rose higher and higher, until it blew quite a storm; and in the night, Mr. Smeaton, hearing a sudden alarm and clamour amongst the crew overhead, ran upon deck in his shirt to ascertain the cause.   It was raining hard, and quite a hurricane was raging.   " It being very dark," he says, "the first thing I saw was the horrible appearance of breakers almost surrounding us; John Bowden, one of the seamen, crying out, 'For God's sake, heave hard at that rope if you mean to save your lives!' I immediately laid hold of the rope, at which he himself was hauling as well as the other seamen, though he was also managing the helm. I not only hauled with all my strength, but called to and encouraged the workmen to do the same thing." The sea was now heard breaking with tremendous violence and noise upon the rocks. In this situation the jibsail was blown to pieces, and, to save the mainsail, it was lowered, when fortunately the vessel obeyed her helm and she rounded off. The night was so dark that nothing of the land could be seen, and the sailors did not know at what part of the coast they were; and in this uncertainty the vessel's head was put round to sea again, the waves occasionally breaking quite over her. At daybreak they found themselves out of sight of land, and the vessel driving towards the Bay of Biscay. Wearing ship, they stood once more for the coast, and before night they sighted the Land's End, but could not then make the shore. Another night and day passed, and, a vessel coming within sight, signals of distress were exhibited, and from her the *Neptune* learned in what direction to steer for the Scilly Islands. The wind coming round, however, they bore up for the Land's End again, passed the Lizard, then Deadman's Point, then Rame Head, and finally, after having been blown about at sea for four days, they came to an anchor in Plymouth Sound, greatly to the joy of their friends, who had begun to despair of their reappearance.

The winter was fully occupied on shore in dressing stones for the next summer's work. Mr. Smeaton himself laid out all the lines on the workroom floor,[1] in

---

[1] Mr. Smeaton had considerable difficulty in finding a room with a floor sufficiently large on which to fit all the moulds together in the order

order to insure the greatest possible accuracy in size and fitting. Nearly four hundred and fifty tons of stone were thus dressed by the time the weather was sufficiently favourable for the commencement of the building. At the same time he bestowed great pains upon experiments, which he himself conducted, for the purpose of determining the best kind of cement to be used in laying the courses of the lighthouse, and eventually fixed upon equal quantities of the lime called *blue lias* and that called *terra puzzolano*, a sufficient supply of which he was fortunate enough to procure from a merchant at Plymouth, who had imported it on adventure, and was willing to sell it cheap. It was also settled to use the finest grout for the intervals between the upright or side joints of the dovetailed part of the work. In the early part of the spring he made several visits of inspection to the quarries where the rough stones were being prepared, in order to satisfy himself as to the progress made. On one of such occasions a severe storm of thunder and lightning occurred at Lostwithiel, by which the church spire was shattered; and this turned his attention to the best mode of protecting his lighthouse against a similar accident. In the mean time he transmitted an account of the storm and the effects of the lightning on Lostwithiel Church to the Royal Society, amongst whose papers it stands recorded.[1] Dr. Franklin had shortly before published his mode of protecting lofty buildings by means of conductors, and Mr. Smeaton eventually determined, for the security of his lighthouse, to adopt this plan.

The building on the rock was fairly begun in the

---

in which they were to be permanently fixed. The engineer applied to the Mayor of Plymouth for the use of the Guildhall for the purpose, but he was refused on the pretence that the chalk-lines would spoil the floor. He was also refused the use of the Assembly-rooms for some similar rea- son. But at length, by taking down a partition in the coopers' workshops, he was eventually enabled to effect his purpose without exposing himself to further refusals from the local magnates of Plymouth.

[1] 'Philosophical Transactions,' vol. 50, p. 198.

summer of 1757, sheers having been erected and the first stone, of two and a quarter tons weight, having been landed and securely set in its place on the morning of Sunday the 12th of June. By the evening of the following day the first course of four stones was safely laid.[1] The work then proceeded from time to time, as the weather permitted; and the second course, of thirteen pieces, was completed by the 30th of the same month. The workmen were occasionally interrupted by ground-swells and heavy seas, which kept them off the rock for days together. At length, on the sixth course being laid, it was found that the building had been raised above the average wash of the sea, and the progress made after that time was much more rapid. From thence the rest of the structure was raised in regular entire courses.

The manner in which the stones were prepared in the yard, arranged in courses, and brought off in the vessels, so that they could be landed in their proper order and fixed in their proper places, was simple and effective. When the separate pieces of which a course was to consist were hewn, they were all brought together in the work-yard, fitted upon the platform in the exact sites they were to occupy in the building, and so marked and numbered that they could readily be restored to their proper relative positions. So much preliminary care having been taken, no difficulty or confusion occurred in the use of the materials, whilst the progress of the building was also greatly accelerated. For the actual details of the manner in which the masonry was proceeded with, we must refer the professional reader to Smeaton's own 'Narrative,' which

---

[1] The sloping form of the rock, to which the foundation of the building was adapted, required but this small number of stones for the lower course; the diameter of the work increasing until it reached the upper level of the rock. Thus the second course consisted of thirteen pieces, the third of twenty-five, and so on.

is remarkably minute, and as a whole exceedingly interesting.[1]

Mr. Smeaton superintended the construction of nearly the entire building. If there was any post of danger from which the men shrank back, he immediately stood forward and took the front place. One morning in the summer of 1757, when heaving up the moorings of the buss preparatory to setting sail for the rock, the links of the buoy-chain came to a considerable strain upon the davit-roll, which was of cast iron, and they began to bend upon the convexity of the roll. To remedy this, Smeaton ordered the carpenter to cut some trenails into short pieces, and split each length into two, with the view of applying the portions betwixt the chain and the roll at the flexure of each link, and so relieve the strain. But some one said that if the chain should break anywhere between the roll and the tackle, the person that applied the pieces of wood would be in danger of being cut in two by the chain or carried overboard along with it. On this Smeaton, making it a rule never to require another to undertake what he was afraid to do himself, at once stepped forward and took

---

[1] The careful manner in which the details of the foundation work were carried on is related by Smeaton at great length. One of his expedients is worthy of notice—the method by which he gave additional firmness to the stones dovetailed into the rock, by oak-wedges and cement inserted between each. To receive the wedges, two grooves were cut in the waist of each stone, from the top to the bottom of the course, an inch in depth and three inches in width. The carpenters dropped into each groove two of the oaken wedges, one upon its head, the other with its point downwards, so that the two wedges in each groove lay heads and points; on which the one was easily driven down upon the other. A couple of wedges were also pitched at the top of each groove; the dormant wedge, or that with the point upward, being held in the hand, while the drift wedge, or that with its point downward, was driven with a hammer. The object of this wedging was to preserve the whole mass steady together, in opposition to the violent agitation of the sea. In addition to this, a couple of holes being bored through every piece of stone, one course was further bound to another by oak trenails, driven stiffly through, and made so fast that they could more easily be torn asunder than pulled out again. "No assignable power," says Smeaton, "less than would by main stress pull these trenails into two, could lift one of these stones from their beds when so fixed, exclusive of their natural weight, as all agitation was prevented by the lateral wedges."

"the post of honour," as he called it, and attended the getting in of the remainder of the chain, link by link, until the operation was completed.

Whilst working at the rock on one occasion, an accident occurred to him which might have been attended with serious consequences, but in which he displayed his usual cheerful courage. The men were about to lay the centre stone of the seventh course on the evening of the 11th of August, when Mr. Smeaton was enjoying the limited promenade afforded by the level platform of stone which had with so much difficulty been raised ; but, making a false step into one of the cavities made for the joggles, and being unable to recover his balance, he fell from the brink of the work down among the rocks on the west side. The tide being low at the time, he speedily got upon his feet and at first supposed himself little hurt, but shortly after he found that one of his thumbs had been put out of joint. He reflected that he was fourteen miles from land, far from a surgeon, and that uncertain winds and waves lay between. He therefore determined to reduce the dislocation at once ; and laying fast hold of the thumb with his other hand, and giving it a violent pull, it snapped into its place again, after which he proceeded to fix the centre stone of the building.

The work now went steadily forward. Occasional damage was done by the heavy seas washing away the stones, tools, and materials ; but these losses were quickly repaired, and by the end of the season the ninth course of stones had been laid complete. The following winter was very tempestuous. The floating lightvessel, stationed about two miles from the rock, was driven from its moorings by the force of the sea, but eventually got safe into harbour. It was the 12th of May before another landing could be effected by Smeaton and his workmen, when he was no less delighted than surprised to find the entire work as he had left it six months before. Not a block had been moved. The cement was found to have set as hard as the stone itself, and the whole of the building which had been raised was one solid mass.

The rock-tackle, with sheers and windlass, having been again fixed, the erection proceeded with compara-

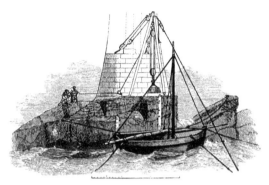

PROGRESS OF THE WORKS TO THE 15TH COURSE.

tively few interruptions until the 24th of September, 1758, when the twenty-fourth course was finished, which completed the solid part of the pillar and formed the floor of the store-room.   The building had now been raised thirty-five feet four inches above its base, or considerably beyond the heavy stroke of the waves.   Above this point were to be formed the requisite apartments for the lighthouse-keepers.   The walls of these were twenty-six inches thick, constructed in circles of hewn blocks, sixteen pieces forming each circle, all joggled and cramped, so as to secure perfect solidity.   The stones were further grooved at the ends, and into the grooves tightly-fitting pieces (rhombs) of Purbeck marble were fixed solid with well-tempered mortar, making the whole perfectly firm and water-tight.   At the end of the season the twenty-ninth course was set, and a temporary house was erected over the work for its protection during the ensuing winter.

While living at Plymouth, Smeaton used to come out upon the Hoe with his telescope, in the early grey of the morning, and stand gazing through it in the direction of the Rock. The Hoe is an elevated promenade, occupying a high ridge of land extending between Mill Bay and the entrance to the harbour, the citadel occupying its eastern end. It forms the sea-front of Plymouth, and overlooks the strikingly beautiful scenery of the Sound. St. Nicholas's Island, strongly fortified, lies immediately in front of it; beyond, rising green from the water's edge, is Mount Edgcumbe Park, with its

SMEATON ON THE HOE. [By P Skelton, and L. Huard]

masses of noble woods backed by green hills. The land juts out in rocky points on either side the bay, some of which are capped with forts and batteries; whilst in the distance now lies the magnificent barricade of the Breakwater, midway between the bluffs of Redding and Staddon Points, boldly interposing between the swell of the Sound and the long ocean waves rolling in from the Atlantic. From the Hoe the Spanish Armada was first descried making for the English coast. It was the lookout of Drake, as it now was of Smeaton, but with a

far different purpose. After a rough night at sea, he
had no eye for the picturesque beauties of the Sound :
his sole thought was of his lighthouse ; for though he had
done all that human care, forethought, and skill could do
to root his column firmly upon that perilous rock, he was
not yet altogether free from anxiety as to the security
of the foundation. There were still many who persisted
in asserting that no building erected of stone could possibly stand upon the Eddystone ; and again and again
the engineer, in the dim grey of the morning, would
come out and peer through his telescope at his deep-sea
lamp-post. Sometimes he had to wait long, until he
could see a tall white pillar of spray shoot up into the
air. Thank God! it was still safe. Then, as the light
grew, he could discern his building, temporary house
and all, standing firm amidst the waters ; and, thus far
satisfied, he could proceed to his workshops, his mind
relieved for the day.

At the end of the third year's operations the engineer returned to London to proceed with the designs
for the iron rails of the balcony, the cast and wrought
iron and copper works, as well as the glass for the
lantern, all of which were, like the rest of the work,
manufactured under his own eye. The ensuing season
proved so stormy that it was the 5th of July before
the workmen could land upon the rock and recom-
mence their building operations for the year ; but
from this point they proceeded with such rapidity—
the whole of the stones being now in readiness to be
placed—that in thirteen days two entire rooms with
their proper coverings had been erected upon the
column ; and by the 17th of August the last pieces of
the corona were set, and the forty-six courses of masonry
were finished complete. The column was now erected
to its specified height of seventy feet. The last mason's
work done was the cutting out of the words "LAUS
DEO" upon the last stone set over the door of the lan-

tern. Round the upper store-room, upon the course under the ceiling, had been cut, at an earlier period,

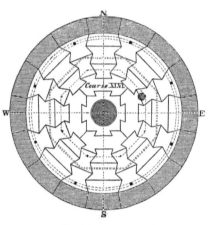

PLAN OF THE 16TH COURSE, SHOWING THE METHOD OF DOVETAILING.

"Except the Lord build the house, they labour in vain that build it." The iron work of the balcony and the lanthern were next erected, and over all the gilt ball, the screws of which Smeaton fixed with his own hands, " that in case," he says, " any of them had not held quite tight and firm, the circumstance might not have been slipped over without my knowledge." Moreover, this piece of work was dangerous as well as delicate, being performed at a height of some hundred and twenty feet above the sea. Smeaton fixed the screws while standing on four boards nailed together, resting on the cupola; his assistant, Roger Cornthwaite, placing himself on the opposite side, so as to balance his weight whilst he proceeded with the operation.

The engineer's work was now so nearly ended, and his anxiety had become so great, that he could not leave it, but took up his abode in the lighthouse, putting his own hands to the finishing of the window-fittings (for skilled workmen were difficult to be had at the lighthouse) and seeing to the minutest details in the execution of the undertaking. At length the lantern was glazed, the lightning-conductor fixed, the rooms were fitted up, and the builder looked upon the work of his hands all finished and complete. The light was first exhibited on the night of the 16th of October, 1759, and the column still stands as firm as on the day on which it was erected. About three years after its completion, one of the most terrible storms ever known raged for days along the south-west coast; and though incalculable ruin was inflicted upon harbours and shipping by the hurricane, all the damage done to the lighthouse was repaired by a little gallipot of putty.

The Eddystone Lighthouse has now withstood the storms of nearly a century,—a solid monument to the genius of its architect and builder. Sometimes, when the sea rolls in with more than ordinary fury from the Atlantic, driven up the Channel by the force of a south-west wind, the lighthouse is enveloped in spray and its light is momentarily obscured. But again it is seen shining clear like a star across the waters, a warning and a guide to the homeward-bound. Occasionally, when struck by a strong wave, the central portion shoots up the perpendicular shaft and leaps quite over the lantern. At other times, a tremendous wave hurls itself upon the lighthouse, as if to force it from its foundation. The report of the shock to one within is like that of a cannon: the windows rattle, the doors slam, and the building vibrates and trembles to its very base. But the tremor felt throughout the lighthouse in such a case, instead of being a sign of weakness, is the strongest proof of the unity and close connection of the fabric in all its parts.[1]

Many a heart has leapt with gladness at the cry of

---

[1] At first the men appointed as lightkeepers were much alarmed by the fury of the waves during storms. The year after the light was exhibited, the sea raged so furiously that for twelve days together it dashed over the lighthouse so that the men could not open the door of the lantern or any other. In a letter addressed to Mr. Jessop by the man who visited the rock after such a storm, he says: "The house did shake as if a man had been up in a great tree. The old men were almost frightened out of their lives, wishing they had never seen the place, and cursing those that first persuaded them to go there. The fear seized them in the back, but rubbing them with oil of turpentine gave them relief." Since then, custom has altogether banished fear from the minds of the lighthouse-keepers. The men became so attached to their home, that Mr. Smeaton mentions the case of one of them who was even accustomed to give up to his companions his turn for going on shore!

"The Eddystone in sight!" sung out from the maintop.
Homeward-bound ships, from far-off ports, no longer
avoid the dreaded rock, but eagerly run for its light
as the harbinger of safety. It might even seem as if
Providence had placed the reef so far out at sea as the
foundation for a beacon such as this, leaving it to man's
skill and labour to finish His work. On entering the
English Channel from the west and the south, the cau-
tious navigator feels his way by early soundings on the
great bank which extends from the Channel into the
Atlantic, and these are repeated at fixed intervals until
land is in sight. Every fathom nearer shore increases
a ship's risks, especially in nights when, to use the
seaman's phrase, it is "as dark as a pocket." The men
are on the look-out, peering anxiously into the dark,
straining the eye to catch the glimmer of a light, and
when it is known that "the Eddystone is in sight!"
a thrill runs through the ship, which can only be
appreciated by those who have felt or witnessed it after
long months of weary voyaging. Its gleam across
the waters has thus been a source of joy and given a
sense of deep relief to thousands; for the beaming of
a clear light from one known and fixed spot is infallible
in its truthfulness, and a safer guide for the seaman than
the bearings of many hazy and ill-defined headlands.

By means of similar lights, of different arrangements
and of various colours, fixed and revolving, erected upon
rocks, islands, and headlands, the British Channel is
now lit up along its whole extent, and is as safe
to navigate in the darkest night as in the brightest sun-
shine. The chief danger is from fogs, which alike hide
the lights by night and the land by day. Some of the
homeward-bound ships entering the Channel from North
American ports first make the St. Agnes Light, on the
Scilly Isles, revolving once in a minute, at a height of
138 feet above high water. But most Atlantic ships
keep further south, in consequence of the nature of the

soundings about the Scilly Isles; and hence they oftener make the Lizard Lights first, which are visible about twenty miles off.   These are two in number, standing on the bold headland forming the most southerly point of the English coast, against which the sea beats with tremendous fury in south-westerly gales.   From this point the coast retires, and in the bend lie Falmouth (with a revolving light on St. Anthony's Point), Fowey, the Looes, and Plymouth Sound and Harbour; the coast-line again trending southward until it juts out into the sea in the bold craggy bluffs of Bolt Head and Start Point, on the last of which is another house with two lights,—one revolving, for the Channel, and another, fixed, to direct vessels inshore clear of the. Skerries shoal.   But between the Lizard and Start Point, which form the two extremities of this bend in the land of Cornwall and Devonshire, there lies the Eddystone Rock and Lighthouse, standing fourteen miles out from the shore, almost directly in front of Plymouth Sound and in the line of coasting vessels steaming or beating up Channel.   From this point it gradually con- tracts, and the way becomes lighted on both sides to the Downs.   On the south are seen the three Casquet Lights on the Jersey side; and on the north the two fixed lights on Portland Bill.   The next is St. Catherine's, a brilliant fixed light on the extreme south point of the Isle of Wight.   Next are the lights exhibited at different heights on the Nab, and then the single fixed light exhi- bited on the Owers vessel.   Beachy Head, on the same line, exhibits a powerful revolving light 285 feet above high water, its interval of greatest brilliancy occurring every two minutes.   Then comes Dungeness, exhibiting a fixed red light of great power, situated at the extre- mity of the low point of Dungeness Beach.   Next are seen Folkestone, and then Dover, harbour lights; whilst on the south are the flash light, recently stationed on the Varne Bank; and, further up Channel, on the French

coast, is seen the brilliant revolving light on Cape
Grisnez. The Channel is passed with the two South
Foreland Lights, one higher than the other, on the left;
and the Downs are entered with the South Sandhead
floating light on the right; and when the Gull and the
North Sandhead floating lights have been passed on
the one hand, and the North Foreland on the other,
then the Tongue, the Prince's Channel, and the Girdler,
are passed. The Nore Light comes next in sight;
and from thence it is as easy for the navigator to pilot
his ship up the Thames as for a foot-passenger to thread
his way along the streets of London. Such, in a few
words, is the admirable manner in which our coasts are
lighted up for the guidance of the mariner, and such are
among the benefits to navigation which have followed
close upon Smeaton's great enterprise—the building of
the Eddystone Lighthouse.

THE LIGHT AT THE NORE. [By R. P. Leitch]

CHAPTER V.

MR. SMEATON'S EXTENSIVE EMPLOYMENT AS A CIVIL ENGINEER.

THE completion of the Eddystone Lighthouse was regarded as a matter of much interest, and it excited so
eager a curiosity on the part of the public, that for
some time Mr. Smeaton's rooms at Gray's Inn were the
resort of numerous visitors, who called there for the
purpose of inspecting the model of his extraordinary
building.   This at length so broke in upon his time, that
he found it necessary to depute his wife to attend to these
curious persons and explain to them the details of the
model.   But it does not appear that his success led to his
extensive employment on engineering works for several
years, otherwise it is not probable that we should have
found him, in 1764, offering himself a candidate for the
vacant office of receiver for the Derwentwater Estates,
to which he was appointed about the end of that year.
There was as yet, indeed, but small demand for constructive skill.   The roads were still in a very bad state, bridges
were much wanted in most districts, and little had been
done to provide harbour accommodation beyond what
nature had effected ; but the country was too poor or too
spiritless to undertake their improvement on any comprehensive scale.   The industrial enterprise of England
had not yet begun ; and the country was content to jog
along in its old paths, displaying its energies principally
in warfare by land and sea.   The victory of Wolfe on
the heights of Abraham occurred in the same year that
Smeaton completed his lighthouse on the Eddystone, and
doubtless excited a far more general interest.   It is true
the trade and commerce of the country were making

progress,[1] though they had to labour under heavy imposts and serious restrictions. The public expenditure was great, provisions were dear in proportion to wages, and food-riots were frequent. Under these circumstances domestic improvements, involving any unusual outlay, were of a very limited character. When Smeaton was called upon to examine an undrained district, or a dangerous and inaccessible harbour, or a decaying bridge, he had little difficulty in advising what was best to be done; and his reports were searching, explicit, and almost exhaustive. But then arose the invariable impediment. The requisite improvements could not be executed without money, and money was scarce and could not be raised. Hence the greater number of his reports, though containing much excellent and carefully-considered advice, fell dead upon the minds of those to whom they were addressed; and no action was taken to carry them into effect until the country had become richer, and a new race of capitalists, engineers, and contractors had sprung into existence.

One of the earliest subjects on which Mr. Smeaton was consulted, was the opening up of river navigations. In 1760 he reported to the magistrates of Dumfries as to the improvement of the Nith; but his advice—to form a navigable canal rather than deepen and straighten the river at a much greater cost—was not carried out for want of funds. He was also consulted as to the lockage of the Wear, the opening up of the navigation of the Chelmer to Chelmsford, of the Don above Doncaster, of the Devon in Clackmannanshire from Melloch Foot to the Forth, of the Tetney Haven navigation near Louth, and the improvement of the river Lea, which has been a fertile source of contention amongst

---

[1] It may, however, be questioned whether the trade of England did make progress during the twelve years ending 1762; for we find that, although the value of the cargoes exported increased about a million sterling during that period, the quantity exported was less by 60,000 tons.—See 'Chalmers's Estimate,' p. 131.

engineers down even to our own day; but it does not appear that any works of importance followed the elaborate advice which he gave on those subjects. The first large engineering undertaking which he conducted was in his own county, where he was employed in making extensive repairs of the dams and locks on the river Calder in Yorkshire; and he carried out many important improvements in that navigation, the planning of which required much skill and judgment, in consequence of the rapid floods which swept down in rainy seasons from Blackstone Edge. At the same time he was consulted as to the Aire navigation from Leeds to its junction with the Ouse, which he succeeded in greatly improving.

Another subject on which he was early and often consulted was the recovery of the flooded lands in the Lincoln Fens, and in the low-lying lands near Doncaster and Hull, in Yorkshire. The river Witham, between Lincoln and Boston, was still a source of constant grief and loss to the farmers along its banks. It had become choked up by neglect, so that not only had the navigation of the river become almost lost, but a large extent of otherwise valuable land was constantly laid under water. In reporting on this subject in 1761, Mr. Smeaton was associated with Mr. John Grundy and Mr. Langley Edwards; and the result of their joint examination was an elaborate report, accompanied by plans, in which they clearly pointed out the causes of the existing evils and the best mode of remedying them. For the purpose of improving the outfall, they recommended the cutting of an entirely new river, about twelve and a half miles in length, from a place called Chapel Hill to a little above Boston. They also at the same time recommended a plan for the drainage of Wildmore and West Fens by a new cut and sluice in place of the old Anthony's Gout, with sundry other improvements which they set forth in detail. But the total estimated cost being upwards of 40,000l., which was then considered a " mint of money " for a comparatively poor county to raise, the recommendation of the consulting engineers produced no result; and the greater part of the lands remained drowned until they were effectually cleared of their surplus water by Mr. Rennie, about half a century later. Mr. Smeaton was also consulted, in 1762, about the improvement of the Fossdyke, an old cut joining the Trent and the Witham, which had been allowed to fall into decay; but only a few pottering improvements were made, in lieu of the thorough measure of general drainage which he so strongly recommended. After the lapse of twenty years Mr. Smeaton was again called in, and further advised the proprietors on the subject; but although he then submitted a much more limited scheme, it was still beyond the capability of the county to undertake it.

At a still later period he was consulted as to the drainage of the North Level of the Fens, and the improved outfall of the river Nene at Wisbeach. In his report on this subject, he went at great length into the probable causes of the flooding of the fens, and from these he reasoned out the improvements necessary for their effectual remedy. The principal measure which he proposed was, to build a powerful outfall sluice upon the mouth of the Nene. In this report he brought the observations which he had made while on his journey through the Low Countries to bear upon the case; and he argued that, as the outfall channels at Middlesburgh and Ostend were kept wholly open by sluices, the same method would equally apply at Wisbeach. But, like his predecessor Vermuyden, Mr. Smeaton did not seem sufficiently to have taken into account the different circumstances of the two tracts of country; and it is perhaps fortunate that his plans were not carried out, as subsequent experience has shown that, if executed, they would most probably have proved a failure.

Considerable success, however, attended his operations in improving the drainage of the Isle of Axholme, originally executed by Vermuyden. The lower lying lands in the district had fallen into a wretched state through neglect of the river outfalls; and when Smeaton was consulted as to a remedy, he advised the diversion of the old river Torne, which was carried out, and it was also widened and deepened. The result was on the whole highly satisfactory; though, many years after, we find Mr. Rennie describing the drainage as still very imperfect, and urgently demanding an effectual remedy. It would occupy too much space to detail the works of a similar kind on which Mr. Smeaton was consulted; but we may content ourselves with merely mentioning the more important, which were these : the drainage of the lands adjacent to the river Went, in Yorkshire; the Earl of Kinnoul's lands lying along the Almond and the Tay, in Perthshire; the Adling Fleet Level, at the junction of the Ouse and the Trent; Hotham Carrs, near Market Weighton; the Lewes Laughton Level, in Sussex; the Potterick Carr Fen, near Doncaster; the Torksey Bridge Fen, near Gainsborough; and the Holderness Level, near Hull. These works, though of a highly useful character, possess but small interest in a narrative, and we therefore proceed to describe the undertakings of a different character on which our engineer was about the same time employed.

Having fully proved his mastery of the art of construction, and his skill in overcoming the difficulties arising from insecure foundations, by his erection of the Eddystone Lighthouse, he was frequently called upon for advice as to the repairs of old bridges, as well as the erection of new ones. Thus, in 1762, we find him consulted as to the repairs of Bristol Old Bridge; and in the following year he was called upon by the Corporation of London to advise them as to the best means of improving, widening, and enlarging Old London Bridge. Although considerable alterations and improvements had been made in it, the structure was in a very rickety state and a source

of constantly recurring alarm to the public. When
Labelye's New Westminster Bridge was opened for traffic
in 1749, the defects of the old structure became more
apparent than ever. The Corporation even went so far
as to entertain a project for rebuilding it. The city surveyor, however, after examining the foundations of the
piers in 1754, declared them still to be good, and capable
of lasting for ages! His report relieved the public
anxiety for a time, and the old patching process went on
as before. The bridge was still overhung with houses
on either side, and the roadway between them was very
narrow and dark. Labelye's opinion was then taken as

OLD LONDON BRIDGE BEFORE THE ALTERATION OF 1758.

[After the Painting by Samuel Scott.]

to the improvement of the structure, and he recommended
the removal of the starlings, which so blocked up the
waterway as to cause a fall of nearly five feet during
the greater part of every tide. He also advised the removal of some of the piers, as had been recommended
by Sir Christopher Wren, and throwing several of the
arches together. The discussions of the Common Council,
however, ended in the proposal to erect a new bridge
at Blackfriars, and the removal at the same time of the
houses from the old bridge, both of which measures

were eventually carried out. The great middle pier was also removed, and the two adjoining locks were thrown into one by turning a new arch, which occupied the whole space.

OLD LONDON BRIDGE AFTER THE REMOVAL OF THE HOUSES.

[By E. M. Wimperis]

It was now found, however, that the increased scour of the water passing under the new archway placed the adjoining piers in great peril, by washing away the bed of the river under their foundations. The apprehensions of danger were such that but few persons would pass either over or under the bridge, and the Corporation becoming alarmed, at this juncture sent in all haste for Mr. Smeaton. He was then living at his house at Austhorpe, from whence he was summoned by express to town. On his arrival, he proceeded to survey the bridge and examine the foundations which were giving way. His advice to the Corporation was, to buy back immediately the stones of the City gates, which had recently been taken down and the materials sold, and throw them into the river outside the starlings, for the purpose of protecting them against the scour of the river. Another object of this measure, as explained in Smeaton's reports, was to restore the old dam by again raising a barrier of stones across the water-way, and thus increase the head of the current under the other arches, which was necessary for the purpose of driving the wheels by means of which a considerable part of the water required for the supply of the City was still raised. Mr. Smeaton's recommendations were adopted as the most advisable course to be pursued under the circumstances; and horses, carts, and barges were at once set to work, and the stones were tumbled into the stream at the base of the tottering piers. By these means the destruction of the foundations was temporarily stayed, and the process of patching up the old bridge went on from time to time for sixty years more, until it was at length effectually remedied by the erection of the present structure.

In connection with the works at Old London Bridge, Mr. Smeaton also furnished a design for a new pumping-engine, which was placed in the fifth arch, and worked by the rise and fall of the tide. Before the invention of the steam-engine, this was an economical though an irregular method of obtaining motive power. The same tides that lifted great ships up the river and let them down again twice in each day, then drove pumping-engines and even flour-mills—the driving-wheels turning one way as the tide rose and another as it fell.[1] This power was, however, shortly superseded by the still more economical power of steam: for the steam-engine, though involving a considerable expenditure of coal, proved cheaper in the end, because it was so much more certain, regular, and expeditious than the natural power of the tides.

The bridges erected after Mr. Smeaton's original designs, were those of Perth, Coldstream, and Banff; the only one which he erected in England being at Hexham, in Northumberland, which proved a failure. He was consulted about the new bridge at Perth as early as the year 1763, when he visited the place, fixed upon the best site for the structure, and afterwards furnished the

---

[1] 'Encyclopedia Metropolitana,' vol. vii., p. 139.

design which was carried into effect. The river Tay being subject to sudden floods—in one of which a former bridge had been swept away—it was necessary to take every precaution with the foundations, which were got in by means of coffer-dams. That is, a row of piles was driven into the bed of the river, on which a quantity of " gravel and even mould earth mixed together " was thrown in all round the piles, with a view to render the enclosed space impervious to water. Pumping power was then applied, and the bed of the river laid dry within the coffer-dam thus formed, after which the gravel or clay was dug out to a proper depth, until a solid foundation was secured for the piers. Piles were driven into the earth under the intended foundation-frame, and the building proceeded upward in the usual way.[1] The bridge is a handsome structure, consisting of seven principal arches, and is about 900 feet in length, including the approaches. It was completed and opened for traffic in 1772, and has proved of great service to the locality.

Smeaton's employment at Perth on this occasion introduced him to a considerable amount of engineering business in the North. He was consulted at Edinburgh respecting the improved supply of water for that city, and at Glasgow about the security of its old bridge. But the most important work on which he was employed in Scotland, about this time, was the designing and construction of the Forth and Clyde Canal for connecting the navigation of the eastern and western seas. The success of the Duke of Bridgewater's Canal had directed public attention in all parts of the kingdom to the formation of similar lines of internal communication; and the movement had also extended to Scotland.

[1] It may be worthy of remark that John Gwin, the person recommended by Mr. Smeaton to conduct the trial borings for the foundations, took with him two experienced men from England to conduct the works, stipulating that they should each receive wages at the rate of 14s. a week.

James Watt, then carrying on a small business as a
mathematical instrument maker in Glasgow, had been
employed to survey a "ditch canal," of a very limited
capacity, by a round-about route, through the Perthshire
lochs; but his genius being as yet unrecognised, the
projectors thought it desirable to call in an engineer of
higher standing, and Smeaton was consulted by them in
1764. He had before been employed to examine the
Grand Trunk line, as surveyed by Brindley, and his
report on the subject was regarded as a very able one.
Brindley was also advised with respecting the Forth and
Clyde scheme, but his time was so much occupied by the
projects which he was engaged in carrying out in the
western counties of England, that he could not undertake the working survey; and it was accordingly placed
in the hands of Mr. Smeaton. As early as the year
1764, we find him reporting upon the several schemes
which had been proposed for connecting the Forth with
the Clyde, and advocating the plan which in his judgment was the best calculated to carry out the intentions
of the projectors. He declared himself strongly in
favour of forming the most direct line across the country
between the two Friths, of such a capacity as to accommodate vessels of large burden. Lord Dundas, the
leading promoter of the scheme, adopting the view put
forward by Mr. Smeaton, took the requisite steps to
obtain an Act authorizing the construction of the Forth
and Clyde Canal, which passed accordingly, and the
works were commenced in 1768. The canal runs almost
parallel with the line of the wall of Antoninus, built by
the Romans to restrain the incursions of the Caledonian
tribes, some vestiges of which are said to be still trace-
able near Port Dundas, the point at which the main
canal joins the Clyde, a few miles below Glasgow. It
is about 38 miles in length, and includes 39 locks, with
a rise of 156 feet from the sea to the summit level. It
was one of the most difficult works of the kind which

had, up to that time, been constructed in the kingdom; the engineer having to encounter rocks and quicksands; the canal in some places passing over deep rivers, in others along embankments more than twenty feet high. It crosses many roads and rivulets, and two rivers, the Luggie and the Kelvin,—the bridge over the latter being 275 feet in length and 68 feet in height. The depth of the canal was 8 feet, and vessels of 19 feet beam and 68 feet keel were capable of easily passing along the navigation between the east and west coasts. Although the total cost of the undertaking was estimated at only about 150,000*l.*, and the important uses of the navigation were unquestionable, the greatest difficulty was experienced in raising the requisite funds; and long before the canal could be opened to the Clyde, the works came to a complete stand-still. Twenty years passed before the money could be raised to finish them, and this was only effected by the aid of a public grant. At length the canal was opened in 1790, having been finished by Mr. Whitworth (one of Brindley's pupils), and the opening of the communication between the eastern and western seas was celebrated with great rejoicings,—the Chairman of the Canal Committee symbolically performing the feat by launching a hogshead of water brought from the Forth into the Clyde.

Mr. Smeaton was next employed to build a bridge across the Tweed at Coldstream. He furnished several designs, and that eventually selected by the trustees was executed under his superintendence. It consisted of five principal arches of the segment of a circle, the centre one being 60 feet 8 inches from pier to pier; the two next, 60 feet 5 inches; and the two land or side arches, 58 feet. The design presents no features worthy of special notice, nor was any unusual difficulty experienced in getting in the foundations. The piers were founded on piles driven deep into the bottom of the river; and the building, where beneath the level of the stream, was carried on, as at Perth, within coffer-dams. To give additional protection to the piers during winter time, when heavy floods sweep down the valley of the Tweed, they were surrounded by strong sheet-piling,[1] as well as by rubble slopes pointing up stream. The bridge was finished at a total cost of about 6000*l.*, and was opened for carriage traffic in October, 1766, having been rather more than three years in building.

COLDSTREAM BRIDGE.   [By E. M. Wimperis, after a drawing by J S. Smiles.]

Whilst engaged on his engineering business in Scotland, Mr. Smeaton formed the acquaintance of Dr. Roebuck, the enterprising but eventually unfortunate projector of the Carron Iron Works near Falkirk. That gentleman was one of the first who attempted to develope the iron trade of Scotland, since become so important. He was then engaged in the double task of carrying on

---

[1] Sheet-piling consists of a row of timbers driven firmly side by side into the earth, and is used for the protection of foundation-walls or piers from the effects of water. Cast-iron is now employed in many cases for the same purpose, instead of timber.

iron works at Carron and working coal mines at Borrow-stonness. Dr. Roebuck was a man full of expedients, and possessed an uncommon knowledge of mechanics for his time. Smeaton was a kindred spirit, whom he very early sought out and invited to his house at Kinneil, near Borrowstonness, for the purpose of consulting him as to the pumping machinery of his mines, and the vari-ous arrangements of his iron manufactory at Carron. Dr. Roebuck was one of the first to employ coal in iron-smelting on a large scale, and for that purpose he required the aid of the most powerful blowing apparatus that could be procured. Mr. Smeaton succeeded in con-triving and fixing for him, about the year 1768, a highly effective machine of this kind, driven by a water-wheel.[1] He also supplied the same Company with a design for a double-boring mill for cylinders and guns,—the manu-facture of carronades, or " smashers," having been a very early branch of the business at the Carron Works. At the same time he pointed out how the water power of the little river Carron might be so concentrated and increased by damming, as to work the apparatus he contrived with the greatest possible effect. Smeaton was afterwards repeatedly consulted by the Carron Com-pany as to the several manufactures carried on at the works—such as the making of shot-moulds, the best form of slide-carriages for guns, the construction of furnaces, and such like matters, of which the plans and descriptive details are to be found in his published reports.[2]

Another fine bridge, of which Smeaton furnished the design in the year 1772, was that subsequently erected over the river Deveron, near the town of Banff in Scot-land. It is of seven arches, segments of circles, and is of the total length of 410 feet between the abutments, with

---

[1] The author endeavoured to ob-tain an inspection of this long-disused apparatus, for the purposes of this work, in the autumn of 1858 ; but the reply of the manager was, " Na,

na, it canna be allooed—we canna be fashed wi' straingers here."

[2] ' Reports of the late John Smea-ton, F.R.S.' In 3 vols. London, 1812. Vol. i., pp. 359-412.

a roadway twenty feet wide over all. The design is similar in most respects to those of the bridges previously erected by the same engineer at Perth and Coldstream; and the beauty of its situation, in the immediate vicinity of Duff House, the mansion of the Earl of Fife, and its noble surrounding grounds, renders it an object of even greater pictorial interest.[1]

The only peculiarity to be noted in the designs of Smeaton's bridges, is the circular perforations left in the spandrels of the arches, somewhat after the method adopted by Edwards at Pont-y-Pridd, and in several Continental bridges. This had the effect of lightening the weight which pressed upon the piers and their foundations, and was doubtless an advantage. He also invariably adopted segmental or elliptical in preference to semi-circular arches, probably because of the less cost of bridges after the former design. Much ability was displayed by our engineer in the designing of his centres, which have been much admired for their strength as well as economy of material.

Smeaton was much less successful in the construction of his only English bridge than he was with his Scotch ones. He was called upon to furnish the design for a structure across the Tyne at Hexham, in 1777, and a very handsome bridge of nine arches was erected after it under the superintendence of Mr. Pickernell, the resident engineer. It had scarcely been finished ere a subsidence in the foundations of one of the piers took place, which was attempted to be remedied by sheet-piling and filling up the cavities in the river's bed with rough rubble-stones. But it appeared that the foundations had been imperfectly laid from the beginning. In the spring of 1782 a violent spate or flood swept down the Tyne, and in the course of a few hours Smeaton's beautiful Hexham Bridge lay a wreck in the

---

[1] See engraving at p. 2.

bottom of the river. Writing to Pickernell, he said,—
" All our honours are now in the dust ! It cannot now
be said that in the course of thirty years' practice, and
engaged in some of the most difficult enterprises, not
one of Smeaton's works has failed ! Hexham Bridge is
a melancholy instance to the contrary." Thus the same
engineer who had founded a lighthouse far out at sea, so
firmly as to bid defiance to the utmost fury of the waves,
was baffled by an inland stream. " The news came to
me," he says, " like a thunderbolt, as it was a stroke I
least expected, and even yet can scarcely form a practical belief as to its reality. There is, however, one
consolation that attends this great misfortune, and that
is, that I cannot see that anybody is really to blame, or
that anybody is blamed ; as we all did our best, according
to what appeared ; and all the experience I have gained
is, not to attempt to build a bridge upon a gravel bottom
in a river subject to such violent rapidity." The fault
committed seems to have been, that Smeaton was satisfied with setting his piers upon a crust of gravel slightly
beneath the bottom level of the river ; and that the
increased scour of the stream under the arches, caused
by the contraction of the water-way, had washed away
the bottom, and thus undermined the work. But the
founding of piers in deep rivers was as yet very imperfectly understood ; and the art was not brought to its
perfection until the time of Rennie, who went down
through the bed of the river, far beneath all possible
scour, until he had reached a solid foundation, which he
also piled, and on that secure basis he planted the strong
masonry of his piers.

    Among his various works, Smeaton was also employed
in the designing of harbours. With the exception, however, of Ramsgate, these were for the most part confined
to the improvement of the existing accommodation. At
St. Ives, in Cornwall, where he formed his first harbour,
in 1766, nature had provided a convenient haven

enclosed in a bay between two headlands, one of which was formed by " the Island," and the other by Penolver Point, as shown in the annexed plan. It was thus well

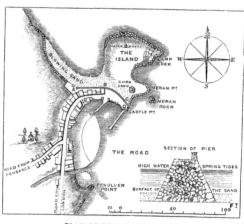

PLAN OF ST. IVES HARBOUR.

protected from the north, west, and south, and from the prevalent storms along that coast, which mostly blow from a south-westerly direction. All that was wanted to give shelter for shipping from the remaining quarters, the east and north-east, was the provision of a pier running nearly south from Castle Point. The works were carried out after Smeaton's design; and as the port is the seat of considerable trade, arising from the pilchard fishery and the mining operations of the country inland, the facilities thereby provided for shipping, and the protection to navigation along that coast, proved of great advantage to the district.

Our engineer was also consulted respecting numerous other harbours : Whitehaven, Workington, and Bristol, on the west coast; Christchurch, Rye, and Dover, on the south ; and Yarmouth, Lynn, Scarborough, and Sunderland, on the east ; but in nearly every case want of money prevented the improvements suggested by him from being fully carried out. This was pre-eminently the case at Bristol, where the merchants gave him an unanimous " vote of thanks " for his report and plan for keeping the ships at the quay constantly afloat by docking the river, and also for enlarging the harbour by a new canal through Cannon's Marsh. But nothing was

ST. IVES HARBOUR. [By E. M. Wimperis.]

done ; the Bristol vessels continued to lie upon the mud
and get "hogged," and a considerable time elapsed before the commercial interest became alive to the necessity of improving the conveniences of the harbour. This
was eventually accomplished by William Jessop, a pupil
of Smeaton's ; but not until Liverpool had taken the
lead of Bristol among the western ports, in respect of
the convenient accommodation which it provided for
shipping, as well as its more ready connection with the
best markets.

The principal harbour works actually executed by
Mr. Smeaton were those of Ramsgate. The proximity
of this harbour to the Downs and the mouth of the
Thames rendered it of considerable importance ; and its
improvement for purposes of trade, as well as for the
shelter of distressed vessels in stormy weather, was long
regarded as a matter of almost national importance.
The neighbourhood of Sandwich was first proposed for
a harbour of refuge as early as the reign of Queen

Elizabeth, and the subject was revived in succeeding reigns. In 1737, Labelye, the architect of Westminster Bridge, was called upon to investigate the subject; and ten years later, a committee of the House of Commons, after taking full evidence and obtaining every information, reported that " a safe and commodious harbour may be made into the Downs near Sandown Castle, fit for the reception and security of large merchantmen and ships of war, which would also be of great advantage to

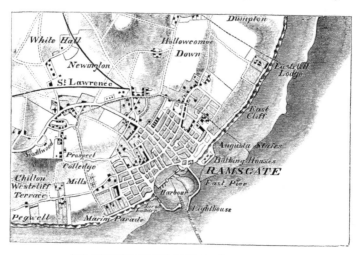

MAP OF RAMSGATE AND HARBOUR. [Ordnance Survey.]

the naval power of Great Britain." The estimated cost of the proposed harbour was, however, considered too formidable, although it was under half a million; and the project lay dormant until a violent storm occurred in the Downs in 1748, by which a great number of ships were forced from their anchors and driven on shore. Several vessels, however, found safety in the little haven at Ramsgate, which was then only used by fishermen, the whole extent of its harbour accommodation consisting merely of a rough rubble pier. This circumstance seems to have had the effect of directing attention to Ramsgate as the proper place for a har-

RAMSGATE HARBOUR.

[By Percival Skelton, after his original drawing.]

bour of refuge for vessels in distress from bad weather in the Downs. The Legislature was petitioned on the subject, and an Act was passed in 1749, enabling a harbour to be constructed at Ramsgate. A large number of plans were sent in, from which the Trustees made selections, adopting the east stone pier of one amateur, and the west wooden one of another. The plan of the east pier was made by one of the trustees, and that of the west pier by a captain resident at Margate. Whilst the works were in progress, the Harbour Trustees proposed to reduce its area, and consequently the extent of accommodation for shipping. On this decision becoming known, the shipping interest memorialised Parliament on the subject, in 1755, and an inspection of the works was ordered, during which they were entirely suspended, and remained in that state during the next six years. Differences arose between the officers appointed by the Government and the Harbour Trustees as to the plan most proper to be carried out. At length the trustees gave way, and that part of the works which had been executed with a view to the contraction of the harbour was taken up, and the piers proceeded in the direction originally intended. It was, however, a matter of great vexation to observe that even while the construction of the piers was in progress, and especially when they were carried out so far as to bend towards each other, with the object of affording the requisite protection to the shipping within them, large quantities of sand and silt began to collect in the harbour, threatening to choke it up altogether. This accumulation of silt went on notwithstanding every effort made to remove it.

At this juncture, in 1774, Mr. Smeaton was called upon to advise the Harbour Board as to the steps most proper to be taken in the matter. After a careful examination, he ascertained that no less than 268,700 cubic yards of sand and mud had already silted up, every tide bringing in a fresh quantity and depositing it in the still water of the harbour, which was without any natural scour to carry it away. He accordingly recommended a plan for accomplishing this object by means of sluices, supplied by an artificial backwater. He pointed out that Ramsgate Harbour, having a sound bottom of chalk, was well adapted for the execution of this scheme, and that provided the silt could be thus scoured out, the tide, running cross-ways upon the harbour's mouth, would easily carry it away. Mr. Smeaton accordingly accompanied his report with a plan showing the details of his design. He proposed to enclose two spaces of four acres each, and to provide them with nine draw-gates : four upon the westernmost, and five upon the easternmost basin, the whole being pointed in three different directions : two towards the curve of the western pier, four towards the harbour's mouth, and three towards the curve in the eastern pier. To give the sluices all possible effect, he proposed to construct a caisson, shaped something like the pier of a bridge, which, being floated to its place, and then sunk, might be used to direct the current to the right hand or the left according to circumstances. Several experiments having been made with a lighter filled with water and scuttled when the tide was out, the efficacy of the scouring process was thus ascertained. It was finally resolved to adopt the general features of Mr. Smeaton's plan, though it was not carried out in the exact manner designed by him. But it was shortly found that the process of sluicing endangered the foundations of the piers. Our engineer was accordingly again called in, when he recommended further improvements, including a new dock, the first stone of which was laid in July, 1784. In the course of the excavations numerous springs were tapped, which broke through the pavement with which the dock had been laid, and Portland blocks were then substituted ; but this not proving effectual, the engineer was again sent for, and from that time forward the execution of the further works in connection with the harbour was placed entirely in his hands. The dock was rebuilt, a timber floor laid in the most complete manner throughout, and an additional thickness given to the walls ; the east pier was rebuilt of stone, and carried out into deep water to a further extent of 350 feet. In carrying out the elongated pier, Smeaton first employed the diving-bell in building the foundations, making use of a square iron chest weighing about half a ton. It was 4 feet 6 inches in height and length,

and 3 feet wide, affording room for two men to work in
it; and they were provided with a constant supply of
fresh air by means of a forcing pump placed in a
boat which floated above them.   The works, when
finished, were found to answer remarkably well.   The
harbour included an area of forty-two acres, the piers
extending 1310 feet into the sea, the opening between
the pier-heads being 200 feet in width.   The inner basin
is used as a wet dock, and also contains a dry dock
for the repair of ships.   With its many defects, and
its limited depth, the harbour is nevertheless the best
upon that coast, and in stormy weather affords a refuge
to vessels of considerable draught of water that run for
protection there at tide time.

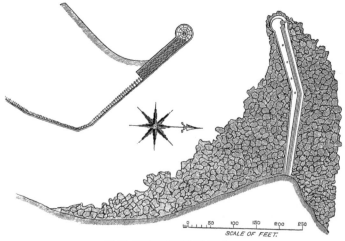

SCALE OF FEET.

PLAN OF EYEMOUTH HARBOUR.

Besides the harbours constructed or improved by
him at different points of the English coast, Smeaton
was frequently employed during his Scotch journeys in
inspecting the northern harbours and advising the local
authorities as to means of increasing their security
and accommodation.   Thus the harbour at Aberdeen
was altered after his plans in 1770, and a greater
depth of water was secured over the bar and in the

channel of the river Dee, by the erection of the old North Pier, and other additions which served their purpose until the enlarged trade of the town required the more ample accommodation hereafter to be described in the Life of Telford. He also inspected and reported on the harbours of Dundee and Dunbar, then of very limited capacity, and several improvements of a minor character were carried out by his advice. The small harbours of Portpatrick on the west, and

EYEMOUTH HARBOUR. [By R. P. Leitch.]

Eyemouth on the east coast, were constructed after his plans; and in his report on Scarborough Pier, dated August, 1781, he states that they had " given entire satisfaction." Both of these harbours were in a great measure formed by nature, and the improvement of them demanded comparatively small skill on the part of the engineer. He had merely to follow the direction of the rocks, which provided a natural foundation for his piers at both places. Of his little harbour at Eyemouth he was somewhat proud, as it was one of the

first he constructed, and very effectually answered its
purpose at a comparatively small outlay of money. It
lies at the corner of a bay, opposite St. Abb's Head, on
the coast of Berwickshire, and is almost landlocked,
excepting from the north. Smeaton accordingly carried
his north pier into deep water for the purpose of protecting the harbour's mouth from that quarter, as well
as enlarging the accommodation of the haven. The
harbour was thus rendered perfectly safe in all winds,
and proved of great convenience and safety to the
fishing-craft by which it is chiefly frequented.

It would occupy too much space to refer in detail to
the various other public works on which Mr. Smeaton
was employed in the course of his professional career.
There was scarcely a crazy old bridge in the kingdom
on which he was not called upon to report. He was
consulted respecting canal projects almost until the close
of his life : amongst others, on the improvement of the
Birmingham Canal, the Ure Canal, the Dublin Grand
Canal, and various other schemes of the same sort. He
was the principal authority on lighthouses, and, amongst
others, he erected two on Spurn Point, at the entrance
to the Humber, between the years 1771-6, which were
lighted by coal-fires down to a comparatively recent
period. The Government consulted him respecting their
dockyards at Plymouth and Portsmouth. Water companies consulted him as to water supply, and landowners
and coalowners as to the best method of draining their
lands or working their mines. He was called upon to
design many weirs, sluices, and dams, and his dam on
the Coquet, north of Newcastle, was considered one of
the most complete works of its kind.

He was ready to supply a design of any new machine, from a ship's pump or a fire-bucket to a turning-
lathe or a steam-engine. His machinery was neatly
designed, and he was very particular as to its careful
execution and finish   The water-pumping engine which
he erected for Lord Irwin, at Temple Newsam, near his

own house at Austhorpe, to pump the water for the supply of the mansion, is an admirable piece of workmanship, and continues at this day in good working condition. His advice was especially sought on subjects connected with mill-work, water-pumping, and engineering of every description—flour-mills and powder-mills, wind-mills and water-mills, fulling-mills and flint-mills, blade-mills and forge hammer-mills. From a list left by him in his own handwriting, it appears that he designed and erected forty-three water-mills of various kinds, besides numerous wind-mills. Water-power was then used for nearly all purposes for which steam is now applied: such as grinding flour, sawing wood, boring and hammering iron, fulling cloth, rolling copper, and driving all kinds of machinery. Smeaton also bestowed much patient study on the development of the infant powers of the steam-engine. In order to investigate the subject by experiment, he expressly erected a model engine, after Newcomen's principle, near his house at Austhorpe; and by improving it in all its arrangements he succeeded in rendering it as complete as it was possible to make it; his Chacewater engine of 150-horse power being regarded as the finest and most powerful of its kind which had until then been erected. In this field of invention, however, he found himself distanced by Watt, the superior merit of whose condensing-engine —notwithstanding the time and labour Smeaton had bestowed on the improvement of Newcomen's—he generously acknowledged, frankly admitting, after he had inspected Watt's invention, that "the old engine, even when made to do its best, was now driven from every place where fuel could be considered of any value." The fame of Smeaton, therefore, does not rest upon his improvements in this machine, though what he accomplished in bringing out the full powers of Newcomen's engine cannot fail to elicit the admiration of the practical mechanic.

SMEATON'S HOUSE AT AUSTHORPE.

[By Percival Skelton, after an original Drawing by T. Sutcliffe, Leeds.]

CHAPTER VI.

SMEATON'S PRIVATE LIFE—DEATH AND CHARACTER.

WHILST Mr. Smeaton was thus extensively employed as an engineer throughout the three kingdoms, his home continued to be at Austhorpe, near Leeds, where he had been born. The mechanical experiments of his boyhood had been conducted there, as were also those of his maturer years. His father had allowed him the privilege of a workshop in an outhouse, which he long continued to enjoy; after which, when the house had become his settled home, he erected a shop, study, and observatory, all in one, for his own special use. The building was in the form of a square tower, four stories high, standing apart from his dwelling, on the opposite side of the yard, as represented in the above engraving. The ground floor contained his forge; the first floor his lathe; the second his models; the third was his drawing-room and study; and the fourth was a sort of lumber-room and attic. From the little turreted staircase on the top, a door opened on to the leads. A vane was fixed on the summit, which worked the hands of a dial upon the ceiling of his drawing-room, so that by raising his head he could at any moment ascertain precisely which way the wind blew.

When he entered his sanctum, strict orders were given that he was not to be disturbed on any account. No one was permitted to ascend the circular staircase that led to his study. When he heard a footstep below, he would call out and inquire what was wanted. His black-smith, Waddington, was not allowed even to announce himself, but was ordered on such occasions to wait in the lower apartment until Mr. Smeaton came down; and as the smith was equally paid for his time, whether he was sitting there or blowing his forge, it was much the same to him.

When not engaged in drawing plans or writing reports, much of the engineer's time was occupied with astronomical studies and observations. Even in the height of his professional career, and when fully em-ployed, he continued to indulge in this solitary pleasure, and for many years was a regular contributor of papers on astronomical subjects to the Royal Society, of which he was a Fellow.[1] The instruments with which he

---

[1] The following are the papers read by him before the Royal Society, in addition to those previously men-tioned :—' Discourse concerning the Menstrual Parallax, arising from the mutual gravitation of the earth and moon, its influence on the observation of the sun and planets, with a me-thod of observing it ;' read before the Royal Society May 12th, 1768.—' De-scription of a new method of observ-ing the heavenly bodies out of the meridian ;' read May 16th, 1768.—' Observation of a Solar Eclipse, made at the Observatory at Austhorpe ;' read June 4th, 1769.—' A description of a new hygrometer, by Mr. J. Smeaton, F.R.S. ;' read March 21st, 1771.—' An experimental examina-tion of the quantity and proportion of mechanic power necessary to be em-ployed in giving different degrees of

was accustomed to illustrate his papers were of the most
beautiful workmanship, all made by his own hands,
which had by no means lost their cunning.   Indeed, he
was nowhere so happy as in his workshop amongst his
tools, except, it might be, at his own fireside, where he
was all but worshipped.

His contrivances of tools were endless, and he was
perpetually inventing and making new ones.   There
are large quantities of these interesting relics still in
existence in the possession of the son of his blacksmith,
who lives in the neighbourhood.   When the author
lately made inquiry after them, they were found laid in
a heap in an open shed, covered with dirt and rust.
One article, after having been well scrubbed with a
broom, at length displayed the form of a jack-plane, the
tool with which Smeaton himself had worked.   Picked
out from the heap were also found his drill, the bow
formed of a thick piece of cane ; his trace, his T square,
his augers, his gouges, and his engraving tools.   There
was no end of curiously arranged dividers ; pulleys in
large numbers, and of various sizes ; cog-wheels ; brass
hemispheres ; and all manner of measured, drilled,
framed, and jointed brass-work.   His lathe is still in the
possession of Mr. Mathers, engineer, Hunslet ;[i] but many

---

velocity to heavy bodies from a state
of rest;' read April 25th, 1776.—
'New fundamental experiments on
the collision of bodies;' read April
18th, 1782.—'Observations on the
graduation of astronomical instruments;' read November 17th, 1785.—
'Account of an observation of the
right ascension and declination of
Mercury out of the meridian, near his
greatest elongation, September, 1786,
made by Mr. John Smeaton, with an
equatorial micrometer of his own invention and workmanship, accompanied with an investigation of a method of allowing for refraction in such
kind of observations;' read June 27th,
1787.—'Description of an improve-

ment in the application of the quad-
rant of altitude to a celestial globe,
for the resolution of problems de-
pendent on azimuth and altitude;'
read November 20th, 1788.—'Description of a new hygrometer;' read
before the same Society.

[i] The lathe stands on three legs,
which are fastened together in such a
way that they, as well as the rest of
the framework, are still as firm as if
they had been only just made, and
yet the machine has been in use ever
since Smeaton made it.   The fly-
wheel is of dark walnut-wood, and
slightly inclines from the perpendicular, by which the driving-cord is
allowed to be crossed and to play

of the other interesting remains of the great engineer are equally worthy of preservation. To mechanics, there

is a meaning in every one of them. They do not resemble existing tools, but you can see at once that each was made for a reason; and one can almost detect what the contriver was thinking about when he made them so different from those we are accustomed to see. Even in the most trifling matters, such as the kind of wood or metal used, the direction of the fibre of the wood, and such like, each detail has been carefully studied.

SMEATON'S LATHE.

Much even of the household furniture seems to have been employed in their fabrication, possibly to the occasional amazement of the ladies in Smeaton's house over the way. We are informed that so much "rubbish," as it was termed, was found in that square tower at his death, that a fire was kindled in the yard, and a vast quantity of papers, letters, books, plans, tools, and scraps of all kinds, were remorselessly burnt.

We have said that Smeaton was a born mechanic; and a mechanic he remained to the last. He contrived and constructed for the pure love of it. Among the traditions which survive about him at Whitkirk, is this, that when new gates were erected at the entrances to Temple Newsam Park, near his house at Austhorpe, he volunteered to supply the designs, and they were made and hung after his plans. The people of the neighbourhood, however, think his most wonderful work

---

with a greater amount of friction on the other wheels. The metal-work is of brass, iron, and steel, all nicely finished; and the whole is very compact, curious, and thoroughly Smeaton-like.

is the ingenious hydraulic ram, by means of which the
water is still raised in the grounds of Temple Newsam.
His pursuits in his workshop, and at his desk, were
varied by visits to his blacksmith's shop. One of his
principal objects, on such occasions, was to experiment
upon a boiler,—the lower part copper and the upper
part lead,—which he had fitted up in an adjoining building, for the purpose of ascertaining the evaporative
power of different kinds of fuel, and other points connected with the then little understood question of steam
power. He was on very familiar terms with the smith,
and if he thought him not very handy about a piece of
work he was engaged upon, he would take the tools himself and point out how it should be done. One of the
maxims which he frequently quoted to his smith was,
" Never let a file come where a hammer can go."

When getting work done in other parts of the country,
if a workman appeared to him unhandy, or at a loss how
to proceed, he would pass him on one side, take up the
tools, and finish the piece of work himself. " You know,
Sir," observed the son of Smeaton's blacksmith, still
living, " workmen didn't know much about drawings
at that time a-day, and so when Mr. Smeaton wanted
any queer-fangled thing making, he'd cut one piece out
o' wood, and say to my father, ' Now, lad, go make me
this.' And so on for ever so many pieces; and then
he'd stick all those pieces o' wood together, and say,
' Now, lad, thou knows how thou made each part, go
mak it now all in a piece.' And I've heard my father
say, 'at he's often been cap't to know how he could tell
so soon when owt ailed it, for before ever he set his foot
at t' bottom of his twisting steps, or before my father
could get sight of his face, if t' iron had been wrong,
thear'd been an angry word o' some sort, but t' varry
next words were, ' Why, my lad, thou s'ud a' made it so
and so : now go mak another.' "

Mr. Smeaton's professional engagements necessarily

called him frequently to London, where he spent part of every year, occupying chambers in Gray's Inn. He had joined his friend Mr. Holmes, in 1771, in the proprietorship of the works for supplying Deptford and Greenwich with water, which also required his presence in town, and he devoted considerable attention to the requisite mechanical arrangements. On the occasion of his visits to London, it was a source of great pleasure to him to attend the meetings of the Royal Society, as well as to cultivate a friendship with the distinguished members of the Royal Society Club.[1] He was also a frequent witness before committees of both Houses of Parliament[2] in support of bills for authorising the construction of bridges, canals, and water-works; and was accustomed on such occasions to give his evidence in a modest, simple, and straightforward manner, which is calculated to win confidence and respect far more than that glib and unscrupulous style which has since become the fashion. Moreover, he was known to be a most conscientious man, and that he would not express an opinion on any subject until he had thoroughly mastered it.

During the time spent by Mr. Smeaton in town, he was accustomed to meet once a week, on Friday evenings, in a sort of club, a few friends of the same calling,—canal-makers, bridge-builders, and others of the

---

[1] James Watt writes: "When I was in London in 1785, I was received very kindly by Mr. Cavendish and Dr. Blagden, and my old friend Smeaton, who has recovered his health, and seems hearty. I dined at a turtle feast with them, and the select Club of the Royal Society; and never was turtle eaten with greater sobriety and temperance, or more good fellowship."—'Rise and Progress of the Royal Society Club.' 1860.

[2] It is stated in a recent work, edited by the learned Recorder of Birmingham, M. D. Hill, Esq., entitled 'Our Exemplars,' that "Smeaton was for several years an active member of Parliament, and many useful bills are the result of his exertions. . . . . His speeches were always heard with attention, and carried conviction to the minds of his auditors." This must, however, be a mistake, as Smeaton was never in Parliament, except for the purpose of giving engineering evidence before committees; and, instead of being eloquent, Mr. Playfair says he was very embarrassed even in his ordinary conversation.

class then beginning to be known by the generic term of Engineers. The place of meeting was the Queen's Head Tavern in Holborn; and after they had come together a few times, the members declared themselves a Society, and kept a register of membership,—free social conversation on matters relating to their business being the object of their meetings. Some personal disagreement, however, occurring, through the offensive behaviour of one of the members, Mr. Smeaton withdrew from the club, which came to an end in 1792. Mr. Holmes says of him, that though of a very kindly and genial nature, he was occasionally abrupt, and, to those who did not know him, apparently harsh in his manner; and that he would sometimes break out hastily when anything was said that did not tally with his ideas, not being disposed to yield upon any point on which he argued until his mind was convinced by sound reasoning.[1]

Mr. Smeaton earned a fair income by the practice of his profession; but he was no worshipper of money. Though he had an insatiable appetite for work, and was occupied in useful pursuits from youth to old age, his pecuniary wants were most moderate. Those were not the days when great fortunes were to be made by engineering; and Mr. Smeaton was satisfied to be paid two guineas for a full day's work. Moreover, he refused new engagements rather than imperfectly perform what he had already undertaken. He also limited his professional employment, that he might be enabled to devote a certain portion of his time to self-improvement and scientific investigation. The maxim which governed his life was, that "the abilities of the individual were a debt due to the common stock of public well-being." This high-minded principle, on which he faithfully acted, kept him free from sordid self-aggrandisement, and he

---

[1] Mr. Holmes's 'Short Narrative,' p. 15.

had no difficulty in resisting the most tempting offers which were made to attract him from his own settled course. When pressed on one occasion to undertake some new business, and the prospect of a lucrative recompense was held out to him, he called in the old woman who took charge of his chambers at Gray's Inn, and pointing to her said, "Her attendance suffices for all my wants." If urgently called by duty, he was ready with his help; but he would not be bought. When the Princess Dashkoff urged him to go to Russia and enter the service of the Empress, she held out to him very tempting promises of reward. But he refused: no money would induce him to leave his home, his friends, and his pursuits in England; and, though not rich, he had enough and to spare. "Sir," exclaimed the Princess, unable to withhold her admiration, "I honour you! You may have your equal in abilities perhaps; but in character you stand alone. The English minister, Sir Robert Walpole, was mistaken, and my Sovereign has the misfortune to find one Man who has *not* his price." [1]

Influenced by the same spirit, Mr. Smeaton, towards the close of his life, believing that he should be rendering a service to his country by publishing an account of the various works in which he had been engaged as an engineer, endeavoured to avoid as much business as he consistently could to devote himself to that work, and eventually determined to retire altogether from the profession; [2] but the only portion that he lived to complete was

---

[1] Letter written by Mrs. Dixon, daughter of the engineer, to the Committee of Civil Engineers, dated 30th October, 1797, relative to the life and character of her deceased father.— Smeaton's 'Reports,' vol. i., p. 28.

[2] A year before his death, Mr. Smeaton formally took leave of the profession in the following circular:—

"Mr. Smeaton begs leave to inform his Friends and the Public in general, that having applied himself for a great number of Years to the business of a Civil Engineer, his wishes are now to dedicate the chief part of his remaining Time to the Description of the several Works performed under his Direction. The Account he lately published of the Building of Eddystone Lighthouse of Stone has been so favourably received, that he is persuaded he cannot be of more service to the Public, or show a greater Sense of his Gratitude, than to continue to employ himself in the way now specified. He therefore flatters himself, that in not yielding to the many applica-

his Narrative of the construction of the Eddystone Light-house.   Indeed, he states that he found the task of de-scribing this work even more difficult than that of erecting it, and he consequently seems to have become inordinately impressed with a sense of the importance of literary com-position.   He very naïvely observes in the Preface : " I am convinced that to write a book tolerably well is not a light or an easy matter; for, as I have proceeded in this work, I have been less and less satisfied with the execu-tion.   In truth, I have found much more difficulty in writing than I did in building, as well as a greater length of time and application of mind to be employed.   I am indeed now older by thirty-five years than I was when I first entered on that enterprise, and therefore my faculties are less active and vigorous; but when I consider that I have been employed full seven years, at every opportu-nity, in forwarding this book, having all the original draughts and materials to go upon, and that the produc-tion of these original materials as well as the building itself were despatched in half that time, I am almost tempted to subscribe to the sentiment adopted by Mr. Pope, that ' Nature's chief masterpiece is writing well.' It is true that I have not been bred to literature, but it is equally true that I was no more bred to mechanics : we must therefore conclude that the same mind has in reality a much greater facility in some subjects than in others."

Smeaton's story of the Eddystone Lighthouse is, how-ever, told in a very effective manner.   It possesses an interest almost dramatic, exhibiting a contest between a strong, skilled, and determined man, and the most tre-mendous forces of nature.   It is truly observed by the late Lord Ellesmere, in his ' Essays on Engineering,' that bloody battles have been won, and campaigns conducted to a successful issue, with less of personal exposure to

tions made to him lately for further Under-takings, but confining himself in future to the Objects above mentioned, and to such occasional Consultations as will not take up

much Time, he shall not incur the Dis-approbation of his Friends.
" Gray's Inn, 6th October, 1791."

physical danger on the part of the commander-in-chief than was constantly encountered by Smeaton during the greater part of those years in which the lighthouse was in course of erection. In all works of danger he himself led the way—was the first to spring upon the rock and the last to leave it; and by his own example he inspired with courage the humble workmen engaged in carrying out his plans, who, like himself, were unaccustomed to the special terrors of the scene.

The portrait prefixed to this volume gives a good representation of Mr. Smeaton's countenance, the expression of which was gentle, yet shrewd. In person he was of a middle stature, broad and strong made, and possessed originally of a vigorous constitution. In his manners he was simple, plain, and unassuming. He had the bluntness and straightforwardness of speech which usually mark the north-countryman, and never acquired that suavity and polish which are more common amongst educated men in our southern districts. He spoke in the dialect of his native county, and was not ashamed to admit it.[1] Yet he mixed in good society when in town, though his diffidence, as well as his reluctance to bestow too much time on social enjoyment, caused him to contract his circle as his professional engagements increased. His daughter has related the anecdote of his meeting on one occasion with the Duke and Duchess of Queensberry, which led to a pleasant intercourse with that family. Mr. Smeaton was walking with his wife in Ranelagh Gardens—the fashionable place of resort at that time—when he observed an elderly lady and gentleman fix their marked attention upon him. At length they came up, and the lady, who proved to be the eccentric Duchess of Queensberry, said

---

[1] In the Preface to his Eddystone Narrative he says : "As I speak and write a provincial language, and was not bred to letters, I am greatly obliged to my friends in the country for perusing and abundantly correcting my manuscript."

to Mr. Smeaton, "Sir, I do not know who you are or what you are; but so strongly do you resemble my poor dear Gay (the poet), that we *must* be acquainted. You shall go home and sup with us; and if the minds of the two men accord, as do the countenances, you will find two cheerful old folks, who can love you well; and I think (or you are a hypocrite) you can as well deserve it." Mr. Smeaton and his wife accepted the invitation, and it proved the commencement of one of his most pleasant London friendships. It happened that the Duke and Duchess had a great love of card-playing, which Smeaton detested. But his good-nature would not permit him to hold aloof when asked to take a hand. He played, however, like a boy, his attention never following the game. On one occasion, when it was Pope Joan, and the stake in "Pope" had accumulated to a considerable sum, it became Mr. Smeaton's turn by the deal to double it. Regardless of his cards, he took up a scrap of paper, made some calculations on it, and laid it on the table. The Duchess eagerly asked what it was. He replied, "Your Grace will recollect that the field in which my house at Austhorpe stands may be about five acres, three roods, and seven perches, which, at thirty years' purchase, will be just my stake; and if your Grace will make a Duke of me, I presume the winner will not dislike my mortgage." The hint thus given in a joke was kindly taken, and from that time they never played but for the merest trifle.

In his own home he was beloved and revered. His wife died in 1784, after which his two daughters kept house for him until his own death. The eldest has left on record a charming picture of his domestic character, which we cannot do better than transcribe:—"Though communicative on most subjects," she says, "and stored with ample and liberal observations on others, of himself he never spoke. In nothing does he seem to have stood more single than in being devoid of that egotism which

more or less affects the world.　It required some address,
even in his family, to draw him into conversation directly
relating to himself, his pursuits, or his success.　Self-
opinion, self-interest, and self-indulgence, seemed alike
tempered in him by a modesty inseparable from merit—
a moderation in pecuniary ambition, a habit of intense
application, and a temperance strict beyond the common
standard. . . . . Devoted to his family with an
affection so lively, a manner at once so cheerful and
serene, that it is impossible to say whether the charm of
conversation, the simplicity of instruction, or the gentle-
ness with which it was conveyed, most endeared his
home—a home in which from infancy we cannot recollect
to have seen a trace of dissatisfaction or a word of asperity
to any one.　Yet with all this he was absolute!　And
it is for casuistry, or education, or rule, to explain his
authority ; it was an authority as impossible to dispute
as to define."

Mrs. Dixon illustrates the benevolence of her father's
character by referring to a painful and trying event
in his life.　Mr. Smeaton had befriended a young
man whom he had formerly employed as a clerk, and
successfully exerted himself to procure for him a situation of trust and responsibility, further becoming bound,
jointly with another gentleman, in a considerable sum.
The young man fell into bad habits : his expenses
outran his income ; he committed a forgery to meet the
deficiency, and he was detected, apprehended, and given
up to justice.　The same post brought Mr. Smeaton the
intelligence of the young man's ruin, the claim for the
amount of the forfeited bond, and the refusal of the other
person to pay the moiety.　Mrs. Smeaton's health being
delicate at the time, her husband suppressed all appearance of emotion ; nor, until all was put in train for settlement, did a word or look betray the exquisite distress
which these painful circumstances had caused him.　He
even exerted himself to save the prisoner's life, in which

he eventually succeeded, and he did all that he afterwards could to soothe the remorse of the wretched youth who had betrayed him.[1]

Of Mr. Smeaton's intellectual powers it would be difficult to speak too highly.   James Watt always mentioned him in terms of sincere admiration, speaking of him as "father Smeaton."   Writing to Sir Joseph Banks, he said : " In justice to him we should observe that he lived before Rennie, and before there were one-tenth of the artists there are now.   *Suum cuique;* his example and precepts have made us all engineers."   Even after the great works of the railway era, and the variety of practical ability which they called forth and fostered, Robert Stephenson pronounced Smeaton to be the engineer of the highest intellectual eminence that had yet appeared in England.   Speaking of him to the author in 1858, he observed, " Smeaton is the greatest philosopher in our profession this country has yet produced. He was indeed a great man, possessing a truly Baconian mind, for he was an incessant experimenter.[2]   The principles of mechanics were never so clearly exhibited as in his writings, more especially with respect to resistance, gravity, the power of water and wind to turn mills, and so on.   His mind was as clear as crystal, and his demonstrations will be found mathematically conclusive.   To

---

[1] The engineer's daughter, who has related these beautiful features in his character, became the wife of Jeremiah Dixon, Esq., at one time mayor of Leeds, afterwards of Fell Foot, Windermere, and an active county magistrate.   She possessed much of the force of character and benevolence of disposition which distinguished her father; and was regarded as a woman of great practical ability.   She survived her husband many years, and during her lifetime built and endowed a free-school for girls at Staveley, about a mile from her residence, which is now, and has been ever since its establishment, of very great benefit to the population of the neighbour-

hood.   Mrs. Dixon was also an artist of some merit, and painted in oils; the altar-piece and decorated Ten Commandments now in Staveley church being of her execution.

[2] One of Smeaton's rules was, never to trust to deductions drawn from theory in any case where one could have an opportunity for actual experiment. "In my own practice," he said, "almost every successive case would have required an independent theory of its own.   In my intercourse with mankind I have always found those who would thrust theory into practical matters to be, at bottom, men of no judgment, and pure quacks."

this day there are no writings so valuable as his in the
highest walks of scientific engineering ; and when young
men ask me, as they frequently do, what they should
read, I invariably say, Go to Smeaton's philosophical
papers ; read them, master them thoroughly, and nothing
will be of greater service to you. Smeaton was indeed
a very great man."

From what we have said, it will be obvious that
Smeaton was, throughout his whole career, a most industrious man,—indeed, industry was the necessity and
habit of his life. His daughter describes him as having
been incessantly occupied from six years old to sixty.
He was a great economist of time, and laid it out in
such a way as to obtain from its use the greatest amount
of valuable result. When at home, his forenoons were
devoted to writing reports, and the various business
arising out of his professional engagements ; and his
afternoons were occupied by the pursuits in which he
took most pleasure,—working at his forge or in his
workshop, making mechanical experiments, or preparing his papers on scientific subjects for the Royal
Society. Though naturally possessed of an excellent
constitution, and capable of enduring much fatigue, it is
to be feared that he taxed his brain too much, and " o'er
informed his tenement of clay," by continuous and
intense application to study during his long periods
of seclusion at Austhorpe. His robust frame became
fragile, and his strength was further impaired by the
abstinence which he was subsequently compelled to adopt.
Moreover, it appears that brain disease was hereditary
in his family, and he long apprehended the stroke which
eventually terminated his life. This only made him the
more eager to employ to the greatest advantage the time
which it might yet be permitted him to live : and he
dreaded above all things the blight of his mental powers
—to use his own words, " lingering over the dregs after
the spirit had evaporated "—chiefly as depriving him of

the means of doing further good. The last public measure on which he was professionally engaged in London, was the passing of the Bill through Parliament for the construction of the Birmingham and Worcester Canal. It was very strongly opposed, and its support in Committee cost him much application, thought, and anxiety His friends saw him visibly breaking down, and apprehended that the powers of his vigorous mind were beginning to fail. The bill passed by a small majority, and Mr. Smeaton went down to his home at Austhorpe for repose. But shortly after, when walking in his garden, he was struck with palsy. Happily his faculties returned to him, and he expressed his thankfulness to the Almighty that his intellect had been spared. He was very resigned and cheerful, and took pleasure in seeing the usual social occupation of the family going on about him. He would, however, complain of his growing slowness of apprehension, and excuse it with a smile, saying, " It could not be otherwise : the shadow must lengthen as the sun goes down." Some phenomena relating to the moon formed the subject of conversation one evening, when it shone very bright full into his room. Fixing his eyes upon it, he said, " How often have I looked up to it with inquiry and wonder, and thought of the period when I shall have the vast and privileged views of an hereafter, and all will be comprehension and pleasure ! " He even continued to dictate letters to his friends ; and in one of these, addressed to Mr. Holmes, after describing his health and feelings, he said : " In consequence of the foregoing, I conclude myself nine-tenths dead, and the greatest favour the Almighty can do me (as I think) will be to complete the other part ; but as it is likely to be a lingering illness, it is only in His power to say when that is likely to happen." His suffering, however, did not last long ; and after the lapse of about a month from the writing of this letter, the engineer's spirit found repose.   He died on the 28th of October, 1792, in the 68th year of his age; and was buried with his fore-fathers in the old parish church of Whitkirk, where a tablet with the following inscription was erected to his memory :—

SACRED TO THE MEMORY

OF JOHN SMEATON, F.R.S.

A Man whom God had endowed with the most extraordinary abilities, which
he indefatigably exerted for the benefit of Mankind in works of science
And Philosophical research :

More especially as an Engineer and Mechanic.   His principal work, the Edystone Lighthouse, erected on a rock in the open sea, (where one had been washed away by the violence of a storm, and another had been consumed by the rage of fire,) secure in its own stability and the wise precautions for its safety, seems not unlikely to convey to distant ages, as it does to every Nation of the Globe, the Name of its constructor.

He was born at Austhorpe, June 8, 1724.
And departed this Life October 28, 1792.

Also Sacred to the Memory of
ANN, the Wife of the said JOHN SMEATON, F.R.S.,
who died January 17th, 1784.

Their two surviving Daughters,
Duly imprest with sentiments of Love and Respect
For the kindest and tenderest of Parents,
Pay this tribute to their Memory.

SMEATON'S BURIAL PLACE IN WHITKIRK CHURCH.

[By C. Cattermole, after an original Sketch by T. Sutcliffe, Leeds.]

John Rennie, F.R.S.

Engraved by W. Holl, after the portrait in Crayons
by Archibald Skirving

Published by John Murray, Albemarle Street, 1861.

LIFE OF JOHN RENNIE.

VIEW OF THE BREAKWATER FROM MOUNT EDGCUMBE.

[By Percival Skelton, after his original Drawing.]

LIFE OF JOHN RENNIE.

CHAPTER I.

SCOTLAND AT THE MIDDLE OF LAST CENTURY.

JOHN RENNIE, the architect of the three great London bridges, the engineer of the Plymouth Breakwater, of the London and East India Docks, and various other works of national importance, was born at the farm-steading of Phantassie, in East Lothian, on the 7th of June, 1761. His father was the owner of the small estate of the above

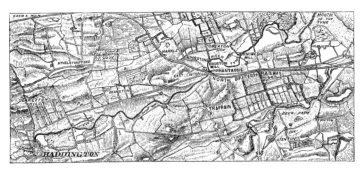

RENNIE'S NATIVE DISTRICT [Ordnance Survey.]

name, situated about midway between Haddington and Dunbar, at the foot of the gently-sloping hills which rise from it towards the south, the village of East Linton lying close at hand on the further bank of the little river Tyne. The property had been in the family for generations, and Mr. Rennie had the reputation of being one of the best farmers in the neighbourhood. But the art of agriculture, like every thing else in Scotland, was in an incredibly backward state, compared with either England or even Ireland, at the time when our engineer was born.

The traveller through the Lothians—which now exhibit perhaps the finest agriculture in the world, where every inch of ground is turned to profitable account, and the fields are cultivated to the very hedge-roots—will scarcely believe that less than a century ago these districts were not much removed from the state in which nature had left them. In the interior there was little to be seen but bleak moors and quaking bogs. The chief part of each farm consisted of " out-field " or unenclosed land, no better than moorland, from which even the hardy black cattle could scarcely gather herbage enough to keep them from starving in winter time. The " in-field " was an enclosed patch of ill-cultivated ground, on which oats and " bear" or barley were grown; but the principal crop was weeds.

Of the small quantity of corn raised in the country nine-tenths were grown within five miles of the coast;[1] and of wheat very little was raised—not a blade north of the Lothians. When the first crop of that grain was to be seen on a field near Edinburgh, people flocked to look upon it as a wonder. Clover, turnips, and potatoes had not yet been introduced, and no cattle were fattened : it was with difficulty they could be kept alive. Mr. Rennie, the engineer's father, was one of the first to introduce turnips as a regular farmer's crop. All loads were as yet carried on horseback; but where the farm was too small, or the crofter too poor, to keep a horse, his own or his wife's back bore the load. The horse brought peats from the bog and coals from the pit, and carried the crops to market. Sacks filled with manure were also sent a-field on horseback; but the uses of manure were so little understood, that if a stream was

---

[1] Professor Forbes's ' Considerations on the Present State of Scotland,' p. 14.

near, it was thrown in and floated away, and in summer it was burnt.[1] The towns were for the most part collections of thatched mud cottages,[2] giving scant shelter to a miserable population. The whole country was poor, desponding, gaunt, and almost haggard. The common people were badly fed and wretchedly clothed ; those in the country living in despicable huts with their cattle.[3] The poor crofters were barely able to exist. Lord Kaimes says of the Scotch tenantry of the early part of last century, that they were so benumbed by oppression and poverty[4] that the most able instructors in husbandry could have made nothing of them. A writer in the Scotch 'Farmer's Magazine' sums up his account of the country at that time in these words : " Except in a few instances, it was little better than a barren waste."[5]

What will scarcely be credited, now that the industry of Scotland has become thoroughly educated by a century's discipline of work, was the inconceivable listlessness and laziness of the people at that

<hr />

[1] 'Farmer's Magazine,' No. xxxiv., p. 200.

[2] It is stated in MacDiarmid's Picture of Dumfries' that at the middle of the century no lime was used in building, "except a little shell-lime, made of cockle-shells, which was burned at Colvend, and brought to Dumfries in bags." And, "in 1740, when Provost Bell (the chief magistrate or mayor of that town) built his house, the under storey was built of clay, and the upper storeys with lime brought from Whitehaven in dry-ware casks."

[3] The Rev. Dr. Playfair in 'Statistical Account of Scotland.' First edition. Vol. I., p. 513.

[4] Bad although the condition of Scotland was at the beginning of last century, there were many who believed that it would be made *worse* by the carrying of the Act of Union.

The Earl of Wigton was one of these. Possessing large estates in the county of Stirling, and desirous of taking every precaution against the impending ruin, he disposed to his tenants, on condition that they continued to pay him their then rents, low though they were, his extensive estates in the parishes of Denny, Kirkintulloch, and Cumbernauld, retaining only a few fields round the family mansion.[1] Fletcher of Saltoun equally feared the ruinous results of the Union, though he was less precipitate than the Earl of Wigton. We need scarcely say how completely all those apprehensions were falsified by the actual results.

[5] 'Farmer's Magazine,' 1803. No. xiii., p. 101.

<hr />

[1] Farmer's Magazine, 1808, No. xxxiv., p. 193.

period.[1] They left the bog unreclaimed and the swamp undrained. They would not even be at the trouble to enclose lands easily capable of cultivation. There was no class possessed of any enterprise or wealth. A middle rank could scarcely be said to exist, or any condition between that of the starving peasantry and the impoverished proprietary, whose available means were principally expended on hard drinking.[2] Mr. Brown, an East Lothian farmer, said of the latter class, that they were still too proud, and perhaps too ignorant, to interest themselves about the amelioration of their own domains.[3] The educated class—strictly so called—was as yet extremely small, and displayed a general indifferentism on all subjects of social, political, or religious interest, which some regarded as philosophic, but which was only an exhibition in another form of the prevalent national indolence. An idea of the general poverty may be formed from the fact that about the middle of the century the whole circulating medium of the Edinburgh banks was only 200,000*l.*, which was found amply sufficient for the requirements of trade and commerce, which had scarcely yet sprung into existence.[4] Even in East Lothian, which was probably in advance of the other Scotch counties, the ordinary wage of a day labourer was only fivepence in

---

[1] Miss Craik, in describing the difficulties which her father (William Craik, of Arbigland) had to contend against in introducing agricultural improvements in the county of Kirkcudbright, about the middle of last century, says: "For many years the indolent obstinacy of the lower class of people was almost unconquerable. Amongst other instances of their laziness, I have heard him say that, upon his first introduction of the mode of dressing the grain at night which had been thrashed during the day, all the servants in the neighbourhood refused to adopt the measure, and even threatened to destroy the houses of their employers by fire if they continued to insist upon the business. My father speedily perceived that a forcible remedy was required for the evil. He gave them their choice of removing the thrashed grain in the evening, or becoming inhabitants of Kirkcudbright jail; they preferred the former alternative, and open murmurings were no longer heard."— 'Farmer's Magazine,' No. xlvi. (June, 1811), p. 155. Art.: 'Account of William Craik, Esq., of Arbigland.'

[2] See the 'Autobiography of Dr. Alexander Carlyle,' *passim*.

[3] Brown on 'Rural Affairs,' Vol. I., p. 58. [4] Ibid.

winter and sixpence in summer.[1] The food of the working class was almost wholly vegetable, and even that was insufficient in quantity. The little butcher's meat consumed by the better class was salted beef and mutton, which was stored up at Ladner Time, betwixt Michaelmas and Martinmas, for the year's consumption. Mr. Buchan Hepburn says the sheriff of the county of East Lothian informed him that he remembered when not a single bullock was slaughtered in the butcher-market at Haddington for a whole year, except at the above period; and when Sir David Kinloch, of Gil-merton, sold ten wedders to an Edinburgh butcher, he stipulated for three several terms to take them away, to prevent the Edinburgh market from being overstocked with fresh butcher's meat![2]

The rest of Scotland was in no better state: in some parts it was even worse. The now rich and fertile county of Ayr, which glories in the name of "the garden of Scotland," was for the most part a wild and dreary common, with here and there a poor, bare, homely hut, where the farmer and his family were lodged.[3] There were no enclosures of land, except one or two about a gentleman's seat, and black cattle roamed at large over the face of the country.[4] More deplorable still was the

---

[1] G. Buchan Hepburn's 'General View of the Agriculture and Economy of East Lothian.' Edinburgh, 1794. P. 95.

[2] Ibid., p. 55.

[3] The Rev. Mr. Robertson, in the 'Statistical Account of Scotland.'

[4] When it was attempted, in 1723, to form enclosures in the adjoining county of Kirkcudbright, for the purpose of preventing the black cattle from straying, the poor people, who had squatted or were small tenants on the land, were turned out, and mobs assembled at different points and levelled the enclosures. "It is not pleasant," says a Kirkcudbright chronicler, "to represent the wretched

state of individuals as times then went in Scotland. The tenants in general lived very meanly, on kail, groats, milk, graddon ground in querns turned by the hand, the grain being dried in a pot, together with a crock ewe now and then about Mar-tinmas. They were clothed very plainly, and their habitations were most uncomfortable. Their general wear was of cloth, made of waulked plaiding, black and white wool mixed, very coarse, and the cloth rarely dyed. Their hose (when they wore them) were made of white plaiding cloth, sewed together; with single-soled shoes, and a black or blue bonnet—none having hats but the lairds, who

condition of those counties which immediately bordered the wild Highland districts, the inhabitants of which regarded the Lowlands as their lawful prey.   The only method by which security of a certain sort could be obtained for their property was by the payment of blackmail to some of the principal caterans; though this was not sufficient to protect them against the lesser marauders.   Regular contracts were drawn up between proprietors in the counties of Perth, Stirling, and Dumbarton, and the Macgregors, in which it was stipulated that if less than seven cattle were stolen—which peccadillo was styled *picking*—no redress should be required; but if the number stolen exceeded seven—such amount of theft being termed *lifting*—then the Macgregors became bound to recover.   This blackmail was regularly levied as far south as Campsie—then within six miles of Glasgow, but now almost forming part of it—down to within a few months of the outbreak of the rebellion of 1745.[1]   Under such circumstances agricultural improvement was impossible.   Another evil was, that the lawless habits of their neighbours tended to make the Lowland farmers almost as ferocious as the Highlanders themselves.   Feuds were of constant occurrence between neighbouring baronies, and even contiguous parishes; and the county fairs, which were tacitly recognised as the occasions for settling quarrels, were the scenes of as

---

thought themselves very well dressed for going to church on Sunday with a black kelt-coat of their wife's making. The distresses and poverty felt in the country continued till about the year 1735.   During these times, when potatoes were not generally raised (having been only introduced into the stewartry in 1725), there was, for the most part, a great scarcity of food, bordering on famine; for, in the whole of Kirkcudbright and Dumfries, there was not as much victual produced as was necessary for the supply of the inhabitants; and the chief part of what was required for

that purpose was brought from the Sandbeds of Esk, in tumbling cars, to Dumfries; and when the waters were high by reason of spates, and there being no bridges so that these cars could not come with the meal, I have seen the tradesmen's wives in the streets of Dumfries crying, because there was none to be got."— Letter of John Maxwell, in Appendix to MacDiarmid's 'Picture of Dumfries.'   Edinburgh, 1832.

[1] 'Farmer's Magazine:' 'Account of the Husbandry of Stirlingshire,' No. xxxiv., p. 198.

bloody fights as were ever known in Ireland, even in its worst days.

The country was as yet almost without roads, so that communication between one town and another was exceedingly difficult, especially in winter. The old track between Haddington and Edinburgh still exists as it was left when the new system of turnpike roads was introduced in Scotland. It is now used only by fox-hunters riding to cover, but it continues to bear out the description of a local writer : " Nothing," he says, " can be a greater contrast with the roads of modern times. In some places, where there was space for taking room, it was not spared. There might be seen four or five or more tracks, all collateral to one another, as each in its turn had been abandoned and another chosen, and all at last equally impassable. In wet weather they became mere sloughs, in which the carts or carriages had to slumper through in a half-swimming state, whilst in time of drought it was a continued jolting out of one hole into another." [1]

Such being the state of the highways, it will be evident that very little traffic could be conducted in vehicles of any sort. Single horse traffickers, called cadgers, plied between country towns and villages, supplying the inhabitants with salt, fish, earthenware, and articles of clothing, which they carried in sacks or creels hung across the horse's back. Even the trade between Edinburgh and Glasgow was carried on in the same primitive way. So limited was the consumption of the comparatively small population of Glasgow about the middle of last century, that most of the butter, cheese, and poultry raised within six miles of that city was carried by cadgers to Edinburgh in panniers on horse-back. On one occasion, a load of ducks, brought from Campsie to Edinburgh for sale in the Grassmarket,

---

[1] George Robertson's 'Rural Recollections,' p. 38.

finding themselves at liberty, rose upon wing and flew westward. Some of them were afterwards found at Linlithgow, and others succeeded in reaching their native "dub" at Campsie, some forty-five miles distant.[1]

It was long before travelling by coach was introduced in Scotland. When Smollett went from Glasgow to Edinburgh in 1739, on his way to London, there was neither coach, cart, nor waggon on the road. He accordingly accompanied the carriers as far as Newcastle, "sitting upon a pack-saddle between two baskets, one of which," he says, "contained my goods in a knapsack." The first vehicle which plied between the two chief cities of Scotland was not started until 1749. It was called "The Edinburgh and Glasgow Caravan," and performed the journey of forty-four miles in two days; but the packhorse continued to be the principal means of communication between the two places. Ten years later another vehicle was started, which was named "The Fly," because of its extraordinary speed, and it contrived to make the journey in rather less than a day and a half.[2] When a coach with four horses was put on between Haddington and Edinburgh, it took a full winter's day to perform the journey of sixteen miles. The effort was to reach Musselburgh in time for dinner, and go into town in the evening.[3]

In some parts of the country—as in Spain to this day —the beds of rivers served the double purpose of a river in wet, and a road in dry weather. When a common carrier began to ply between Selkirk and Edinburgh, a distance of only thirty-eight miles, he occupied a fortnight in performing the double journey. Part of the road lay along Gala Water, and in summer the carrier drove his rude cart along the bed of the stream; in winter the route was of course altogether impassable.

---

[1] 'Farmer's Magazine,' No. xxxiv., p. 200.
[2] Robertson's 'Rural Recollections.'
[3] G. Buchan Hepburn's 'Account of East Lothian.' 1794.

The townsmen of this adventurous individual, on the morning of his way-going, were accustomed to turn out and take leave of him, wishing him a safe return from his perilous journey.

The great post-road between London and Edinburgh passed close in front of the house at Phantassie in which John Rennie was born ; but even that was little better than the tracks we have already described. It passed westward over Pencrake, and followed the ridge of the Garleton Hills towards Edinburgh. The old travellers had no aversion to hill tops, rather preferring them because the ground was firmer to tread on, and they could see better about them. This line of high road avoided the county town, which, lying in a hollow, was unapproachable across the low grounds in wet weather ; and, of all things, swamps and quagmires were then most dreaded. A portion of this old post-road was visible until within the last few years, upon the high ground about a mile to the north of Haddington. In some places it was very narrow and deep, not unlike an old broad ditch, much waterworn, and strewn with loose stones. Along this line of way Sir John Cope passed with his army, in 1745, to protect Edinburgh against the Highland rebels ; and it is related that, on marching northward to intercept them, he was compelled to halt for several days, waiting for a hundred horse-loads of bread required for the victualling of his army.

In 1750, a project was set on foot for improving the high road through East Lothian, and a Turnpike Act was obtained for the purpose—the first Act of the kind obtained north of the Tweed.[1] The inhabitants of the town of Haddington complained loudly of the oppression practised on them, by making them pay toll for every bit of coal they burned ; though before the road was made it was a good day's work for a man and

---

[1] G. Buchan Hepburn's 'Account,' p. 151.

horse to fetch a load of "divot" from Gladsmuir, or of coal from the nearest colliery, only some four miles distant. By the year 1763 this post-road must have been made practicable for wheeled vehicles; for in that year the one stage-coach, which for a time formed the sole communication of the kind between London and all Scotland, began to run; and John Rennie, when a boy, was familiar with the sight of the uncouth vehicle lumbering along the road past his door. It "set out" from Edinburgh only once a month, the journey to London occupying from twelve to eighteen days, according to the state of the roads.

Such, however, had not always been the miserable condition of Scotland. The fine old bridges which exist in different parts of the country alone serve to show that at some early period a degree of civilization and prosperity had prevailed, from which it had gradually fallen. Professor Innes has clearly pointed this out in a recent work :[1] "When we consider," he says, "the long and united efforts required in the early state of the arts for throwing a bridge over any considerable river, the early occurrence of bridges may well be admitted as one of the best tests of civilization and national prosperity." As in England itself, the original reclamation of lands, the improvement of agriculture, the making of roads, and the building of bridges throughout the Lowlands of Scotland, were for the most part due to the old churchmen; and when their ecclesiastical organization was destroyed the country again relapsed into the state from which they had raised it, and it lay in ruins almost until our own day, when it has again been rescued from barrenness, even more effectually than before, by the combined influences of education and industry.

The same " Brothers of the Bridge," who erected so many fine old bridges across the rivers of England, were

---

[1] Cosmo Innes's 'Sketches of Early Scottish History.' 1861.

equally busy beyond the Tweed, providing those essential means of intercourse for the community. Thus we find bridges early erected across most of the rapid rivers in the Lowlands, especially in those places where the ecclesiastical foundations were the richest; and to this day the magnificent old abbey or cathedral of the neighbourhood—in some corner of which the Presbyterian Church holds its worship—serves to remind one of the contemporaneous origin of both classes of structures. Thus, as early as the thirteenth century, there was a bridge over the Tay at Perth; bridges over the Esk at Brechin and Marykirk; one over the Dee at Kincardine O'Neil; one at Aberdeen; and one at the mouth of Glenmuick. The fine old bridge over the Dee, at Aberdeen, is still standing : it consists of seven arches, and, as usual, the name of a bishop—Gawin Dunbar—is connected with its erection. There is another old bridge over the Don near the same city, said to have been built by Bishop Cheyne in the time of Robert the Bruce—the famous " Brig of Balgonie," celebrated in Lord Byron's stanzas as " Balgownie Brig's black wa'." It consists of a spacious Gothic arch, resting upon the rock on either side. There was even an old bridge over the rapid Spey at Orkhill. Then at Glasgow there was a fine bridge over the Clyde, which used, in old times, to be called "the Great Bridge of Glasgow," said to have been built by Bishop Rae in 1345. Though the bridge was only twelve feet wide, it consisted of eight arches; somewhat similar to the ancient fabric which still spans the Forth under the guns of Stirling Castle. This last-mentioned bridge was, until recent times, a structure of great importance, affording almost the only access into the northern parts of Scotland for wheeled carriages.

But the art of bridge-building in Scotland, as in England, seems for a long time to have been almost entirely lost; and until Smeaton was employed to erect the bridges of Coldstream, Perth, and Banff, next to nothing was done to improve this essential part of the communications of the country. Where attempts were made by local builders to erect such structures, they very rarely stood the force of a winter's, or even a summer's, flood. " I remember," says John Maxwell, " the falling of the Bridge of Buittle, which was built by John Frew in 1722, and fell in the succeeding summer, while I was in Buittle garden seeing my father's servants gathering nettles." [1] A similar fate befell the few attempts that were made about the same time to maintain the lines of communication by replacing the old bridges where they had gone to ruin, or substituting new ones in place of fords.

The mechanical arts had indeed fallen into the very lowest state. All kinds of tools were of the most imperfect description. The implements used in agriculture were extremely rude. They were mostly made by the farmer himself, in the roughest possible style, without the assistance of any mechanic. But a plough, which was regarded as a complicated machine, was reserved for the blacksmith. It was made of young birch trees, and, if the tradesman was expert, it was completed in the course of a winter's evening.[2] This rude implement scratched, without difficulty, the surface of old crofts, but made sorry work in out-fields, where the sward was tough and stones were large and numerous. Lord Kaimes said of the harrows used in his time, that they were more fitted to raise laughter than to raise mould. Machinery of an improved kind had not yet been introduced in any department of labour. Its first application, as might be expected, was in agriculture, then the leading, and indeed almost the only, branch of industry in Scotland; and its introduction will be found to be both curious and interesting in its bearing upon the subject of our present memoir.

---

[1] Appendix to ' Picture of Dumfries.' By John MacDiarmid. Edinburgh, 1832.

[2] ' Farmer's Magazine,' No. xxxiv., p. 199.

CHAPTER II.

RENNIE'S MASTER—ANDREW MEIKLE.

ANDREW FLETCHER, of Saltoun, fled into Holland during the political troubles in the reign of Charles II., and during his residence there he was particularly struck by the expert methods employed by the Dutch in winnowing corn and shealing barley. The chaff was then ordinarily separated from the corn by means of wind upon a knoll, or a draught of air blowing through the barndoor; and barley was shealed by pounding the grains with water in the hollow of a stone, until by that means the husks were rubbed off. Fletcher saw that there was a great waste of labour in these processes, and during his residence abroad he determined to introduce the Dutch methods into his own country. Writing home to his brother, he desired him to send out to Holland one James Meikle, an ingenious country wright of Wester Keith,[1] for the purpose of learning the above arts and importing the requisite machinery into Scotland. After a stay of about two months in that country, Meikle returned home, bringing with him a winnowing-machine, commonly called a pair of fanners, and the ironwork requisite for a barley-mill. These were safely transported to Leith, and afterwards conveyed to Saltoun, where the barley-mill was erected and set to work; and for many years it was the only machine of the kind in the British dominions, so slow were people in those days to copy the

---

[1] It would seem that the ancestors of Meikle were held in esteem as ingenious workmen for generations; the Scots Parliament having, in 1686, passed a special Act for the encourage- | ment of John Meikle, founder, who, it appears, was the first person to introduce the art of iron-founding into Scotland.

improvements of their neighbours. "Saltoun barley" was the name by which dressed pot-barley then became known, and it continued to preserve the name long after barley-mills had come into general use. James Meikle was equally successful in setting his fanners to work; but they had a good deal of superstitious prejudice to encounter, the country people looking upon the grain cleaned by them with suspicion, as procured by "artificially-created wind." The clergy even argued that "winds were raised by God alone, and it was irreligious in man to attempt to raise wind for himself, and by efforts of his own;" and one clergyman even refused the holy communion to those of his parishioners who thus irreverently raised "Devil's wind." The readers of 'Old Mortality' will remember Mause Headrigg's indignation when it was proposed that her "son Cuddie should work in the barn wi' a new-fangled machine for dightin' the corn frae the chaff, thus impiously thwarting the will of Divine Providence by raising wind for your leddyship's ain particular use by human art, instead of soliciting it by prayer, or waiting patiently for whatever dispensation of wind Providence was pleased to send upon the shealing-hill." Scott, however, was obviously guilty of an anachronism in this passage, for the first pair of fanners was not set up at Saltoun until the year 1720— long after the period of Cuddie Headrigg's supposed trial—and it was not until seventeen years later that another winnowing-machine was set up in the neighbouring shire of Roxburgh, and employed as an ordinary agency in farming operations.

Andrew Meikle was the only surviving son of Fletcher's millwright, and like him was an ingenious mechanic. He had married and settled at Houston Mill, on Mr. Rennie's Phantassie estate, where he combined the occupations of small farmer, miller, and millwright. He had himself fitted up the machinery of the mill, of which he was the tenant; and adjoining it was his mill-

BOUSTON MILL.

[By E. M. Wimperis, after a Drawing by J. S. Smiles]

wright's shop, where he carried on his small business in connection with mill-work—the demands of the district being as yet of an extremely limited character. But the march of social improvement had by this time fairly begun in East Lothian. The public spirit displayed by Fletcher of Saltoun was imitated by his neighbours. But probably the gentleman who gave the greatest impulse to agricultural progress in the county, which shortly after extended itself over Scotland, was Mr. Cockburn of Ormiston, to whom belongs the honour of adopting the system of long leases. He early became convinced that the surest way of stimulating the industry of the farmer was to give him a substantial interest in the improvement of the land which he farmed. One of his tenants having enclosed his fields with hedges and ditches at his own cost—the first farmer in Scotland who adopted the practice [1]—his landlord, to encourage his spirit of improvement, granted him a lease of his farm for nineteen years, renewable at the expiry of that term for a like period.

The results were found so satisfactory, that Mr. Cockburn was induced to extend the practice, and before long it became generally adopted throughout the county. From this point, then, agriculture advanced with extraordinary rapidity. The more thriving farmers sent their sons into England—a practice long since reversed—to learn the best methods of farming: they employed better implements and improved methods of culture; their landlords, further to encourage them, built more commodious steadings and farmhouses; and they were greatly helped in this course by the unusual facilities for obtaining credit which persons of standing and property possessed, on the general extension, from about the middle of last century, of what is called the Scotch system of banking.[2] These measures very shortly put an entirely new face upon the country. The distinction of "in-field" and "out-field" altogether ceased. Farms became completely enclosed, and sheep and black cattle were no longer allowed to roam at large. Fields were thrown together, and small holdings consolidated into large ones. The moorland and the bog were reclaimed and converted into fruitful farms. A single instance, of some historical interest, may be given. When the Royal army lay upon the field of Prestonpans in 1745, their front was "protected by a deep bog," across which Robert Anderson, a young gentleman of the county, who knew every foot of the ground, contrived to lead the Pretender's army by a path known only to himself. That bog, like so many others, has long since been reclaimed by drainage and cultivation, and now forms part of one of the most fertile farms in the Lothians.

---

[1] Brown on 'Rural Affairs.'
[2] See Adam Smith's 'Wealth of Nations,' Book II., Chap. 2.

Such was the improving state of affairs in East Lothian when Andrew Meikle began business at Houston Mill. There were as yet very few mills in the district; but his reputation as a mechanic and his skill in millwork were such, that he was usually employed on any new erection of the sort, travelling also into the adjoining counties of Edinburgh and Berwick to repair or fit up mills. Being an ingenious and thoughtful man, he eagerly turned his attention to the improvement of agricultural machinery, more especially of that connected with the thrashing, winnowing, dressing, and grinding of grain. Thus, as early as the year 1768, we find him taking out a patent—one of the very first taken out by any Scotch mechanic—for a new machine contrived by him for dressing and cleansing corn.[1] It was a combination of the riddle and fanners; and though of no great novelty, it showed the direction in which his inventive faculties were thus early at work. But Meikle's most important invention was made at a considerably later period of his life; and in the interval he devoted himself to the ordinary business of his humble calling as a country miller and millwright.

Nothing caused so much loss and vexation to the farmer in former times as the operation of separating the corn from the straw. In some countries it was trodden out by cattle, as in the old Scriptural times; hence the proverb, "Thou shalt not muzzle the ox that treadeth out the corn." Sledges or trail-carts were also used for the same purpose; but the most common instrument employed was the flail. By either of these methods, however, the process of thrashing was slowly performed, whilst a considerable portion of the grain was damaged or lost. Many attempts had been made before Meikle's

---

[1] Patent No. 896. The name of Robert Mackell (employed with James Watt in the survey of the " ditch canal " through Perthshire—see Life of Smeaton) was associated with that of Meikle in this patent; Mackell probably finding the money, and Meikle the brains.

time to invent a machine which should satisfactorily perform this operation; but without effect. An East Lothian gentleman, named Michael Menzies, contrived one upon the principle of the flail, arranging a number of flails so as to be worked by a water-wheel; but they were soon broken to pieces by the force with which they went. Another experiment was made in 1758 by a Stirlingshire farmer, named Leckie, who invented a machine on the principle of the horizontal flax-mill. It consisted of a vertical shaft, with four cross-arms fixed in a box, and when set in motion the arms beat off the grain from the straw when let down upon them by hand. Though this machine succeeded very well in thrashing oats, it cut off the heads of every other kind of corn presented to it. Similar attempts were made about the same time by farmers in the south, more especially by Mr. Ilderton at Alnwick, Mr. Smart at Wark, and Mr. Oxley at Flodden, about 1772-3. The machine employed by these gentlemen was composed of a large drum, about six feet in diameter, resembling a sugar hogshead, round which were placed a number of fluted rollers, which pressed inwards upon the drum by means of springs. The corn, in passing the cylinder and rollers, was no doubt rubbed out; but a large proportion of it being bruised and damaged by the operation, this plan too was eventually abandoned. Mr. Oxley is said to have afterwards tried the plan of stripping the corn from the straw by means of a scutcher; but the machine constructed with this object did not answer, and it was also laid aside.

Mr. Kinloch,[1] of Gilmerton in East Lothian, had however seen the last-mentioned machine at work, and he conceived the idea of improving it. He accordingly had a model made, in which he contrived that the drum, mounted with four pieces of fluted wood, should work

---

[1] Afterwards Sir Francis Kinloch.

upon springs, pressing with less force upon the corn in the process of rubbing it out. This model was shown to Meikle, with whom Mr. Kinloch had many conversations on the subject; and at the millwright's suggestion several improvements were made in it, one of which was the substitution of smooth feeding rollers for fluted ones. When the model had been completed, Mr. Kinloch sent it to Houston Mill to be tried by the power of Meikle's water-wheel. On being set to work, however, it was driven in pieces in a few minutes; and the same fate befell a larger machine after the same model, which Mr. Kinloch got made for one of his tenants a few years later.

The best result of Mr. Kinloch's experiments was, that they had the effect of directing the inventive mind of Andrew Meikle to the subject. After several years' thinking and planning, about the year 1776 he constructed a thrashing-machine, consisting of a number of flails fixed in a strong beam moved by a crank, which beat out the corn on two platforms, one on each side of the beam. Although the performance of this machine before some East Lothian farmers who went to see it at work was on the whole satisfactory, it did not come up to Meikle's expectations; and on one of the gentlemen observing that the flails and platforms probably would not bear the force of the stroke, the inventor replied, that in case the machine did not answer, he intended to try a method of beating out the corn by means of fixed scutchers or beaters.[1] Accordingly he proceeded to work out this idea in practice, and after a few years he succeeded in perfecting his invention on this principle, which was entirely new. These scutchers, shod with iron, were fixed upon a strong beam or cylinder, which revolved with great velocity, and in the process

---

[1] 'A Reply to an Address to the Public, but more particularly to the Landed Interest of Great Britain and Ireland, on the subject of the Thrashing Machine.' By John Shirreff. Edinburgh, 1811.

of so revolving beat off the corn instead of rubbing
it off by pressure, as had been attempted by former
contrivers. By dint of study and perseverance, he succeeded at length in perfecting his machine; to which
he added solid fluted feeding rollers, and afterwards a
machine for shaking the straw, fanners for winnowing
the corn, and other improvements. Meikle is said to
have been superintending a mill job at Leith at the time
he was engaged in working out the contrivance in his
mind. He was accustomed to walk there and back within
the same day while the job was in hand, or a distance of
about forty miles. He studied the subject during his
journey, and would occasionally stop while travelling to
draw a rapid diagram upon the road with his walking-
stick. It is related of him that on one occasion, whilst
very much engrossed with the subject of his thrashing-
mill, he had, absorbed by his calculations, wandered
considerably from the right path. He stopped short
suddenly, and hastily sketching his plan on the road,
exclaimed, "I have got it! I have got it!" Archimedes himself, when he cried "Eureka," could not have
been more delighted than our millwright was at the
happy upshot of his deliberations.

The first machine erected on Meikle's new principle
was put up in 1787 for Mr. Stein of Kilbeggie, in
Clackmannanshire, who had great difficulty in procuring
a sufficient number of barnsmen for thrashing straw to
litter the large stock of cattle he had on hand; but the
novelty of the experiment, and the doubt entertained
by Mr. Stein as to the efficacy of the proposed machine,
induced him to require, as a condition, that if it did not
answer the intended purpose, Meikle was not to receive
any payment for it. The result, however, proved quite
satisfactory, and the thrashing-machine at Kilbeggie,
which was driven by water-power, long continued in
good working order. The next he erected was for Mr.
George Rennie, at Phantassie, in the same year; and by

this time he had so perfected his machine as to enable it to be driven by water, wind, or horses. That at Phantassie was worked by the latter power. In 1788 Meikle took out a patent for his invention, describing himself in the specification as "engineer and machinist."[1]

ANDREW MEIKLE. [By T. D. Scott after Ruddock.]

The thrashing-machine proved to be one of the greatest boons ever conferred upon the husbandman, effecting an immense saving of labour as well as of corn. By its means from seventy to eighty bushels of oats, and from thirty to fifty bushels of wheat, might be thrashed and cleaned in an hour; and it is calculated to have effected a saving, as compared with the flail, of one-hundredth part of the whole corn thrashed, or equal to a value of not less than two millions sterling in Great Britain alone.

---

[1] Patent No. 1645 : "Machine for separating Corn from Straw."

In the course of twenty years from the date of the patent, about three hundred and fifty thrashing-mills were erected in East Lothian alone, at an estimated outlay of nearly forty thousand pounds; and, shortly after, it became generally adopted in England, and indeed all over the civilized world. We regret, however, to add, that Meikle did not reap those pecuniary advantages from his invention which a less modest and more pushing man would have done. Pirates fell upon him on all sides and deprived him of the fruits of his ingenuity, even denying him any originality whatever. When growing old and infirm, Sir John Sinclair bestirred himself to raise a subscription in his behalf; and a sum of 1500l. was collected, which was invested for his benefit. Mr. Dempster, M.P., wrote to Sir John, when on his charitable mission in 1809, "Should your tour in East Lothian procure a suitable reward to the inventor of the thrashing-machine, it will redound much to your and the country's honour: our heathen ancestors would have assigned a place in heaven to Meikle." [1] Mr. Smeaton knew Meikle intimately, and frequently met him in consultation respecting the arrangements of the Dalry Mills, near Edinburgh, and other works; and he was accustomed to say of him, that if he had possessed but one-half the address of other people, he would have rivalled all his contemporaries, and stood forth as one of the first mechanical engineers in the kingdom.

Among the various improvements which this ingenious mechanic introduced in mill-work, were those in the sails of windmills. Before his time, these machines were liable to serious accidents on the occurrence of a sudden gale, or a shift in the direction of the wind. By Meikle's contrivance, the machinery was so arranged that the whole sails might be taken in or let out in half a minute, according as the wind required, by a person merely

---

[1] 'Memoirs of Sir John Sinclair,' vol. ii., p. 99.

pulling a rope within the mill. The machinery was at the same time kept in more uniform motion, and all danger from sudden squalls completely avoided. His improvements in water-wheels were also important, and on one occasion proved effectual in carrying out an improvement of a remarkable character in the county of Perth. This was neither more nor less than washing away into the river Forth some two thousand acres of peat moss, and thus laying bare an equivalent surface of arable land, now amongst the most valuable in the Carse of Stirling. The Kincardine Moss was situated between the rivers Teith and Forth. It was seven feet in depth, laid upon a bottom of rich clay. In 1766 Lord Kaimes, who had entered into possession of the Blair Drummond estate, to which it belonged, determined if possible to improve the tract; and it occurred to him that the easiest plan would be to wash the moss entirely away. But how was this to be done? The river Teith, which was the only available stream at hand, was employed to drive a corn-mill. But Lord Kaimes saw that it would answer his intended purpose if he could get possession of it. He accordingly made an arrangement by which he became owner of the mill, which he pulled down, and then turned the mill-stream in upon the moss. Labourers were set to work to cut away the stuff, which was thrown into the current, and much of it thus washed away. But the process was slow; and the clearing of the land had not advanced very far by the year 1783, when Lord Kaimes's son, Mr. Home Drummond, entered into possession of the estate. A thousand acres still remained, which he determined to get rid of, if possible, in a more summary manner than his predecessor had done. He consulted several engineers—amongst others Mr. Whitworth, a pupil of Brindley's—who recommended one plan; but George Meikle, a millwright at Alloa, the son of Andrew, proposed another, the invention of his father; and Mr. Whitworth, with much

I 2

candour and liberality, at once acknowledging its superiority to his own, urged Mr. Drummond to adopt it.
The invention consisted of a newly-contrived wheel,
28 feet in diameter and 10 feet broad, for raising
water in a simple, economical, and powerful manner, at
the rate of from 40 to 60 hogsheads a minute; and it
was necessary so to raise it about 17 feet, in order to
reach the higher parts of the land. The machinery
on being erected was set to work, and with such good
results, that in the course of a very few years the four
miles of barren moss was completely washed away, and
the district was shortly after covered with thriving farmsteads, as it remains to this day.

Meikle was a thorough mechanical inventor, and,
wherever he could, he endeavoured to save labour
by means of machinery. Stories are still told in the
neighbourhood in which he lived of the contrivances
he adopted with this object in his own household, some
of which were of an amusing character. One day a
woman came to the mill to get some barley ground, and
was desired to sit down in the cottage hard by until
it was ready. With the first sound of the mill-wheels
the cradle and churn at her side began to rock and to
churn, as if influenced by some supernatural agency. No
one was in the house besides herself at the time, and she
rushed from it frightened almost out of her wits. Such
incidents as these brought an ill name on Andrew, and
the neighbours declared of him that he was "no canny."
He was often sent for to great distances for the purpose
of repairing pumps or setting mills to rights. On one
occasion, when he undertook to supply a gentleman's
house with water, so many country mechanics had tried
it before and failed, that the butler would not believe
Meikle when he told him he would send in the water
next day. Meikle, however, told him to get everything ready. "It will be time enough to get ready,"
said the incredulous butler, "when we see the water."

Meikle quietly pocketed the affront, but set his machinery to work early next morning; and when the butler got out of bed, he found himself up to his knees in water, so successfully had the engineer performed his promise.

Meikle lived to an extreme old age, and was cheerful to the last. He was a capital player on the Northumbrian bagpipes. The instrument he played on was made by himself, the chanter being formed out of a deer's shank-bone. When ninety years old, at the family gatherings on " Auld Hansel Monday," his six sons and their numerous families danced about him to his music. He died in 1811, in his ninety-second year, and was buried in Prestonkirk churchyard, close by Houston Mill, where a simple monument is erected to his memory.[1]

Such was the master who first trained and disciplined the skill of John Rennie, and implanted in his mind an enthusiasm for mechanical excellence. Another of his apprentices was a man who exercised almost as great an influence on the progress of mechanics, through the number of first-rate workmen whom he trained, as did Rennie himself in the art of engineering. We allude to Peter Nicholson, an admirable mechanic and draughts-man, the author of numerous works on carpentry and architecture, which to this day are amongst the best of their kind. We now pursue the career of Andrew Meikle's most distinguished pupil.

---

[1] It is remarkable that Scotch biography should be altogether silent respecting this ingenious and useful workman. In the most elaborate of the Scotch biographical collections—that of Robert Chambers, in four large volumes—not a word occurs relating to Meikle. An article is devoted to Mickle, the translator of another man's invention in the shape of a poem, the 'Lusiad;' but the name of the inventor of the thrashing-machine is not even mentioned; affording a singular illustration of the neglect which this department of biography has heretofore experienced, though it has been by men such as Meikle that this country has in a great measure been made what it is.

CHAPTER III.

YOUNG RENNIE AT SCHOOL, WORKSHOP, AND COLLEGE.

FARMER RENNIE died in the old house at Phantassie
in the year 1766, leaving a family of nine children, four
of whom were sons and five daughters. George, the
eldest, was then seventeen years old. He was discreet,
intelligent, and shrewd beyond his years, and from that

RENNIE'S BIRTHPLACE, PHANTASSIE.

[By E. M. Wimperis, after a Drawing by J. S. Smiles.]

time forward he managed the farm and acted as the
head of the family. The year before his father's death
he had made a tour through Berwickshire, for the purpose of observing the improved methods of farming
introduced by some of the leading gentry of that county,

and he returned to Phantassie full of valuable practical information. The agricultural improvements which he was shortly afterwards instrumental in introducing into East Lothian were of a highly important character; his farm came to be regarded as a model, and his reputation as a skilled agriculturist extended far beyond the bounds of his own country, insomuch that he was resorted to for advice as to farming matters by distinguished visitors from all parts of Europe.[1]

Of the other sons, William, the second, went to sea: he was taken prisoner during the first American war, and was sent to Boston, where he died. The third, James, studied medicine at Edinburgh, and entered the army as an assistant-surgeon. The regiment to which he belonged was shortly after sent to India: he served in the celebrated campaign of General Harris against Tippoo Saib, and was killed whilst dressing the wound of his commanding-officer when under fire at the siege of Seringapatam. John, the future engineer, was the youngest son, and he was only five years old at the death of his father. He was accordingly brought up mainly under the direction of his mother, a woman possessed of many excellent practical qualities, amongst which her strong common sense was not the least valuable.

The boy early displayed his strong inclination for mechanical pursuits. When about six years old, his best loved toys were his knife, hammer, chisel, and saw, by means of which he indulged his love of construction. He preferred this kind of work to all other amusements, taking but small pleasure in the ordinary sports of boys of his own age. His greatest delight was in frequenting the smith's and carpenter's shops in the neighbouring village of Linton, watching the men use their tools, and

---

[1] Amongst Mr. Rennie's other illustrious visitors in his later years was the Grand Duke Nicholas (afterwards Emperor) of Russia, who stayed several nights at Phantassie, and during the time was present at the celebration of a "hind's wedding."

trying his own hand when they would let him. But his favourite resort was Andrew Meikle's millwright's shop, down by the river Tyne, only a few fields off. When he began to go to the parish school, then at Prestonkirk, he had to pass Meikle's shop daily, going and coming; and he either crossed the river by the planks fixed a little below the mill, or by the miller's boat when the waters were high. But the temptations of the millwright's workshop while passing to school in the mornings not unfrequently proved too great for him to resist, and he played truant; the delinquency being only discovered by the state of his fingers and clothes on his return home, when an interdict was laid against his "idling" away his time at Andrew Meikle's shop.

The millwright, on his part, had taken a strong liking for the boy, whose tastes were so congenial to his own. Besides, he was somewhat proud of his landlady's son frequenting his house, and was not disposed to discourage his visits. On the contrary, he let him have the run of his workshop, and allowed him to make his miniature water-mills and windmills with tools of his own. The river which flowed in front of Houston Mill was often swollen by spates or floods, which descended from the Lammermoors with great force; and on such occasions young Rennie took pleasure in watching the flow of the waters, and following the floating stacks, field-gates, and other farm wreck along the stream, down to where the Tyne joined the sea at Tyningham, about four miles below. Amongst his earliest pieces of workmanship was a fleet of miniature ships. But not finding tools to suit his purposes, he contrived, by working at the forge, to make them for himself; then he constructed his fleet, and launched his ships, to the admiration and astonishment of his playfellows. This was when he was about ten years old. Shortly after, by the advice and assistance of his friend Meikle, who took as much pride in his performances as if they had been

his own, Rennie made a model of a windmill, another of
a fire-engine (or steam-engine), and a third of Vellore's
pile-engine, displaying upon them a considerable amount
of manual dexterity; some of these early efforts of the
boy's genius being still preserved.

Though young Rennie thus employed so much of his
time on this amateur work in the millwright's shop, he
was not permitted to neglect his ordinary education at
the parish school. That of Prestonkirk was kept by a
Mr. Richardson, who seems to have well taught his pupil
in the ordinary branches of education; but by the time
he had reached twelve years of age he seems almost to
have exhausted his master's store of knowledge, and his
mother then thought the time had arrived to remove him
to a seminary of a higher order.[1]   He was accordingly

---

[1] Though a poor country, as we
have seen, Scotland was already rich
in parish and burgh schools; the
steady action of which upon the rising
generation was probably the chief
cause of that extraordinary improvement in all its branches of industry
to which we have above alluded.
John Knox—himself a native of East
Lothian—explicitly set forth in his
first 'Book of Discipline'—"That every
several kirke have ane schoolmaister
appointed," "able to teach grammar
and the Latin tongue, if the town be
of any reputation;" and if an upland
town, then a reader was to be appointed, or the minister himself must
attend to the instruction of the children and youth of the parish. It
was also enjoined that "provision be
made for the attendance of those that
be poore, and not able by themselves
nor their friends to be sustained at
letters;" "for this," it was added,
"must be carefully provided, that no
father, of what estate or condition
that ever he may be, use his children
at his own fantasie, especially in their
youthhead; but all must be compelled
to bring up their children in learning
and virtue." During the troubles in
which Scotland was involved, almost
down to the Revolution of 1688,
although attempts were made to establish a school in every parish, they
seem to have been attended with
comparatively small results; but at
length, in 1696, the Scottish Parliament was enabled, with the concur-
rence of William of Orange, to put in
force the Act of that year, which is
regarded as the charter of the parish-
school system in Scotland. It is there
ordained "that there be a school settled and established, and a schoolmaster appointed in every parish not
already provided, by advice of the
heritors and minister of the parish."
The consequence was, that the parish
schools of Scotland, working steadily
upon the rising generation, all of
whom passed under the hands of the
parish teachers during the preceding
half-century, had been training a
population whose intelligence was
greatly in advance of their material
condition as a people; and it is in
this circumstance, we apprehend, that
the true explanation is to be found of
the rapid start forward which the
whole country now took, dating more
particularly from the year 1745.
Agriculture was naturally the first
branch of industry to exhibit signs of
decided improvement; to be speedily
followed by like advances in manufac-

taken from the parish school at that age, but his friends had not made up their minds as to the steps they were to adopt with reference to his further education. The boy, however, found abundant employment for himself with his tools, and went on model-making; but feeling that he was only playing at work, he became restless and impatient, and entreated his mother that he might be allowed to go to Andrew Meikle's to learn to be a millwright. This was agreed to, and he was sent to Meikle's accordingly, where he worked for two years, during which period he learnt one of the most valuable parts of education—the use of his hands. He seemed to overflow with energy, and was ready to work at anything — at smith's work, carpenter's work, or millwork; taking most pleasure in the latter, in which he shortly acquired considerable expertness. Having the advantage of books—limited though the literature of mechanism was in those days—he studied the theory as well as the practice of mechanics, and the powers of his mind became steadily strengthened and developed with application and self-culture.

At the end of two years his friends determined to send him to the burgh school of Dunbar, one of that valuable class of seminaries directed and maintained by the magistracy, which have been established for the last hundred years and more in nearly every town of any importance in Scotland.[1] Dunbar High School was

---

tures, commerce, and shipping. Indeed, from that time, the country never looked back, but her progress went on at a constantly accelerating rate, issuing in results as marvellous as they have probably been unequalled.

[1] The origin of what are technically termed "Grammar Schools" in Scotland, is involved in considerable obscurity. They are, for the most part, of ancient foundation, and are supposed to have been endowed by generous individuals, who vested in some public body, usually the borough cor-poration, sums of money for the purpose of educating the youth of the towns in which they are established. The money or property so devised was legally termed a "mortification." Many of such bequests were made in the remote times when Scotland was a Catholic nation. John Knox himself was educated at the Grammar School of Haddington, near to which town he was born and brought up, and there he says he learnt the elements of the Latin language.

then a seminary of considerable celebrity. Mr. Gibson, the mathematical master, was an excellent teacher, full of love and enthusiasm for his profession; and it was principally for the benefit of his discipline and instruction that young Rennie was placed under his charge. The youth, on entering this school, possessed the advantage of being fully impressed with a sense of the practical value of intellectual culture. His two years' service in Meikle's workshop, while it trained his physical powers had also sharpened his appetite for knowledge, and he entered upon his second course of instruction at Dunbar with the disciplined powers almost of a grown man. He had also this advantage, that he prosecuted his studies there with a definite aim and purpose, and with a determinate desire to master certain special branches of education required for the successful pursuit of his intended business. Accordingly, we are not surprised to find that in the course of a few months he outstripped all his schoolfellows and took the first place in the school. A curious record of his proficiency as a scholar is to be found in a work by one Mr. David Loch, Inspector-General of Fisheries, published in 1779. It was his duty to hold a court of the herring skippers of Dunbar, then the principal fishing-station on the east coast; and it appears that at one of his visits to the town he attended an examination of the burgh schools, and was so much pleased with the proficiency of the pupils that he makes special mention of it in his book.[1]

---

[1] After speaking of the teachers of Latin, English, and arithmetic, he goes on to say: " But Mr. Gibson, teacher of mathematics, afforded a more conspicuous proof of his abilities, by the precision and clearness of his manner in stating the questions which he put to the scholars; and their correct and spirited answers to his propositions, and their clear demonstrations of his problems, afforded the highest satisfaction to a numerous audience. And here I must notice in a particular manner the singular proficiency of a young man of the name of Rennie: he was intended for a millwright, and was breeding to that business under the famous Mr. Meikle, at Linton, East Lothian; he had not then attended Mr. Gibson for the mathematics, &c., much more than six months, but on his examination he discovered such amazing power of genius, that one would have imagined him a second Newton. No problem was too hard for him to de-

Rennie remained with Mr. Gibson for about two years. During that period he went as far in mathematics and natural philosophy as his teacher could carry him, after which he again proposed to return to Meikle's workshop. But at this time the mathematical master was promoted to a higher charge—the rectorship of the High School of Perth—and a question arose as to the appointment of his successor. The loss to the town was felt to be great, and Mr. Gibson was pressed by the magistrates to point out some person whom he thought suitable for the office. The only one he could think of was his favourite pupil; and though not yet quite seventeen years old, he strongly recommended John Rennie to accept the appointment. The young man, how-ever, already beginning to be conscious of his powers, had formed more extensive views of life, and could not entertain the idea of settling down as the "dominie" of a burgh school, respectable and responsible though that office must be held to be. He accordingly declined the honour which the magistrates proposed to confer upon him, but agreed to take charge of the mathematical classes until Mr. Gibson's successor could be appointed. He continued to carry on the classes for about six weeks, and conducted them so satisfactorily that it was matter of much regret when he left the school and returned to his family at Phantassie for the purpose of prosecuting his intended profession.

At home he pursued the study of his favourite branches of instruction, more particularly mathematics, mechanics, and natural philosophy, frequenting the work-shop of his friend Meikle, assisting him with his plans,

---

monstrate. With a clear head, a de-cent address, and a distinct delivery, his master could not propose a ques-tion, either in natural or experimental philosophy, to which he gave not a clear and ready solution, and also the reasons of the connection between causes and effects, the power of gravi-tation, &c., in a masterly and con-vincing manner, so that every person present admired such an uncommon stock of knowledge amassed at his time of life. If this young man is spared, and continues to prosecute his studies, he will do great honour to his country."

and taking an especial interest in the invention of the
thrashing-machine, which Meikle was at that time employed in bringing to completion. He was also entrusted
to superintend the repairs of corn-mills in cases where
Meikle could not attend to them himself; and he was
sent, on several occasions, to erect machinery at a
considerable distance from Prestonkirk. Rennie thus
gained much valuable experience, at the same time that
he acquired confidence in his own powers; and before
the end of a year he began to undertake millwork on
his own account. His brother George was already well
known as a clever farmer, and this connection helped
the young millwright to as much employment in his
own neighbourhood as he desired. Meikle was also
ready to recommend him in cases where he could not
accept the engagements offered in distant counties;
and hence, as early as 1780, when Rennie was only
nineteen years of age, we find him employed in fitting
up the new mills at Invergowrie, near Dundee. He
designed the machinery as well as the buildings for its
reception, and superintended them to their completion.
His next work was to prepare an estimate and design
for the repairs of Mr. Aitcheson's flour-mills at Bonnington, near Edinburgh. Here he employed cast iron
pinions, instead of the wooden trundles formerly used:
one of the first attempts made to introduce iron into
this portion of the machinery of mills.

These, his first essays in design, were considered very
successful, and they brought him both fame and emolu-
ment. Business flowed in upon him, and before the end
of his nineteenth year he had as much employment
as he could comfortably get through. But he had no
intention of confining himself to the business of a country
millwright, however extensive, aiming at a higher
professional position and a still wider field of work.
Desirous, therefore, of advancing himself in scientific
culture and prosecuting the studies in mechanical philo-

sophy which he had begun at Phantassie and pursued
in the burgh school at Dunbar, he determined to place
himself at the University of Edinburgh, then a seminary of rising celebrity. In taking this step he formed
the resolution—by no means unusual amongst young
men of his country inspired by a laudable desire for
self-improvement—of supporting himself at college entirely by his own labour. He was persuaded that by
diligence and assiduity he would be enabled to earn
enough during the summer months to pay for his
winter's instruction and maintenance; and his habits
being frugal, and his style of living very plain, he was
enabled to prosecute his design without difficulty.

He accordingly matriculated at Edinburgh in November, 1780, and entered the classes of Dr. Robison,
Professor of Natural Philosophy, and of Dr. Black,
Professor of Chemistry; both men of the highest distinction in their respective walks. Robison was an
eminently prepossessing person, frank and lively in
manner, full of fancy and humour, and, though versa-
tile in talent, a profound and vigorous thinker. His
varied experience of life, and the thorough knowledge
which he had acquired of the principles as well as the
practice of the mechanical arts, proved of great use to
him as an instructor of youth. The state of physical
science was then at a very low ebb in this country, and
the labours of Continental philosophers were but little
known even to those who occupied the chairs in our
universities; the results of their elaborate researches
lying concealed in foreign languages, or being known, at
most, to a few inquirers more active and ardent than
their fellows; whilst the general student, mechanic, and
artisan were left to draw their principal information
from the ancient but ordinary springs of observation and
daily experience.

Under Dr. Robison the study of natural philosophy became invested with unusual significance and importance.

The range of his knowledge was most extensive : he was familiar with the whole circle of the accurate sciences, and in imparting information his understanding seemed to work with extraordinary energy and rapidity. The labours of others rose in value under his hands, and new views and ingenious suggestions never failed to enliven his prelections on mechanics, hydrodynamics, astronomy, optics, electricity, and magnetism, the principles of which he unfolded to his pupils in language at once fluent, elegant, and precise. Lord Cockburn remembers him as somewhat remarkable for the humour in which he indulged in the article of dress. " A pigtail so long and thin that it curled far down his back, and a pair of huge blue worsted hose, without soles, and covering the limbs from the heel to the top of the thigh, in which he both walked and lectured, seemed rather to improve his wise elephantine head and majestic person." He delighted in holding familiar intercourse with his pupils, whom he charmed and elevated by his brilliant conversation and his large and lofty views of life and philosophy. Rennie was admitted freely to his delightful social influence, and to the close of his career he was accustomed to look back upon the period which he spent at Edinburgh as amongst the most profitable and instructive in his life.

During his college career Rennie carefully read the works of Emerson, Switzer, Maclaurin, Belidor, and Gravesande, allowing neither pleasure nor society to divert him from his line of study. As a relief from graver topics, he set himself to learn the French and German languages, and was shortly enabled to read both with ease. His recreation was mostly of a solitary character, and, having a little taste for music, he employed some of his leisure time in learning to play upon various instruments. He acquired considerable proficiency on the flute and the violin, and he even went so far as to buy a pair of bagpipes and learn to play upon them, though the selection of such an instrument probably

does not say much for his musical taste. When he had
left Edinburgh, however, and entered seriously upon the
business of life, the extensive nature of his engagements
so completely occupied his time that in a few years, flute,
fiddle, and bagpipes were laid aside altogether.

During the three years that he attended college, our
student was busily occupied in the summer vacation—
extending from the beginning of May to the end of
October in each year—in executing millwork in various
parts of the country. Amongst the undertakings on
which he was thus employed, may be mentioned the
repair or construction of the Kirkaldy and Bonnington
Flour Mills, Proctor's Mill at Glammis, and the Carron
Foundry Mills. When not engaged on distant works,
his brother George's house at Phantassie was his head-
quarters, where he prepared his designs and specifications.
He had the use of the workshop at Houston Mill for
making such machinery as was intended for erection in
the neighbourhood; but when he was employed at some
distant point, the work was executed in the most convenient places he could find for the purpose. There
were as yet no large manufactories in Scotland where
machinery of an important character could be turned out
as a whole; the millwright being under the necessity of
sending one portion to the blacksmith, another to the
founder, another to the brass-smith, and another to the
carpenter,—a state of things involving a great deal of
trouble and often risk of failure, but which was eminently
calculated to familiarize our young engineer with the
details of every description of work required in the
practice of his profession.

His college training having ended in 1783, and being
desirous of acquiring some knowledge of English engineering practice, Rennie set out upon a tour in the
manufacturing districts. Brindley's reputation attracted
him first towards Lancashire for the purpose of inspecting
the works of the Bridgewater Canal. There being no

stage coaches convenient for his purpose, he travelled on horseback, and in this way was enabled readily to diverge from his route for the purpose of visiting any structure more interesting than ordinary. At Lancaster he inspected the handsome bridge across the Lune, then in course of construction by Mr. Harrison, afterwards more celebrated for his fine work of Chester Gaol. At Manchester he examined the works of the Bridgewater Canal; and at Liverpool he visited the docks there in progress.

Proceeding by easy stages to Birmingham, then the centre of the mechanical industry of England, and distinguished for the ingenuity of its workmen and the importance of its manufactures in metal, he took the opportunity of visiting the illustrious Boulton and Watt at Soho. His friend, Dr. Robison, had furnished him with a letter of introduction to James Watt, who received the young engineer kindly and showed him every attention; and a friendship then began which lasted until the close of Watt's life.

The condensing engine had by this time been brought into an efficient working state, and was found capable not only of pumping water—almost the only purpose to which it had formerly been applied—but of driving machinery, though whether with advantageous results was still a matter of doubt. Thus, in November, 1782, Watt wrote to his partner Boulton, " There is now no doubt but that fire-engines will drive mills, but I entertain some doubts whether anything is to be got by them." [1] About the beginning of March, 1783, however, a company was formed in London for the purpose of erecting a large corn-mill, to be driven by one of Boulton and Watt's steam-engines, and the work was in progress at the time Rennie visited Soho. Watt had much conversation with his visitor on the subject of corn-mill machinery, and was gratified to learn the extent and

---

[1] Muirhead's 'Origin and Progress of the Mechanical Inventions of James Watt,' vol. ii., p. 165.

accuracy of his information. He seems to have been
provoked beyond measure by the incompetency of his
own workmen. " Our millwrights," he wrote to his
partner, " have kept working, working, at the corn-mill
ever since you went away, and it is not yet finished ;
but my patience being exhausted, I have told them that
it must be at an end to-morrow, done or undone. There
is no end of millwrights once you give them leave to set
about what they call machinery ; here they have multi-
plied wheels upon wheels until it has now almost as
many as an orrery." [1]

Watt himself had but little knowledge of millwork,
and stood greatly in need of some able and intelligent
millwright to take charge of the fitting up of the Albion
Mills. Young Rennie seemed to him at the time to be
a very likely person ; but, with characteristic caution,
he said nothing of his intentions, but determined to
write privately to his friend Robison upon the subject,
requesting particularly to know his opinion as to the
young man's qualifications for taking the superintendence of such important works. Dr. Robison's answer
was most decided ; his opinion of Rennie's character and
ability was so favourable, and expressed in so confident
a tone, that Watt no longer hesitated ; and he shortly
after wrote to the young engineer, when he had returned
home, inviting him to undertake the supervision of the
proposed mills, so far as concerned the planning and
erection of the requisite machinery.

Watt's invitation found Rennie in full employment
again. He was engaged in designing and erecting
mills and machinery of various kinds. Amongst his
earlier works, we also find him, in 1784, when only
in his twenty-third year, occupied in superintending
the building of his first bridge—the humble forerunner
of a series of structures which have not been surpassed

---

[1] Muirhead, vol. ii., p. 177.

in any age or country. His earliest building of this
kind was erected for the trustees of the county of
Mid-Lothian, across the Water of Leith, near Stevenhouse Mill, about two miles west of Edinburgh. It is
the first bridge on the Edinburgh and Glasgow turnpike
road.

RENNIE'S FIRST BRIDGE.

[By R. P. Leitch, after a Sketch by J. S. Smiles.]

Notwithstanding the extent of his engagements and
the prospect of remunerative employment which was
opening up before him, Rennie regarded the invitation
of Watt as too favourable an opportunity for enlarging
his experience to be neglected, and, after due deliberation, he wrote back accepting the appointment. He
proceeded, however, to finish the works he had in hand;
after which, taking leave of his friends and home at
Phantassie, he set out for Birmingham on the 19th
of September, 1784. He remained there two months,
during which he enjoyed the closest personal intercourse with Watt and Boulton, and was freely admitted
to their works at Soho, which had already become the
most important of their kind in the kingdom. Birmingham was then the centre of the mechanical industry of
England. For many centuries, working in metals had
been the staple trade of the place. Swords were made

K 2

there in the time of the ancient Britons. In the reign
of Henry VIII., Leland found "many smythes in the
town that use to make knives and all manner of cutting
tools, and many loriners that make bittes, and a great
many nailers; so that a great part of the town is maintained by smythes who have their iron and sea-coal out
of Staffordshire."

The artisans of the place thus had the advantage of
the training of many generations; aptitude for handicraft, like every other characteristic of a people, descending from father to son like an inheritance. There
was then no town in England where mechanics were to
be found so capable of satisfactorily executing original
and unaccustomed work, nor has the skill yet departed
from them. Though there are now many districts in
which far more machinery is manufactured than in
Birmingham, the workmen of that place are still superior to most others in executing machinery requiring
manipulative skill and dexterity out of the common
track, and especially in carrying out new designs. The
occupation of the people gave them an air of quickness
and intelligence which was quite new to strangers accustomed to the quieter aspects of rural life. When Hutton
entered Birmingham, he was especially struck by the
vivacity of the persons he met in the streets. "I had,"
he says, "been among dreamers, but now I was among
men awake. Their very step showed alacrity. Every
man seemed to know and prosecute his own affairs."
He also adds, that men whose former disposition was
idleness no sooner breathed the air of Birmingham than
diligence became their characteristic.

Rennie did not stand in need of this infection being
communicated to him, yet he was all the better for his
contact with the population of the town. He made himself familiar with their processes of handicraft, and,
being able to work at the anvil himself, he could fully
appreciate the skill of the Birmingham artisans. The

manufacture of steam-engines at Soho chiefly attracted his notice and his study. He had already made himself acquainted with the principles as well as the mechanical details of the steam-engine, and was ready to suggest improvements, in a very modest way, even to Watt himself, who was still engaged in perfecting his wonderful machine. The partners thought that they saw in him a possible future competitor in their trade; and in the agreement which they entered into with him as to the erection of the Albion Mills, they sought to bind him, in express terms, not only to abstain from interfering in any way with the construction and working of the steam-engines required for the mills, but to prohibit him from executing such work upon his own account at any future period. Though ready to give his word of honour that he would not in any way interfere with Watt's patents, he firmly refused to bind himself to such conditions; being resolved in his own mind not to be debarred from making such improvements in the steam-engine as experience might prove to be desirable. And on this honourable understanding the agreement was concluded; nor did Rennie ever in any way violate it, but retained to the last the friendship and esteem of both Watt and Boulton.

On the 24th of November following, after making himself fully acquainted with the arrangements of the engines by means of which his machinery was to be driven, our engineer set out for London to proceed with the designing of the millwork. It was also necessary that the plans of the building—which had been prepared by Mr. Samuel Wyatt, an architect of reputation in his day—should undergo revision; and, after careful consideration, Rennie made an elaborate report on the subject, recommending various alterations, which were approved by Boulton and Watt, and forthwith ordered to be carried into effect.

CHAPTER IV.

THE ALBION MILLS—MR. RENNIE EXTENSIVELY EMPLOYED AS
AN ENGINEER.

WHEN Rennie arrived in London in 1785, the country
was in a state of serious depression in consequence of the
unsuccessful termination of the American War. Parliament was engaged in defraying the heavy cost of the
recent struggle with the revolted colonies. The people
were ill at ease, and grumbled at the increase of the
debt and taxes. The unruly population of the capital
could with difficulty be kept in order. The police and
local government were most inefficient. Only a few
years before, London had, during the Gordon riots, been
for several days in the hands of the mob, and blackened
ruins in different parts of the city still marked the track
of the rioters. Though the largest city in Europe, the
population was scarcely more than a third of what it is
now ; yet it was thought that it had become so vast as to
be unmanageable. Its northern threshold was at Hicks's
Hall, in Clerkenwell. Somers Town, Camden Town,
and Tyburnia were as yet green fields; and Kensington,
Chelsea, Marylebone, and Bermondsey were outlying
villages. Fields and hedgerows led to the hills of Highgate and Hampstead. The West End of London was
a thinly-inhabited suburb, Fitzroy Square having only
been commenced in 1793. The westernmost building
in Westminster was Millbank, a wide tract of marshy
ground extending opposite Lambeth. Executions were
conducted in Tyburn fields, long since covered with handsome buildings, down to 1783. Oxford Street, from
Princes Street eastward as far as High Street, St. Giles's,

had only a few houses on the north side. "I remember it," says Pennant, "a deep hollow road and full of sloughs, with here and there a ragged house, the lurking-place of cut-throats; insomuch that I was never taken that way by night, in my hackney-coach, to a worthy uncle's who gave me lodgings at his house in George Street, but I went in dread the whole way." Paddington was "in the country," and the communication with it was kept up by means of a daily stage—a lumbering vehicle, driven by its proprietor—which was heavily dragged into the city in the morning, down Gray's Inn Lane, with a rest at the Blue Posts, Holborn Bars, to give passengers an opportunity of doing their shopping. The morning journey was performed in two hours and a half, "quick time," and the return journey in the evening in about three hours.

Heavy coaches still lumbered along the country roads at little more than four miles an hour. A new state of things had, however, been recently inaugurated by the starting of the first mail-coach on Palmer's plan, which began running between London and Bristol on the 24th of August, 1784, and the system was shortly extended to other places. Numerous Acts were passed by Parliament authorising the formation of turnpike-roads and the erection of bridges.[1] The general commerce of the country was also making progress. The application of recent inventions in manufacturing industry gave a stimulus to the general improvement, and this was further helped by a succession of favourable harvests. The India Bill had just been renewed by Pitt, and trade with India was brisk. Besides, a commercial treaty with France was on foot, from which great things were ex-

---

[1] In the interval between 1784 and 1792, not fewer than 302 Acts were passed authorising the construction of new roads and bridges, 64 authorising the formation of canals and harbours, and still more numerous Acts for carrying out measures of drainage, enclosure, paving, and other local improvements—a sufficient indication of the industrial activity of the nation at the time.

pected; although the outbreak of the Revolution, which shortly after took place, put an end for a time to those hopes of fraternity and peaceful trade in which it had originated.   The Government boldly interposed to check smuggling, and Pitt sent a regiment of soldiers to burn the smugglers' boats laid up on Deal beach by the severity of the winter, so that the honest traders might have the full benefit of the treaty with France which Pitt had secured.   Increased trade flowed into the Thames, and ministers and monarch indulged in drawing glowing pictures of prosperity.   When Pennant visited London in 1790, he found the river covered with shipping, presenting a double forest of masts, with a narrow avenue in mid-channel.   The smaller vessels discharged directly at the warehouses along the banks of the river, whilst the India ships of large burden mostly lay down the river as far as Blackwall, and discharged into lighters, which floated up their cargoes to the city wharves. London as yet possessed no public docks,—only a few private ones of very limited extent,—although Pennant speaks of Mr. Perry's dock and ship-yard at Blackwall, on the eastern side of the Isle of Dogs, as "the greatest private dock in all Europe!"   Another was St. Saviour's, denominated by Pennant "the port of Southwark," though it was only thirty feet wide, and used for discharging barges of coal, corn, and other commodities. There was also Execution Dock at Wapping, which witnessed the occasional despatch of seagoing criminals, who were hanged on a gallows at low-water mark, and left there until the tide flowed over their dead bodies.

Among the commercial enterprises to which the increasing speculation of the times gave birth, was the erection of the Albion Mills.   For the more convenient transit of corn and flour, as well as to secure a plentiful supply of water for engine purposes, it was determined to erect the new mills on the banks of the Thames, near the south-east end of Blackfriars Bridge.   Hand-mills, which had in the first place been used for pounding wheat into flour, had long since been displaced by water-mills and windmills; and now a new agency was about to be employed, of greater power than either—the agency of steam. Fire-engines had heretofore been employed almost exclusively in pumping water out of mines; but the possibility of adapting them to the driving of machinery having been suggested to the inventive mind of James Watt, he set himself at once to the solution of the problem, and the result was the engines for the Albion Mills—the most complete and powerful which he had until then turned out of the Soho manufactory. They consisted of two double-acting engines, of the power of 50 horses each, with a pressure of steam of five pounds to the superficial inch—the two engines, when acting together, working with the power of 150 horses. They drove twenty pairs of millstones, each four feet six inches in diameter, twelve of which were usually worked together, each pair grinding ten bushels of wheat per hour, by day and night if necessary. The two engines working together were capable of grinding, dressing, &c., complete, 150 bushels an hour—by far the greatest performance achieved by any mill at that time, and probably not since surpassed, if equalled. But the engine power was also applied to a diversity of other purposes, then altogether novel—such as hoisting and lowering the corn and flour, loading and unloading the barges, and in the processes of fanning, sifting, and dressing—so that the Albion Mills came to be regarded as among the greatest mechanical wonders of the day. The details of these various ingenious arrangements were entirely worked out by Mr. Rennie himself, and they occupied him nearly four years in all, having been commenced in 1784, and set to work in 1788. Mr. Watt was so much satisfied with the result of his employment of Rennie, that he wrote to Dr. Robison, thanking him for his recommendation of his young friend, and speaking in the highest terms of the ability with which he had designed and executed the millwork and set the whole in operation.

THE ALBION MILLS.

Amongst those who visited the new mills and carefully inspected them was Mr. Smeaton, the engineer, who pronounced them to be the most complete, in their arrangement and execution, which had yet been erected in any country; and though naturally an undemonstrative person, he cordially congratulated Mr. Rennie on his success. The completion of the Albion Mills, indeed, marked an important stage in the history of mechanical improvements; and they may be said to have effected an entire revolution in millwork generally. Until then, machinery had been constructed almost entirely of wood, and it was consequently exceedingly clumsy, involving great friction and much waste of power. Mr. Smeaton had introduced an iron wheel at Carron in 1754, and afterwards in a mill at Belper, in Derbyshire —mere rough castings, imperfectly executed, and neither chipped nor filed to any particular form; and Mr. Murdock (James Watt's ingenious assistant) had also employed cast iron work to a limited extent in a mill erected by him in Ayrshire; but these were very inferior specimens of iron work, and exercised no general influence on mechanical improvement. Mr. Rennie's adoption of wrought and cast iron wheels, after a system, was of much greater importance, and was soon adopted generally in all large machinery. The whole of the wheels and shafts of the Albion Mills were of these materials, with the exception of the cogs in some cases, which were of hard wood, working into others of cast iron; and where the pinions were very small, they were of wrought iron. The teeth, both wooden and iron, were accurately formed by chipping and filing to the form of epicycloids. The shafts and axles were of iron and the bearings of brass, all accurately fitted and adjusted, so that the power employed worked to the greatest advantage and at the least possible loss by friction. The machinery of the Albion Mills, as a whole, was regarded as the finest that had been executed to that date, forming a model for future engineers to work by; and although Mr. Rennie exe-cuted many splendid specimens in his after career,[1] he

---

[1] Shortly after the completion of these mills, Mr. Rennie was largely consulted on the subject of machinery of all kinds. The Corporations of London, Edinburgh, Glasgow, Perth, and other places, took his advice as to flour-mills. Agriculturists consulted him about thrashing-mills, millers about grinding-mills, and manufacturers and distillers respecting the better arrangement of their works. He supplied plans for a steel lead-rolling-mill for Messrs. Locke and Co., at Newcastle-on-Tyne; he was called in to remedy the defective boiler-arrangements at Meux and Co.'s brewery; he advised the Government as to the power for working their small-arms manufactory at Enfield, and the Navy Board respecting the apparatus for blowing the forge at Portsmouth. In 1792 he invented the depressing sluice for water-mills, which a Government engineer, a Mr. Lloyd, afterwards brought out (in 1807) as his own invention. The Don Navigation Company's mills at Doncaster were entirely rebuilt after his designs; he sent plans of large flour-mills to one Don Diego at Lisbon, and of the extensive saw-mills erected at Archangel in Russia. In July, 1798, he was called upon to examine the machinery and arrangements at the Royal Mint on Tower Hill. The result was, the construction of an entire new mint, worked by steam-power, with improved rolling, cutting-out, and stamping machinery, after Mr. Rennie's designs. The new machinery was introduced between the years 1806 and 1810. Although it has now been in use for half a century, it continues in as efficient a working state as in the year it was erected. It is still capable of turning out from the metal, in each day of twelve hours, two and a half tons of copper, and a ton each of gold and silver coin. The whole process, as carried out by this apparatus, is extremely beautiful and effi-

himself was accustomed to say that the Albion Mill machinery was the father of them all.

As a commercial enterprise, the mills promised to be perfectly successful : they were kept constantly employed, and were realising a handsome profit to their proprietors, when unhappily they were destroyed by fire on the 3rd of March, 1791, only three years after their completion. Their erection had been viewed with great hostility by the trade, and the projectors were grossly calumniated on the ground that they were establishing a monopoly injurious to the public, which was sufficiently disproved by the fact that the mills were the means of considerably reducing the price of flour while they continued in operation. The circumstances connected with the origin of the fire were never cleared up, and it was generally believed at the time that it was the work of an incendiary. During the night in which the buildings were destroyed, Mr. Rennie, who lived near at hand, felt unaccountably anxious. A presentiment as of some great calamity hung over him, which he could not explain to himself or to others. He went to bed at an early hour, but could not sleep. Several times he went off in a doze, and suddenly woke up, having dreamt that the mills were on fire! He rose, looked out, and all was quiet. He went to bed again, and at last fell into a profound sleep, from which he was roused by the cry of "Fire!" under his windows, and the rumble of the fire-engines on their way to the mills! He dressed hastily, rushed out, and to his dismay found his chef-d'œuvre wrapt in flames, which brightened the midnight sky. The engi-

---

cient. The cutting-out and stamping-machines were the invention of the late Matthew Boulton, of Soho, but the machinery was by Mr. Rennie. On one occasion, in 1819, a million of sovereigns were turned out in eight days! During the great silver coinage in 1826, the eight presses turned out, for nine months, not less than 247,696 pieces per day, the rolling going on day and night, and the stamping for fifteen hours out of every twenty-four. Mr. Rennie also supplied the machinery for the mints at Calcutta and Bombay ; that erected at the former place being capable of turning out 200,000 pieces of silver in every eight hours.

neer was amongst the foremost in his efforts to extinguish the conflagration; but in vain. The fire had made too great progress, and the Albion Mills, Rennie's pride, were burnt completely to the ground, and never rebuilt.

The Albion Mills, however, established Mr. Rennie's reputation as a mechanical engineer, and introduced him to extensive employment. His practical knowledge of masonry and carpentry also served to point him out as a capable man in works of civil engineering, which were in those days usually entrusted to men bred to practical mechanics. There was not as yet any special class trained to this latter profession, the number of persons who followed it being very small; and these were usually determined to it by the strong instinct of constructive genius. Hence the early engineers were mainly self-educated—Smeaton, like Watt, being originally a mathematical instrument maker, Telford a stonemason, and Brindley and Rennie millwrights; force of character and bent of genius enabling each to carve out his career in his own way. The profession of engineering being still in its infancy in England, there was very little previous practice to serve for their guide, and they were called upon in many cases to undertake works of an entirely new character, in which, if they could not find a road, they had to make one. This threw them upon their own resources and compelled them to be inventive: it practised their powers and disciplined their skill, and in course of time the habitual encounter with difficulties brought fully out their character as men as well as their genius as engineers.

When the ruins of the Albion Mills had been cleared away, Mr. Rennie obtained leave from the owners to erect a workshop upon a part of the ground, wherein he continued for the rest of his life to carry on the business of a mechanical engineer. But from an early period the civil branch of the profession occupied a considerable share of his attention, and eventually it became his principal pursuit; though down to the year 1788 he was chiefly occupied in designing and constructing machinery for dye-works, water-works (at London Bridge amongst others), flour-mills, and rolling-mills, in all of which Boulton and Watt's engine was the motive power employed.

Among the friends whom Mr. Rennie's practical abilities attracted about this time, was the eccentric but ingenious Earl Stanhope, who frequently visited his works to see what was going on that was new. His Lordship was one of the busiest mechanical projectors of his time, and England owes him a debt of gratitude for his many valuable inventions, one of the most useful of which was the printing-press which bears his name. He also made important improvements in the process of stereotype printing; in the construction of locks and canals; and among his lighter efforts may be mentioned the contrivance of an ingenious machine for performing arithmetical operations. He especially delighted in the society of clever mechanics, in whose art he took great pleasure; indeed, he was himself a first-rate workman, and it was truly said of him by his father-in-law, the Earl of Chatham, that " Charles Stanhope, as a carpenter, a blacksmith, or millwright, could in any country or in any times preserve his independence and bring up his family in honest and industrious courses, without soliciting the bounty of friends or the charity of strangers." Lord Stanhope even insisted that his children should devote themselves to acquire an industrious calling, as he himself had done,—believing that a time of public calamity was approaching (arising from the extension of French revolutionary principles to England), which would render it necessary for them to depend for their livelihood upon their own personal labour and skill. Indeed a serious difficulty occurred between him and his wife on this very point, which ended in a separation; and the story went abroad that the Earl was crazed.

The application of the power of the steam-engine to the purposes of navigation was one of the subjects in which Lord Stanhope took a more than ordinary interest. As early as the year 1790—before Fulton had applied his mind to the subject—he was in communication with Mr. Rennie as to the best mode of applying this novel power, and in that year he took out his patent for the propulsion of ships by steam; but his plan, though ingenious, was never carried into practical effect.[1] On the 26th of April, 1790, we find Mr. Rennie, in a letter to the Earl, communicating the information which he required as to the cost of applying Boulton and Watt's improved steam-engine to his newly-invented method of propelling ships without sails.   His Lordship had also, it appeared, taken objection to the space occupied by the condensing apparatus, and wished to know whether it could not be dispensed with, so that room might be economized.   To this Mr. Rennie replied that it could, and that high-pressure steam might be employed if necessary; also that the cylinders might be used inclined or vertically, as best suited the space available for their accommodation.   His Lordship proceeded to perfect his invention, and made a trial of its powers in Greenland Dock with a flat-bottomed boat constructed for the purpose; but the vessel not moving with a velocity greater than three miles an hour, the plan was eventually abandoned.

Shortly after the retirement of Mr. Smeaton from the profession, about the end of the year 1791, Mr. Rennie was consulted respecting numerous important canal undertakings projected in different parts of the country; amongst others were a proposed navigation to connect Cambridge with Bury St. Edmunds—another between Andover and Salisbury—and a third between Reading and Bath, which was afterwards carried out by

---

[1] He adopted paddles, placed under the quarters of the vessel, which were made to open and shut like the feet of a duck.

him as the Kennet and Avon Canal. On this, his first
work of civil engineering in England, he bestowed great
pains,—on the survey, the designs for the viaducts and
bridges, as well as on the execution of the works themselves.

The Kennet and Avon Canal commences at Newbury,
at the head of the River Kennet Navigation, passes up
the vale of the Kennet for 16½ miles, by Hungerford to
Crofton, where the summit level begins, which is reached
by 31 locks, rising in all 210 feet. It then proceeds by
Burnslade, Wootton Rivers, and the valley of the Pewsey,
to Devizes; and from Devizes by Foxhanger, Seming-
ton, Bradford, and the vale of the Avon to Bath, joining
that river just above the Old Bath Bridge, where the
navigation from Bristol terminates. The total length of
the canal is 57 miles, the total descent on the west side
of the summit being 404 feet 6 inches, divided into
48 locks. The Kennet is crossed several times,—at
Hungerford by a brick aqueduct of three arches. At
the summit a tunnel 500 yards in extent was necessary, approached by deep cuttings. The strata between
Wootton Rivers and Devizes being mostly open chalk
and sand, great difficulty was experienced in forming a
water-tight bed for the canal, as well as in preventing
slips of the adjacent ground. At that part of the line
which lies between the river Biss and Trowbridge, the
works were carried along the face of a steep slippery hill.
Then near Bradford the cutting is mostly through open
rock, and beyond that through beds of tough clay in-
terspersed with strata of fuller's earth. The water at
these points worked serious mischief, for after a heavy
fall of rain it would filter through the earth, and the
weight of the mass pressing down from above, tended
to force out the soft clay, causing extensive slips. On
one occasion not less than seven acres of land slid
into the canal, forcing the whole down into the river in
the valley below. To remedy this source of mischief,

soughs or small tunnels were carried into the hillside for a considerable distance, at a level much below that of the canal. These again were crossed by other intercepting drains, so that numerous distinct outlets were provided for the water to prevent its reaching the canal works, which were thus made to stand after great difficulties had been overcome and much expense incurred. Besides these works, there were the usual bridges, aqueducts, culverts, &c., all of which were executed in a substantial and satisfactory manner. Among the finest architectural structures forming part of the canal is the aqueduct over the river Avon, about a mile from Limpley Stoke and six miles from Bath, which is greatly admired for the beauty of its elevation; and indeed, wherever there is an aqueduct or a bridge upon the line of this canal, it will be found excellent in workmanship and tasteful in design. As a whole, the navigation was pronounced to be one of the best executed in the kingdom; and the works have stood admirably down to the present time. In a commercial and national point of view the undertaking was of great importance, connecting as it did the navigation of the metropolis with that of Bristol and St. George's Channel, as well as opening up an extensive intermediate district; and it eventually proved highly remunerative to the proprietors.

Another important line of navigation, on which Mr. Rennie was shortly after engaged, was the Rochdale Canal, projected for the purpose of opening up a direct water communication between the manufacturing districts of West Yorkshire and South Lancashire, to avoid the circuitous route of the Leeds and Liverpool Canal. The main line extended from the Duke of Bridgewater's Canal at Manchester, by Rochdale and Todmorden, to the river Calder at Sowerby Bridge, a distance of 31½ miles, with a branch to join the Leeds and Liverpool Canal at Wanless, and other branches to Bury and Bolton. From the rugged nature of the country over

which the canal had to be carried—having to be lifted
from lock to lock over the great mountain-ridge known
as "the backbone of England"—few works have had
greater physical obstacles to encounter than this between Rochdale and Todmorden. A little before the

LOCKS ON THE ROCHDALE CANAL.
[By Percival Skelton, after his original Drawing.]

traveller by railway enters the tunnel near Littleborough,
on his way between Manchester and Leeds, he can
discern the canal mounting up the rocky sides of the
hills until it is lost in the distance; and as he emerges
from the tunnel at its other end, it is again observed
descending from the hill-tops by a flight of locks down
to the level of the railway. In crossing the range at
one place, a stupendous cutting, fifty feet deep, had to
be blasted through hard rock. In other places, where it
climbs along the face of the hill, it is overhung by
precipices. On the Yorkshire side, at Todmorden, the
valley grows narrower and narrower, overhung by steep,
often almost perpendicular, rocks of millstone-grit, with

room, in many parts, for only the water-way, the turn-pike road, and the little river Calder in the bottom of the ravine. At some points, where space allowed, there were mills and manufacturing establishments jealous of their water-supply, which the engineer of the line had carefully to avoid. It was also necessary to provide against the canal being swept away by the winter's floods of the Calder, which rushed down with immense violence from Blackstone Edge. Large reservoirs had to be carefully contrived to store up water against summer droughts for the purposes of the navigation, as well as to compensate the numerous mills along the valley below. One of these, fourteen feet deep, was dug in a bog on Blackstone Edge, and others, of large dimensions, were formed at various points along the hill-route. But as these expedients were of themselves insufficient, powerful steam-engines were also erected to pump back the lockage water into the canal above, as well as into side-ponds near the locks to serve for reservoirs, and thus economize the supply to the greatest extent. No more formidable difficulties, indeed, were encountered by George Stephenson, in constructing the railway passing by tunnel under the same range of hills, than were overcome by Mr. Rennie in carry-ing out the works of this great canal undertaking. The skill and judgment with which he planned them reflected the greatest credit on their designer; and whoever examines the works at this day—even after all that has been accomplished in canal and railway engineering—will admit that the mark of a master's hand is unmistakably stamped upon them. The navi-gation was completed and opened on the 21st of Decem-ber, 1804; and we need scarcely add that it proved of immense service to the trade of Yorkshire and Lanca-shire,—bringing important manufacturing districts into easy and economical connection with each other, enabling cheap fuel to be brought to the doors of the population of the valleys along which it passed, placing them in direct communication with the markets of Manchester and Liverpool, and, through the latter port, opening up a water road to the world at large.

LUNE AQUEDUCT, NEAR LANCASTER.

[By Percival Skelton after his original Drawing]

The Lancaster Canal was another enterprise conducted by Mr. Rennie in the same neighbourhood. A navigable communication between the coal-fields near Wigan and the lime districts about Lancaster, Burton, and Kendal, connecting these towns also with the intervening country as well as with Liverpool, Manchester, and the towns of South Lancashire, had long been regarded as an object of importance. A survey had been made by Mr. Brindley as early as 1772, but nothing further was done until some twenty years later, when a company was formed, with Mr. Rennie as engineer. The line surveyed by him commences near Wigan, and proceeds northward by Chorley, Preston, and Garstang, to Lancaster, where, skirting the east side of the town, and crossing the Lune by a noble aqueduct, it then passes by Haughbridge to its northern terminus at

Kendal; the total length of the main line being 75¾ miles, and the branches two miles more. The aqueduct over the Lune is the principal architectural work on the canal, consisting of five semicircular arches of 75 feet span each; the soffits being 50 feet, and the surface of the canal 62 feet above the average level of the river. The total length of the aqueduct—which forms a prominent feature in the landscape—is 600 feet. The whole is built of hard sandstone, the masonry being in imitation of rockwork, the top surmounted by a handsome Doric cornice and balustrade. It exhibits, in fine combination, the important qualities of strength, durability, and elegance in design; and even at this day it will bear a favourable comparison with the best works of its kind in the kingdom.

Mr. Rennie continued throughout his life to be extensively consulted as a canal engineer.[1] Though navigations were then mostly valued for purposes of internal communication, he seems early to have appreciated the uses of the railway, if not as a substitute for them, at least as an adjunct. Thus, when laying out a new branch of the Grand Trunk Canal at Henley, in the Potteries, he recommended the addition of a short descending railway, connecting the navigation directly

---

[1] The following canal works of Mr. Rennie may be mentioned:—The Aberdeen and Inverurie, 12 miles long, laid out and constructed by him in 1796-7; the Calder Reservoirs and improvement of the Trent and Mersey Canal at Rudyard Valley, near Leek, 1797-8; a branch of the Grand Trunk Canal to Henley, with a railway connecting it with the manufactories. He also made elaborate reports on the Leominster Canal (1798); on the Chelmer and Blackwater Navigation; Somersetshire and Dorsetshire Canal; Horncastle Navigation; River Foss Navigation; Polbrook Canal (1799); Rotherhithe and Croydon; Thames and Medway (1800); and River Lea Navigation (1804). Among the works surveyed by him, but which were not carried out, were these: a canal through the Weald of Kent (1802-3); a ship-canal between the Thames and Portsmouth (1803); a ship-canal between the Medway and Portsmouth (1810); a ship-canal from Chichester Harbour to Chichester (1804); and a ship-canal from Bristol to the English Channel (1811). He was also employed by the Gloucester and Berkeley Canal Company, the Birmingham Canal Company, and the Leeds and Liverpool Canal Company, as their consulting engineer; and various important improvements in these navigations were carried out by his advice and under his superintendence.

with the manufactories at Burslem.   Referring to this
method of communication, he observes that the railroad
" would form a quick and cheap mode of carrying goods.
Indeed," he adds, " I do not know a cheaper or better,
and, in my opinion, it might be substituted with great
advantage for the branch canal in question.   I have
therefore to submit whether, as a matter worthy of the
consideration of the proprietors, this branch might not
be saved, and a railroad substituted in its place."   This
report was written, be it observed, early in 1797, long
before railroads had been introduced; and the suggestion
affords striking evidence of Mr. Rennie's sagacity in
so early detecting and appreciating the advantages of
this new means of communication.

In 1802 Mr. Rennie was requested to examine the
works of the Royal Canal of Ireland.   The origin of
this project was curious.   The Grand Canal had already
been formed to connect the navigation of the Liffey
with that of the Shannon near Banagher; and though
enormous blunders had marred its construction, and its
cost had consequently been excessive, the traffic upon it
was so great as nevertheless to render it exceedingly
profitable to its proprietors.   The managing committee
consisted for the most part of persons of high rank, but
amongst them was a retired shoemaker, who had invested
a very large sum in the undertaking and made himself
exceedingly busy in its concerns.   Offence seems to have
been taken at this person, and his meddling in various
matters without authority caused a rupture between him
and the other members of the committee.   They thwarted
him at every turn, outvoted him, snubbed him, and " sent
him to Coventry."   Vowing revenge, the shoemaker
threw up his seat at the board, and, on parting with his
colleagues, said to them, " You may think me a very
insignificant person, but I will soon show you the contrary.   I will sell out forthwith, start a rival canal, and
carry all the traffic."   The threat was, of course, treated

with contempt, and the shoemaker was laughed out of the board-room. But the indignant man set to work with energy, got up a company, laid down a line of navigation from Dublin to the Shannon near Longford, passing by Mullingar, secured the support of the landed proprietors through whose property the line passed, and succeeded in obtaining an Act of Parliament authorizing the construction of the Royal Canal of Ireland, in an unusually short space of time. The works were commenced with great eclat, but, before they had proceeded far, it was found that the levels were entirely wrong, and there were numerous difficulties to be overcome for which no provision had been made. Then it was that Mr. Rennie was called in, and found the whole concern in confusion; the works at a standstill in many places, in bogs, in cuttings, in embankments, and in limestone rocks, and the proprietors involved in almost endless claims for compensation. He found it necessary to resurvey the whole line and to alter the plans in many essential respects; after which the works proceeded. It proved to be a work of an extraordinary character as regarded the difficulties, mostly unnecessary, which had been encountered in its construction; but as respected the beneficial results to the proprietors, it proved an almost total failure. The shoemaker, no doubt, had his revenge upon his former associates, inflicting great injury upon the Grand Canal by the diversion of much of its traffic; but he accomplished this at a terrible sacrifice to many, and at the almost total loss of his own fortune.

CHAPTER V.

MR. RENNIE'S DRAINAGE OF THE LINCOLN AND CAMBRIDGE FENS.

NOTWITHSTANDING all that had been done for the drainage of the Fens, as described in the early part of this work, large districts of reclaimable lands in Lincoln still lay waste and unprofitable. As early as 1789 Mr. Rennie's attention was drawn to the drowned state of the rich low-lying lands to the south of Ely; and having become impressed with a conviction of the extensive uses to which his friend Watt's steam-engine might be applied, he recommended it for pumping the water from the Botteshaw and Soham Fens, which contained about five thousand acres of what was commonly called "rotten land," because of the rot which infected the sheep depastured upon it. But he found the prejudices in favour of drainage by the old method of windmills, imported from Holland, too strong to be uprooted; and it was not until many years after, that his recommendation was adopted and the steam-engine was applied to pump the water from low-lying swamps which could not otherwise be cleared. The results were so successful that the same agency became generally employed for the purpose, not only in England but in Holland itself, where the forty-five thousand acres of Haarlemer Meer have since been effectually drained by the application of the steam-engine.

One of the most important works of thorough drainage carried out by Mr. Rennie was in that extensive district of South Lincolnshire which extends along the south verge of the Wolds, from near the city of Lincoln eastward to the sea. It included Wildmore Fen, West Fen,

and East Fen, and comprised about seventy-five thousand acres of land which lay under water for the greater
part of every year, and was thus comparatively useless
either for grazing or tillage. The only crop grown
there was tall reeds, which were used as thatch for
houses and barns, and even for churches. The river
Witham, which flows by Lincoln, had been grievously
neglected and allowed to become silted up, its bottom
being in many places considerably above the level of
the land on either side. Hence, bursting of the banks
frequently occurred during floods, causing extensive
inundation of the lower levels, only a small proportion of the flood-waters being able to force their
way to the sea. The wretched state of these lands may
be inferred·from the fact that, about seventy years since,
a thousand acres in Blankney Fen, constituting part of
" the Dales "—now one of the most fertile parts of the
district between Lincoln and Tattershall—were let annually by public auction at Harecastle, and the reserved
bid was only 10*l*. for the entire area![1] It is stated that,
about the middle of last century, there were not two
houses in the whole parish of Dogdyke communicable
with each other for whole winters round except by boat;
this being also the only means by which the Fen-slodgers
could get to church. Hall, the Fen Poet, speaks of
South Kyme, where he was born, as a district in which,
during the winter season, nothing was to be seen—

> " But naked flood for miles and miles."

The entire breadth of Lincolnshire north of Boston often
lay under water for months together :—

> " 'Twixt Frith bank and the wold side bound,
> I question one dry inch of ground.
> From Lincoln all the way to Bourne,
> Had all the tops of banks been one,
> I really think they all would not
> Have made a twenty acre spot."

---

[1] 'Journal of Royal Agricultural Society,' 1847, vol. viii., p. 124.

Until as recent a date as forty years back, the rich and fertile district of Waldersea, about eight thousand acres in extent, was, as its name imports, a sea in winter. Well might Roger Wildrake describe his paternal estate of " Squattlesea Mere " as being in the " moist county of Lincoln ! "

Arthur Young visited this district in 1793, and found the freeholders of the high lands adjoining Wildmore and West Fens depasturing their sheep on the drier parts during the summer months; but large numbers of them were dying of the rot. " Nor is this," he adds, "the only evil, for the number stolen is incredible. They are taken off by whole flocks, as so wild a country (whole acres being covered with thistles and nettles four feet high and more) nurses up a race of people as wild as the fen." The few wretched inhabitants who contrived to live in the neighbourhood for the most part sheltered themselves in huts of rushes or lived in boats. They were constantly liable to be driven out of their cabins by the waters in winter, if they contrived to survive the attacks of the ague to which they were perennially subject.

The East Fen was the worst of all. It was formerly a most desolate region, though it now presents probably the richest grazing land in the kingdom. Being on a lower level than the West and Wildmore Fens, and the natural course of the waters to the sea being through it to Wainfleet Haven, it was in a much more drowned state than those to the westward. About two thousand acres were constantly under water, summer and winter. One portion of it was called Mossberry or Cranberry Fen, from the immense quantities of cranberries upon it. A great part of the remainder of the East Fen consisted of shaking bog, so treacherous and so deep in many places that only a desperate huntsman would venture to follow the fox when he took to it, and then he must needs be well acquainted with the ground.

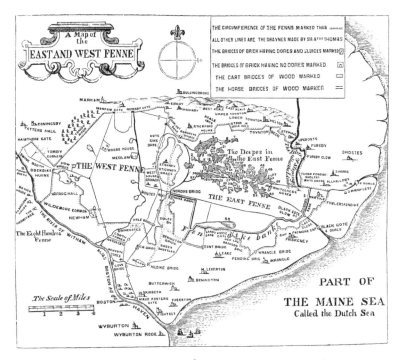

THE LINCOLNSHIRE FENS. [Before their Drainage by Mr. Rennie.]

Matters were in this state when Sir Joseph Banks, then President of the Royal Society, endeavoured to stir up the landowners to undertake the drainage of the district. He was the proprietor of a good estate at Revesby, near Tattershall; and his mansion of Abbot's Lodge, standing on an elevated spot, overlooked the waste of the East and West Fens, of which it commanded an extensive view. Sir Joseph spent a portion of every year at Revesby, as he did at his other mansions, leaving each at special times appointed beforehand, almost with the regularity of clockwork. He was a popular and well-known man, jolly and good-humoured, full of public spirit, and, though a philosopher, not above taking part in the sports and festivities of the neighbourhood in which he resided. While Sir Joseph lived at Revesby he used to keep almost open house,

and a constant succession of visitors came and went,—
some on pleasure, some on friendship, and some on
business. The profuse hospitality of the place was enjoyed not less by the postillions and grooms who drove
thither the baronet's guests than by the visitors themselves; and it was esteemed by the hotel postboys a great
privilege to drive a customer to Revesby. On one occasion, when Mr. Rennie went to dine and sleep at the
Lodge, he took an opportunity of saying to the principal butler that he hoped he would see to his postboy
being kept *sober*, as he wished to leave before breakfast
on the following morning. The butler replied, with
great gravity, that he was sorry he could not oblige
Mr. Rennie, as the same man had left Revesby sober
the last time he was there, but only on condition that he
might be allowed to get drunk the next time he came.
" Therefore," said the butler, " for the honour of the
house, I must keep my word; but I will take care that
you are not delayed for the want of horses and a
postboy." The butler was as good as his word: the
man got drunk, the honour of Revesby was saved, and
Mr. Rennie was enabled to set off in due time next
morning.

From an early period Sir Joseph Banks entertained
the design of carrying out the drainage of the extensive
fen lands lying spread out beneath his hall window,
and making them, if possible, a source of profit to the
owners, as well as of greater comfort and better subsistence for the population. Indeed, the reclamation of
these unhealthy wastes became quite a hobby with him;
and when he could lay hold of any agricultural im-
prover, he would not let him go until he had dragged him
through the Fens, exhibited what they were, and demon-
strated what fertile lands they might be made. When
Arthur Young visited Revesby about 1799, Sir Joseph
immediately started his favourite topic. " He had the
goodness," says Young, in his Report on Lincolnshire,

"to order a boat, and accompanied me into the heart
of East Fen, which had the appearance of a chain of
lakes, bordered by great crops of reed." Sir Joseph
was a man of great public spirit and determination : he
did not allow the matter to sleep, but proceeded to
organize the ways and means of carrying his design
into effect. His county neighbours were very slow to
act, but they gradually became infected by his example,
and his irresistible energy carried them along with him.
The first step taken was to call meetings of the proprietors in the several districts adjoining the drowned
and "rotten lands." Those of Wildmore Fen met at
Horncastle on the 27th of August, 1799, and resolu-
tions were adopted authorizing the employment of Mr.
Rennie to investigate the subject and report to a future
meeting.

Onc reason, amongst others, which weighed with Sir
Joseph Banks in pressing on the measure was the
scarcity of corn, which about that time had risen almost
to a famine price. There was also great difficulty in
obtaining supplies from abroad, in consequence of the
war which was then raging. Sir Joseph entertained
the patriotic opinion that the best way of providing for
the exigency was to extend the area of our English food-
ground by the reclamation of the waste lands ; and hence
his determination to place under tillage, if possible, the
thousands of acres of rich soil, equal to the area of some
English counties, lying under water almost at his own
door. A few years' zealous efforts, aided by the skill
of his engineer, produced such results as amply to
justify his anticipations, and proved his patriotism to
be as wise as his public spirit was beneficent.

The manner in which Mr. Rennie proceeded to work
out the problem presented to him was thoroughly characteristic of the man. Most of the drainage attempted
before his time was of a very partial and inefficient
character. It was enough if the drainers got rid of the

surplus water anyhow, either by turning it into the
nearest river, or sending it in upon a neighbour. What
was done in one season was very often undone, or
undid itself, in the next. The ordinary drainer did not
care to look beyond the land immediately under his own
eyes. Mr. Rennie's practice, on the contrary, was founded
on a large and comprehensive view of the whole subject. He was not bounded by the range of his physical vision, but took into account the whole contour of
the country; the rainfall of the districts through which
his drains were to run, as well as of the central
counties of England, whose waters flowed down upon
the Fens; the requirements of the lands themselves as
regarded irrigation and navigation; and the most effectual method not only of removing the waters from
particular parts, but of providing for their effectual
discharge by proper outfalls into the sea.

What was the problem now to be solved by our
engineer? It was how best to carry out to sea the
surplus waters of a district extending from the eastern
coast to almost the centre of England. Various streams
descending from the Lincolnshire wolds flowed through
the level, whilst the Witham brought down the rainfall
not only of the districts to the north and east of Lincoln, but of a large part of the central counties of
Rutland and Leicester. It was therefore necessary to
provide for the clear passage of these waters, and also to
get rid of the drainage of the Fens themselves, a considerable extent of which lay beneath the level of the
sea at high water. It early occurred to Mr. Rennie
that, as the waters of the interior for the most part came
from a higher level, their discharge might be provided
for by means of distinct drains, and prevented from at
all mingling with those of the lower lying lands. But
would it be possible to "catch" these high land waters
before their descent upon the Fens, and then to carry
them out to sea by means of independent channels?

He thought it would; and with this leading idea in his mind he proceeded to design his plan of a great " catch-water drain," extending along the southern edge of the Lincolnshire wolds.

But there were also the waters of the Fens themselves to be got rid of, and how was this to be accomplished? To ascertain the actual levels of the drowned land, and the depth to which it would be necessary to carry the outfall of his drains into the sea, he made two surveys of the district, the first in October, 1799, and the second in March, 1800,—thus observing the actual condition of the lands both before and after the winter's rains. At the same time he took levels down to the sea outfalls of the existing drains and rivers. He observed that the Wash, into which the Fen waters ran, was shallow and full of shifting sands and silt. He saw that during winter the rivers were loaded with alluvial matter held in suspension, and that at a certain distance from their mouths the force of the inland fresh and the tidal sea waters neutralized each other, and there a sort of stagnant point was formed, at which the alluvium was no longer held in suspension by the force of the current. Hence it became precipitated in the channels of the rivers, and formed banks or bars in the Wash outside their mouths, which proved alike obstructive to drainage and navigation.

It required but little examination to detect the utter inadequacy of the existing outfalls to admit of the discharge of the surplus waters of so extensive a district. The few sluices which had been provided had been badly designed and imperfectly constructed. The levels of the outfalls were too high, and the gowts and sluices too narrow, to accommodate the drainage in flood-times. These outfalls were also liable, in dry summers, to become choked up by the silt settling in the Washes; and when a heavy rain fell, down came the waters from the high lands of the interior, and, unable to find an outlet, they burst the defensive banks of the rivers, and an amount of mischief was thus done which the drainage of all the succeeding summer failed to repair. Accordingly, the next essential part of Mr. Rennie's scheme was the provision of more effectual outfalls; with which object he designed that they should be cut down to the lowest possible level of low water, whilst he arranged that at the points of outlet they should be mounted with strong sluices, opening outwards; so that, whilst the fresh waters should be allowed freely to escape, the sea should be valved back and prevented flowing in upon the land. The third and last point was to provide for the drainage of the Fen districts themselves by means of proper cuts and conduits for the voidance of the Fen waters.

Such were the general conclusions formed by Mr. Rennie after a careful consideration of the circumstances of the case, which he embodied in his report to the Wildmore Fen proprietors[1] as the result of his investigations. The two great features of his plan, it will be observed, were (1) his intercepting or catchwater drains, and (2) his cutting down the outfalls to lower levels than had ever before been proposed. Simple though his system appears, now that its efficacy has been so amply proved by experience, it was regarded at the time as a valuable discovery in the practice of fen-draining, and indeed it was nothing less. There were, however, plenty of detractors, who alleged that it was nothing of the kind. Any boy, they said, who has played at dirt pies in a gutter, knows that if you make an opening sufficiently low to let the whole contained water escape, it will flow away. Very true; yet the thing had never been done until Mr. Rennie proposed it, and, simple though the method was, it cost him many years of arguing, illustration, and enforcement, before he could induce intelligent men in other districts to adopt the

---

[1] Report, dated the 7th of April, 1800.

simple but thoroughly scientific method which he thus invented for the effectual discharge of the drainage of the Fens. And even to this day there are whole districts in which the stubborn obstinacy of ignorant obstructives still continues to stand in the way of its introduction. The Wildmore Fen proprietors, however, had the advantage of being led by a sagacious, clear-seeing man in Sir Joseph Banks, who cordially supported the adoption of the proposed plan with all the weight of his influence, and Mr. Rennie was eventually empowered to carry it into execution.

In laying out the works, he divided them according to their levels, placing Wildmore and West Fen in one plan, and East Fen in another. In the drainage of the former, the outlet was made by Anton's Gowt, about two miles and a half above Boston, and by Maud Foster, a little below that town. But both of these, being found too narrow and shallow, were considerably enlarged and deepened, and provided with double sluices and lifting gates : one set pointing towards the Witham in order to keep out the tides and river-floods; the other to the land, in order to prevent the water in summer from draining too low, and thereby hindering navigation as well as the due irrigation of the lands. An extensive main drain was also cut through the Wildmore and West Fens to the river Witham, about twenty-one miles long and from eighteen to thirty feet wide, the bottom being an inclined plane falling six inches in the mile.

The level of the East Fen being considerably lower than that of the Fens to the westward, it was necessary to provide for its separate drainage, but on precisely the same principles. From the levels which were taken, it appeared that the bottom of "the Deeps," which formed part of the East Fen, was only two feet six inches above the cill of Maud Foster Sluice, thirteen miles distant; whereas its highest parts were but eight feet above the same point, giving a fall of only an inch and eight tenths per mile at low water of neap tides. From some of the more distant parts of the same Fen, sixteen miles from the outfall, there would only have been a fall of five tenths of an inch per mile at low water. It was clear, therefore, that even the higher levels of the East Fen could not be effectually drained by the outfall at Anton's Gowt or Maud Foster; and hence arose the necessity for cutting an entirely separate main drain, with an outfall at a point in the Wash outside the mouth of the river Witham.[1] This east main cut, called the Hobhole Drain, is about eighteen miles long and forty feet wide, diminishing in breadth according to its distance from the outfall; the bottom being an inclined plane falling four inches in the mile towards the sluice at Hobhole in the Wash. This drain is an immense work, defended by broad and lofty embankments extending inland from its mouth, to prevent the contained waters flooding the surrounding lands. It is protected at its sea outlet by a strong sluice, consisting of three openings of fifteen feet each. When the tide rises, the gates, acting like a valve, are forced back and hermetically closed; and when it falls, the drainage waters, which have in the mean time accumulated, force open the gates again, and the waters flow away down to the level of low water. A connection was also formed between the main drains emptying themselves at Maud Foster (three miles higher up the Witham) and the Hobhole Drain, the flow being regulated by a gauge; so that, during heavy floods, not only the low land waters of the East Fen districts were effectually discharged at Hobhole, but also a considerable portion of the drainage of the West and Wildmore Fens.

An essential part of the scheme was the cutting of the catchwater drains, which were carried quite round the

---

[1] See the Map, in Vol. I., of the " Fens as Drained in 1830;" p. 51.

base of the high lands skirting the Fens; beginning
with a six feet bottom, and widening out towards their
embouchures to sixteen feet. The principal work of this
kind commenced near Stickney, and was carried eastward towards Wainfleet, to near the Steepings river. It
was connected at Cowbridge with the main Hobhole
Drain, into which the high land waters brought down
by the catchwater drain were thus carried, without
having been allowed, until reaching that point, to mix
with the Fen drainage at all. It would be tedious to
describe the works more in detail; and perhaps the outline we have given, aided by the maps of the district,
will enable the reader to understand the leading features
of Mr. Rennie's comprehensive design. The works were
necessarily of a very formidable character, the extent of
the main and arterial drains cut during the seven or
eight years they were under execution being upwards
of a hundred miles. They often dragged for want of
funds, and encountered considerable opposition in their
progress; though the wisdom of the project was in all
respects amply justified by the result.[1]

The drainage of Wildmore and West Fens was first
finished, when forty thousand acres of valuable land
were completely reclaimed, and in a few years yielded

---

[1] The following letter, written by
a Lincolnshire gentleman, in January,
1807, appears in the 'Farmer's Magazine' of February in that year:—
"Our fine drainage works begin now
to show themselves, and in the end
will do great credit to Mr. Rennie,
the engineer, as being the most complete drainage that ever was made in
Lincolnshire, and perhaps in England.
I have been a commissioner in many
drainages, but the proprietors never
would suffer us to raise money sufficient to dig deep enough through the
old enclosures into the sea before;
and, notwithstanding the excellency
of Mr. Rennie's plan, we have a party
of uninformed people, headed by a

little parson and magistrate, who keep
publishing letters in the newspapers
to stop the work, and have actually
petitioned Sir Joseph Banks, the lord
of the manor, against it; but he answered them with a refusal, in a most
excellent way. . . . I think Mr. Rennie's great work will promote another
general improvement here, which is,
to deepen and enlarge the river
Witham from the sea, through Boston
and Lincoln, to the Trent, so as to
admit of a communication for large
vessels, as well as laying the water so
much below the surface of the land
as to do away with the engines. We
have got an estimate, and find the
cost may be about 100,000l."

M 2

heavy crops of grain. East Fen was attacked the last, the difficulties presented by its formidable chain of lakes being much the greatest; but the prize also was by far the richest. When the East Fen waters were drained off, the loose black mud settled down into fertile soil. Boats, fish, and wildfowl disappeared, and the plough took their place. After being pared and burnt, the land in the East Fen yielded two and even three crops of oats in succession, of not less than ten quarters to the acre. The cost of executing the drainage had no doubt been very great, amounting to about 580,000*l.* in all, inclusive of expenditure on roads, &c.; against which had to be set the value of the lands reclaimed. In 1814 Mr. Anthony Brown, surveyor and valuer, estimated their improved rental at 110,561*l.*; and allowing five per cent. on the capital expended on the works, we thus find the increased net value of the drained lands to be not less than 81,000*l.* per annum, which, at thirty years' purchase, gives a total increased value of nearly two millions and a half sterling! [1]

It was a matter of great regret to Mr. Rennie that his design was not carried out as respected the improved outfall of the Witham. It was an important part of his original plan that a new and direct channel should be cut for this river from Boston down to deep water at Clayhole, where the tide ebbed out to the main sea level, and there was little probability of the depth being

---

[1] Mr. Brown's estimate was as follows:—

Land reclaimed in the East Fen	..	12,664	acres worth 40s. per ann.			£25,328
„ „ West Fen	..	17,044	„	50s.	„	42,610
„ „ Wildmore	..	10,773	„	42s.	„	22,623
Adjoining low lands improved	..	20,000	„	20s.	„	20,000

Total acreage of improved and drained lands	60,481	Annual value £110,561

Less capital expended on drainage ...	£433,905
„ „ roads, &c...	146,800

£580,705 at 5 per cent.	.. ..	29,035
		£81,526

Which at 30 years' purchase gives £2,445,780.

materially interfered with by silting for many years to come. This new channel would have enabled all the waters—low land as well as high land—to be discharged into the sea with the greatest ease and certainty. It would also have completely restored the navigation of the river, which had become almost entirely lost through the silting up of its old winding channel. But the Witham was under the jurisdiction of the corporation of Boston, who were staggered by the estimated cost of executing the proposed works, though it amounted to only 50,000l. Accordingly, nothing was done to carry out this part of the design, and the channel continued to get gradually worse, until at length it was scarcely possible even for small coasters to reach Boston Quay. As late as the year 1826 the water was so low that little boys were accustomed to amuse themselves by wading across the river below the town even at high water of neap tides. The corporation were at last compelled to bestir themselves to remedy this deplorable state of affairs, and they called in Sir John Rennie to advise them in their emergency. The result was, that as much of the original plan of 1800 [1] was carried out as the state of their funds would permit: the lower part of the channel was straightened, and the result was precisely

---

[1] In his admirable Report, dated the 6th October, 1800, Mr. Rennie pointed out that the lines of direction in which the rivers Welland and Witham entered the Wash tended to the silting-up of the channels of both, and he suggested that the two river outlets should be united in one, and diverted into the centre of the Wash, at Clayhole, which would at the same time greatly increase the depth, and enable a large area of valuable land to be reclaimed for agricultural purposes. This suggestion has since been elaborated by Sir John Rennie, whose plan of 1837, when fully carried out, will have the effect of greatly improving the outfalls of all the rivers entering the Great Wash—the Ouse, the Nene, the Welland, and the Witham—and the drainage of the low level lands depending upon them, comprising above a million of acres, and ultimately gaining from the Wash between 150,000 and 200,000 acres of rich new land, or equal 'to the area of a good-sized county. In the Wash of the Nene, called Sutton Wash, 4000 acres have already been reclaimed after this plan—the land, formerly washed by the sea at every tide, being now covered with rich cornfields and comfortable farmsteads. It was at this point that King John's army was nearly destroyed when crossing the sands before the advancing tide.

that which the engineer had more than thirty years before anticipated. The tide returned to the town, the shoals were removed, and vessels drawing from twelve to fourteen feet water could again come up to Boston Quays at spring tides.

Mr. Rennie was equally successful in carrying out drainage works in other parts of the Fens, and on the same simple but comprehensive principles.[1] He thus drained the low lands of Great Steeping, Thorpe, Wainfleet, All Saints, Forsby, and the districts thereabouts, converting the Steepings river into a catchwater drain, and effectually reclaiming a large acreage of highly valuable land. He was also consulted as to the better drainage of the North Level, the Middle Level, South Holland, and the Great Bedford Level; and his valuable reports on these subjects, though not carried out at the time, for want of the requisite means, or of public spirit on the part of the landowners, laid the foundations of a course of improvement which has gone on until the present day. It is much to be regretted that his grand plan of 1810 for the drainage of the Great Level, by means of more effectual outfalls and a system of intercepting catchwater drains, was not carried out; for there is every reason to believe that it would have proved as completely successful as his drainage of the Fens of Lincolnshire. But the only part of this scheme that was executed in his time was the Eau Brink Cut, for the purpose of securing a more effectual outfall of the river into the Wash near King's Lynn.

The necessity for this work will be more clearly under-

---

[1] Among other important works of the same kind executed by Mr. Rennie, but which it would be tedious to describe in detail, was the reclamation (in 1807) of 23,000 acres of fertile land in the district of Holderness, near Hull. He was extensively employed to embank lands exposed to the sea, and succeeded (in 1812) in effectually protecting the thirty miles of coast extending from Wainfleet to Boston, and thence to the mouths of the rivers Welland and Glen. Two years later (in 1814) he, in like manner, furnished a plan, which was carried out, for protecting the Earl of Lonsdale's valuable marsh land on the south shore of the Solway Frith.

stood when we explain the circumstances under which its construction was recommended. It will be observed from the map of the Fen district (vol. i., p. 51), that the river Ouse flows into the shallows of the Wash near the town of King's Lynn, charged with the waters of the Great Bedford Level as well as those of Huntingdon, Bedford, and Cambridge, and of the high lands of the western parts of Norfolk and Suffolk. Immediately above Lynn the old river made an extensive bend of about five miles in extent, to a point called German's Bridge. This channel was of very irregular breadth and full of great sand beds which were constantly shifting. In some places it was as much as a mile in width, and divided into small streams which varied according as the tidal or the fresh waters were for the time being most powerful. During floods, the flow of the river was so much obstructed that the waters could not possibly get away out to sea during the ebb, so that at the next rise of the tide they were forced back into the interior, and thus caused serious inundations in the surrounding country.[1] The fresh waters were in this way penned up within the land to the extent of about seven feet; and over an extensive plain, such as the Bedford Level, where a few inches of fall makes all the difference between land drained and land drowned, it is clear how seriously this obstruction of the Ouse outfall must have perilled the agricultural operations of the district. Until,

---

[1] When Mr. Rennie was first consulted respecting the drainage of the Great Level, he found that much good land which had been formerly productive had become greatly deteriorated, or altogether lost for purposes of agriculture. Some districts were constantly flooded, and others were so wet that they were rapidly returning to their original state of reeds and sedge. In the neighbourhood of Downham Eau, the harvestmen were, in certain seasons, obliged to stand upon a platform to reap their corn, which was carried to and from the drier parts in boats; and some of the farmers, in like manner, rowed through their orchards in order to gather the fruit from the trees. A large portion of Littleport Fen, in the South Level, was let at 1s. an acre, and, in the summer-time, stock were turned in amongst the reed and "turf-bass," and not seen for days together. In Marshland Fen, the soil was so soft that wooden shoes, or flat boards, were nailed on the horses' feet over the iron ones, to prevent them from sinking into the soil.

therefore, this great impediment to the drainage of the
Level could be removed, it was clear to Mr. Rennie's
mind that no inland works could be of any permanent
advantage.   The remedy which he proposed was, to cut
off the great bend in the Ouse by making a direct new
channel from Eau Brink, near the mouth of Marshland
Drain, to a point in the river a little above the town of
Lynn, as shown in the following plan :—

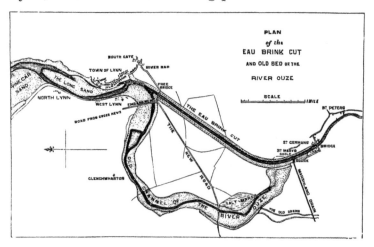

The cut was to be about three miles in length, and of
sufficiently capacious dimensions to contain the whole
body of the river.   By thus shortening the line of the
stream, Mr. Rennie calculated that the channel would be
kept clear of silt by the greater velocity of the current,
and that the fresh waters would at the same time be able
to force their way out to sea without difficulty.   An Act
was accordingly obtained enabling the Eau Brink to
be cut ; but some years passed before any steps were
taken to carry out the works, which were not actually
begun until the year 1817, when Mr. Rennie was formally
appointed the chief engineer.   After about four years'
labour the cut was finished and opened, and its immediate effect was to give great relief to the whole of the
district watered by the Ouse.   An extra fall of not less
than five feet and a half was obtained at St. German's,

by which the surface of the waters throughout the whole of the Middle and South Levels was reduced in proportion. Thus the pressure on all the banks along the rivers of the Level was greatly relieved, whilst inundations were prevented, and the sluices provided for the evacuation of the inland waters were enabled effectually to discharge themselves.

By labours such as these an immense value has been given to otherwise worthless swamps and wastes. The skill of the engineer has enabled the Fen farmers to labour with ever-increasing profit, and to enjoy the fruits of their industry in comparative health and comfort. No wonder they love the land which has been won by toil so protracted and so brave. Unpicturesque though the Fens may be to eyes accustomed to the undulating and hill country of the western districts of England, they nevertheless possess a humble beauty of their own, especially to eyes familiar to them from childhood. The long rows of pollards, with an occasional windmill, stretching along the horizon as in a Dutch landscape—the wide extended flats of dark peaty soil, intersected by dykes and drains, with here and there a green tract covered with sleek cattle—have an air of vastness, and even grandeur, which is sometimes very striking. To this we may add, that the churches of the district, built on sites which were formerly so many oases in the watery desert, loom up in the distance like landmarks, and are often of remarkable beauty of outline.

It has been said of Mr. Rennie that he was the greatest "slayer of dragons" that ever lived,—this title being given in the Fens to persons who, by skill and industry, have perfected works of drainage, and thereby removed the causes of sickness and disease, typified in ancient times as dragons or destroyers.[1] In this sense, certainly, Mr. Rennie is entitled, perhaps more than any other man, to this remarkable appellation.

---

[1] Thompson's 'History of Boston,' 1856, p. 639.

CHAPTER VI.

Mr. Rennie's Bridges.

The bridges erected by our engineer are amongst the finest of his works, and sufficient of themselves to stamp him as one of the greatest masters of his profession. We have already given a representation of his first bridge, erected over the Water of Leith, near Edinburgh, the forerunner of a series of similar structures unrivalled for solidity and strength, contrived with an elegance sometimes ornate, but for the most part of severe and massive simplicity.

Unlike some of his contemporaries, Mr. Rennie did not profess a disregard for theory; for he held that true practice could only be based on true theory. Taken in the sense of mere speculative guessing, however ingenious, he would have nothing to do with it; but as matter of inference and demonstration from fixed principles, he held by theory as his sheet-anchor. His teacher, Professor Robison, had not failed to impress upon him its true uses in the pursuit of science and art; and he never found reason to regret the fidelity with which he carried out his instructions in practice. In 1793 he had the advantage of much close personal intercourse with his old friend the Professor, who paid him a visit at his house in London for the express purpose of conferring with him upon mechanical subjects. In the letter announcing the object of his visit, Dr. Robison candidly avowed that it was in order " that he might extract as much information from him as possible." The Doctor had undertaken to prepare the articles on Mechanics for the third edition of the ' Encyclopedia

Britannica,' and he believed he should be enabled to impart an additional value to his writings by throwing upon
them the light of Rennie's strong practical judgment.
He proposed to take a lodging in the immediate neighbourhood of Rennie's house, then in the Great Surrey
Road, and to board with him during the day ; but Rennie
would not listen to this proposal, and insisted on being
the Professor's entertainer during the period of his visit.

One of the points which he particularly desired to
discuss with Mr. Rennie was the theory of the equilibrium of arches—a subject at that time very imperfectly understood, but which the young engineer had
studied with his usual energy and success. He had
clearly proved that the proper proportion and depth
of the key-stone to that of the extrados (or exterior
curve) should be in proportion to the size and form
of the arch and the materials of which it was composed ; and he had also established the ratio in which the
arch-stones should increase from the key-stone to the
piers or abutments. Up to this time there had been no
rules laid down for the guidance of the engineer or architect, who worked very much in the dark as to principles ;
and it was often a matter entirely of chance whether a
bridge stood or fell when the centres were removed.
According to the views of Hutton and Attwood, the
weight upon the haunches and abutments, to put the
arch in a state of equilibrium so that it should stand, was
unlimited ; whereas Mr. Rennie established the limit to
which the countervailing force or weight on the extrados
should be confined. Hence he adopted the practice of
introducing a flat inverted arch between the extrados of
each two adjoining arches, (at the same time increasing
the width of the abutment,)—the radii of the voussoirs or arch-stones being continued completely through
them. And in order to diminish the masonry, the
lower or foundation course was inclined also,—thus
combining the work more completely together, and

enabling it better to resist the lateral thrust. Dr. Robison had much discussion with Mr. Rennie on these and many other points, and the information he obtained was shortly after worked up into numerous original contributions of great value; amongst which may be mentioned his articles in the 'Encyclopedia' on the Arch, Carpentry, Roof, Waterworks, Resistance of Fluids, and Running of Rivers [1]—on all of which subjects Mr. Rennie had much original information to impart. It may readily be imagined that the evenings devoted by Dr. Robison to conversation and discussion on such topics at Rennie's house were of interest and advantage to both; and when the Doctor returned to his Edinburgh labours, he carried with him the cordial affection and respect of the engineer, who continued to keep up a correspondence with him until the close of his life.

In the early part of his career Mr. Rennie was called upon to furnish designs of many bridges, principally in Scotland, which, however, were not carried out, in most cases because the requisite funds could not be raised to build them. Thus, in 1798, he designed one of eight cast iron arches to span the river Don at Aberdeen. Four years later he was called upon to furnish further designs, when he supplied three several plans, two of granite bridges; but the structures were of too costly a character for the people of Aberdeen then to carry out. The first important bridge which Mr. Rennie was authorised to execute was that across the Tweed at Kelso, and it afforded a very favourable specimen of his skill as an architect. It was designed in 1799 and opened in 1803. It consists of five semi-elliptical arches of 72 feet span, each rising 28 feet, and four piers each 12 feet thick,

---

[1] Dr. Robison was the first contributor to the 'Encyclopedia' who was really a man of science, and whose articles were above the rank of mere compilations. He sought information from all quarters—searched the works of foreign writers, and consulted men of practical eminence, such as Rennie, to whom he could obtain access,—and extraordinary value was thus imparted to his articles.

with a level roadway 23 feet 6 inches wide between the
parapets, and 29 feet above the ordinary surface of the
river. The foundations were securely laid upon the solid
rock in the bed of the Tweed, by means of coffer-dams,
and below the deepest part of the river. The piers and
abutments were ornamented with three-quarter columnar
pilasters of the Roman Doric order, surmounted by a plain
block cornice and balustrade of the same character. The
whole of the masonry was plain rustic coursed work, and
in style and execution it was long regarded as one of the
most handsome and effective structures of its kind. It
may almost be said to have formed the commencement
of a new era of bridge-building in this country. The
semi-elliptical arches, the columnar pilasters on the piers,
the balustrade, and the level roadway, are the same
as in Waterloo Bridge, except as regards size and character ; so that Kelso Bridge may be regarded as the
model of the greater work. We believe it was one of
the first bridges in this country constructed with a *level*

KELSO BRIDGE. [By Percival Skelton.]

MUSSELBURGH BRIDGE.

[By E. M. Wimperis, after a Drawing by J. S. Smiles.]

roadway. Some of the old-fashioned bridges were excessively steep, and to get over them was like climbing the roof of a house. There was a heavy pull on one side and a corresponding descent on the other. The old bridge across the Esk at Musselburgh, forming part of the high road between Edinburgh and London, was of this precipitous character. It was superseded by a handsome and substantial bridge, with an almost level roadway, after a design by Rennie. When the engineer was taking the work off the hands of the contractor, one of the magistrates of the town, who was present, asked a countryman who was passing at the time with his cart how he liked the new brig? "Brig!" said the man, "it's nae brig ava! ye neither ken whan ye're on't, nor whan ye're aff't!"

Mr. Rennie's boldness in design grew with experience, and when consulted as to a bridge near Paxton, over the Whitadder (a rapid stream in Berwickshire), he proposed, in lieu of the old structure, which had been carried away by a flood, a new one of a single arch, of 150 feet span; but unhappily the road trustees could not find the requisite means for carrying it into effect. Another abortive but grand design was proposed by him in 1801. He had been requested by the Secretary of State for Ireland to examine the road through North Wales to Holyhead, with the object of improving the communication with Ireland, which was then in a wretched state. The connection of the opposite shores of the Menai Strait by means of a bridge was considered an indispensable part of any improvement of that route; and Mr. Rennie proposed to accomplish this object by a single great arch of cast-iron 450 feet in span,—the height of its soffit or crown to be 150 feet above high water at spring tides.[1] A similar bridge, of 350 feet span, having its crown 100 feet above the same level, was also proposed by him for the crossing of Conway Ferry. These bridges were to be manufactured after a plan invented by Mr. Rennie in 1791, and communicated by him to Dr. Hutton in 1794; and he was strongly satisfied of its superiority to all others that had been proposed. The designs were alike

---

[1] The great arch of 450 feet was to be supported on two stone piers, each 75 feet thick, the springing to be 100 feet above high water. There were to be arches of stone on the Caernarvon side to the distance of about 156 yards, and on the Anglesea side to the distance of about 284 yards; making the total length of the bridge, exclusive of the wing walls, about 640 yards. The estimated cost of the whole work and approaches was 268,500*l.* The point at which the bridge was recommended to be thrown across was, either opposite Inys-y-Moch island, on which one of the main piers would rest, or at the Swilly rocks, about 800 yards to the eastward; but, on the whole, he preferred the latter site. He also sent in a subsequent design, showing an iron arch on each side of the main one of 350 feet span, in lieu of masonry, with other modifications, by which the dimensions of the main piers were reduced, and the estimate somewhat lessened. Other plans were prepared and submitted, embodying somewhat similar views, the prominent idea in all of them being the spanning of the strait by a great cast-iron arch, the crown of which was to be 150 feet above the sea at high-water. The plans and evidence on the subject are to be found set forth in the 'Reports from Committees of the House of Commons on Holyhead Roads' (1810-22), ordered to be printed 25th July, 1822.

bold and skilful, and it is to be regretted that they were not carried out; for their solidity would not only have proved sufficient for the purposes of a roadway, but probably also of a locomotive railway. In that case, however, we should have been deprived of the after-display of much engineering ability in bridging the straits at Menai and the ferry at Conway. But the plans were thought far too daring for the time, and the expense too great. The whole subject was therefore allowed to sleep for many years, until eventually Telford spanned both these straits with suspension road bridges, and Robert Stephenson afterwards with tubular railway bridges, at a total cost of about a million sterling.

BOSTON BRIDGE. [By Percival Skelton.]

The first bridge constructed by Mr. Rennie in England, and the earliest of his cast iron bridges, was that erected by him over the Witham, in the town of Boston, Lincolnshire, in 1803. It consists of a single arch of iron ribs, forming the segment of a circle, the chord of which is 80 feet. It is simple yet elegant in design; its flatness and width contributing to render it most convenient for the purpose for which it was intended —that of accommodating the street-traffic of one of the most prosperous and busy towns in the Fens.

Mr. Rennie's reputation as an engineer becoming well established by these and other works, he was, during the remainder of his professional career, extensively consulted on this branch of construction;[1] and many solid memorials of his skill in bridge-work are to be found in different parts of the kingdom. But the finest of the buildings of this character which were erected by him are unquestionably those which grace the metropolis itself.

---

[1] Among his minor works may be mentioned the bridge over the stream which issues out of Virginia Water and crosses the Great Western Road (erected in 1805); Darlaston Bridge across the Trent, in Staffordshire (1805); the timber and iron bridge over the estuary of the Welland at Fossdyke Wash, about nine miles below Spalding (1810); the granite bridge of three arches at New Galloway, on the line of the Dumfries and Portpatrick Road (1811); a bridge of five arches across the Cree at Newton Stewart (1812); the cast iron bridge over the Goomtee at Lucknow, erected after his designs in 1814, and frequently referred to in the military operations for the relief of that city a few years ago; Wellington Bridge, over the Aire, at Leeds (1817); Isleworth Bridge (1819); a bridge of three elliptical arches of 75 feet span each, at Bridge of Earn, Perthshire (1819); Cramond Bridge, of eight semi-circular arches of 50 feet span, with the roadway 42 feet above the river (1819); and Ken Bridge, New Galloway, of five stone arches, the centre 90 feet span (1820). An adventure of some peril attended Mr. Rennie's erection of the bridge at Newton Stewart. He happened to visit the works on one occasion during a heavy flood, which swept down the valley with great fury; and the passage of the ferry was thus completely interrupted. Mr. Rennie and his son (the present Sir John) were consequently unable to cross over to Newton Stewart, on the further side of the river, and they were under the necessity of spending the night in a miserable public-house on the eastern bank. About 11 P.M. the violence of the storm had somewhat abated, and the moon came out, though obscured by the clouds which drifted across her face. Mr. Rennie went out at that late hour to look at the bridge works, and even to try whether he might not reach the other side by crossing the timber platform by means of which the works were being carried on. There was a gangway of only two planks from pier to pier on the eastern side, and this he safely crossed. The torrent was still raging furiously beneath, shaking the frail timbers of the scaffolding. As Mr. Rennie was about to place his foot on the plank which led to the third pier, his son observed the framework tremble, and pulled his father back, just in time to see the whole swept into the stream with a tremendous crash. Fortunately the planking still stood across which they had passed, and they succeeded in retracing their steps in safety. The bridge was finished and opened during the summer of 1814.

The project of erecting a new bridge to connect the
Strand, near Somerset House, with the Surrey side of the
Thames at Lambeth, was started by a Bridge Company
in 1809—a year distinguished for the prevalence of one
of those joint-stock fevers which periodically seize the
moneyed classes of this country. The first plan considered was the production of Mr. George Dodds, a well-
known engineer of the time. The managing committee
were not satisfied with the design, and referred it to
Mr. Rennie and Mr. Jessop for their opinion. It was
found to be for the most part a copy of M. Peyronnet's
celebrated bridge of Neuilly, with modifications rendered
necessary by the difference of situation and the greater
width of the river to be spanned. It showed a bridge
of nine arches of 130 feet span; each being a com-
pound curve, the interior an ellipsis, and the face or
exterior a segment of a circle, as in the bridge at
Neuilly.[1] The reporting engineers pointed out various

---

[1] In their report on this design, Mr.
Rennie and his colleague observed:
"We should not have thought it necessary to quote the production of a
foreign country for the sake of showing the practicability of constructing
arches of 130 feet span, had we not
been led to it by the exact similarity
of the designs, and by the principle
which is therein adopted of the com-
pound curve; because our own country affords examples of greater boldness in the construction of arches
than that of Neuilly. There is a
bridge over the river Taff, in the
county of Glamorgan, of upwards of
135 feet span, with a rise not exceeding 32 feet, and what is more remarkable is, that the depth of the
arch-stones is only 30 inches; so that
in fact that bridge far exceeds in
boldness of design that of Neuilly."
[See our Memoir of William Edwards
in Vol. I. of this work.] After some
observations as to the importance and
necessity of making a bridge in such
a situation, at the bend of the river,
with as large arches as possible, to

accommodate the navigation and present as little obstruction as possible to
the rise and fall of the water, they
proceed: "We confess we do not
wholly approve of M. Peyronnet's
construction as adapted for the intended situation. It is complicated
in its form, and, we think, wanting
in effect. The equilibrium of the
arches has not been sufficiently attended to; for when the centres of
the bridge at Neuilly were struck,
the top of the arches sank to a degree
far beyond anything that has come to
our knowledge, whilst the haunches
retired or rose up, so that the bridge
as it now stands is very different in
form from what it was originally designed. No such change of shape
took place in the bridge over the
Taff (Pont-y-Prydd); the sinking after
the centres were struck did not amount
to one-half of that at Neuilly, although
the one was designed and built under
the direction of the first engineer of
France, without regard to expense,
whilst the other was designed and
built by a country mason with par-

objections to Mr. Dodds's design, as well as to the plan proposed by him for founding the piers; and they showed that his estimate of cost was altogether insufficient.   The result was, that no further steps were taken with Mr. Dodds's plan; but when the Act authorising the construction of the bridge had been obtained, the committee again applied to Mr. Rennie; and on this occasion they requested him to furnish them with the design of a suitable structure.[1]   The first step which he took was to prepare an entirely fresh chart of the river and the adjacent shores, after a careful and accurate survey made by Mr. Francis Giles.   In preparing his plan, he kept in view the architectural elegance of the structure as well as its utility; and while he designed it so as to enhance the beauty of the fine river front of Somerset House, by contriving that the face of the northern abutment should be on a line with its noble terrace, he laid out the roadway so that it should be as nearly upon a level with the great thoroughfare of the Strand as possible,—the rise from that street to the summit on the bridge being only 1 in 250, or about two feet in all. Two designs were prepared—one of seven equal arches, the other of nine; and the latter being finally approved by the committee as the less costly, it was ordered to be carried into effect.

The structure as executed is an elegant and substantial bridge of nine arches of 120 feet span, with piers 20 feet thick; the arches being plain semi-ellipses, with their soffits or crowns 30 feet above high-water of ordinary spring tides.   Over the points of each pier are placed

---

simonious economy.   Our opinion therefore is, that the arches of the bridge over the Thames should either be plain ellipses, without the slanting off in the haunches so as to deceive the eye by an apparent flatness which does not in reality exist, or they should be of a flat segment of a circle formed in such a manner as to give the re-

quisite room for the passage of the current and barges under it."

[1] In June, 1810, we find him accepting the direction of the new bridge at 1000l. a year for himself and assistants, or 7l. 7s. a day and expenses; but on no account were any of his people to have to do with the payment or receipt of moneys.

two three-quarter Doric column pilasters, after the design
of the temple of Segesta in Sicily. These pilasters are
5 feet 8½ inches diameter at the base, and 4 feet 4 inches
at the under side of the capital, forming recesses in the
roadway 17 feet wide and 5 feet deep. The depth of
the arch-stones at the crown is 4 feet 6 inches, and they
increase regularly to 10 feet at the haunches. Between
each pair of arches, at the level of 19 feet above the
springing, there is an inverted arch, the stones of which

SECTION OF WATERLOO BRIDGE.

are 4 feet 6 inches deep at the crown, and decrease
regularly on each side as they unite and abut against the
extrados or backs of the voussoirs of the main haunches.
The abutments are 40 feet in thickness at the base, and
decrease to 30 feet at the springing. The cope of the
arches and piers is surmounted by a Grecian Doric
block-cornice and entablature, upon which is placed a
balustrade parapet 5 feet high. The total width of the
bridge from outside to outside of the parapets is 45 feet.
The footpaths on each side are 7 feet wide, and the
roadway for carriages 28 feet. There are four sets of
landing-stairs—two to each abutment; and the arrangement of this part of the work has been much admired,
on account of its convenience for public uses as well as
its architectural elegance.

In the construction of this bridge, there are four features of distinctive importance to be noted :—1st. The
employment of coffer-dams in founding piers in a great
tidal river—an altogether new use of that engineering
expedient, though now become customary. 2nd. The
ingenious method employed for constructing, floating,
and fixing the centres ; since followed by other engineers
in works of like magnitude. 3rd. The introduction and
working of granite stone to an extent before unknown,

and in much larger and more substantial pieces of masonry than had previously been practised. 4th. The adoption of elliptical stone arches of an unusual width, though afterwards greatly surpassed by the same engineer in his New London Bridge.

Mr. Rennie invariably took the greatest pains in securing the most solid foundations possible for all his structures, and especially of his river works, laying them far below the scour of the river, at a depth beyond all probable reach of injury from that cause. - The practice adopted in founding the piers of the early bridges across the Thames, was to dredge the bottom to a level surface, and build the foundations on the bed of the river, protecting them outside by rubble, by starlings, or by sheet-piling. Mr. Dodds had proposed to follow the method employed by Labelye at Westminster Bridge, of founding the piers by means of caissons; but Mr. Rennie insisted on the total insufficiency of this plan, and that the most effectual method was by means of coffer-dams. This would no doubt be more costly in the first instance, but vastly more secure; and he foresaw that the inevitable removal of the piers of Old London Bridge, by increasing the current of the river, would severely test the foundations of all the bridges higher up the stream—which proved to be the case. Having already extensively employed coffer-dams in getting in the foundations of the London and East India Dock walls, he had no doubt as to their success in this case; and they were adopted accordingly.[1]

---

[1] The coffer-dams in which the foundations of the abutments were built, were formed by driving two rows of piles 13 by 6½ inches each, with a counter or abutting pile at every 12 feet, 12 by 12, driven in the form of an ellipsis, and strongly cemented together, at low-water and high-water levels, by double horizontal walings or bracks, having a space of about 8 inches clear between them for the intermediate or half piles. The whole were driven close together from 15 to 20 feet deep into the ground, well caulked, so as to be water-tight, and all connected firmly together by strong wrought iron bars and bolts, besides shores and intermediate braces. The spaces between the two rows of piles were then rammed close with well-tempered clay, so that they formed, as it were,

Mr. Rennie also introduced a practice of some novelty and importance in the centering upon which the arches of the bridge were built. He adopted the braced prin-

CENTERING OF ARCH, WATERLOO BRIDGE. [After E. Blore.]

ciple. The centres spanning the whole width of the arch were composed of eight ribs each, formed in one piece, resting upon the same number of solid wedges,

a solid vat or tub impermeable to water; and within these, when pumped clear of water, the excavation was made to the proper depth, and in the space so dug out the building operations proceeded. The coffer-dams for the piers were formed in a similar manner, with modifications according to circumstances. By this means the bed of the river, where the piers were to be erected, was exposed and dug out to the proper depth, and the foundations were commenced from a level nine feet at least below low water mark. The foundations there rested upon timber piles from 20 to 22 feet long, driven into the solid bed of the river. Upon the heads of these piles half-timber planking was spiked, and on this the solid masonry was built—every stone being fitted, mortared, and laid with studious accuracy and precision. The whole work was done with such solidity that, after the lapse of fifty years, the foundations have not yielded by a straw's breadth at any point.

supported by inclined tressels placed upon longitudinal bearers, firmly fixed to the offsets of the piers and abutments. At the intersecting point of the bearers or braces in each rib, there was a cast iron box with two holes or openings in it, so that the butt-ends rested firmly against the metal; and to prevent them from acting like so many wedges to tear the rib to pieces when the vertical weight of the arch began to act upon them, pieces of hard wood were driven firmly into the holes above described, to check the effect of the bearers or strutts of the ribs; and this arrangement proved completely successful. The eight ribs were firmly connected together by braces and ties, so as to form one compact frame, and the curve or form of the arch was accurately adjusted by means of transverse timbers, 12 inches wide and 6 inches thick, laid across the whole of the ribs, set out to the exact form of the curve by ordinates from the main or longitudinal axis of the ellipsis; and in proportion as the voussoirs or arch-stones were carried up from the adjoining piers, the weight which had been laid upon the top of the centre to keep it in equilibrium according to the form of the arch during construction, was gradually removed as it advanced towards completion. When the arch was about two-thirds completed, a small portion of it was closed with the centre, and the remaining part of each side was brought forward regularly by offsets to the crown until the whole was finished. Each key-stone was accurately fitted to its respective place, and the last portion of each, for the space of about eighteen inches, was driven home by a heavy wooden ram or pile-engine, so as almost to raise the crown of the arch from the centre.

About ten days after the main arches had been completed, and the inverts and spandrel walls between them carried up to the proper height, the arches were gently slackened, to the extent of about two inches, so as to bring each to its bearing to a cer-

tain extent. This was effected by driving back the
wedges upon which the ribs of the centres rested, by
means of heavy wooden rams attached to them, so
that they could swing backwards and forwards with
great facility when any external force was applied to
them; and this was done by ropes worked by hand-
labour. After the first striking or slackening, the
arches were allowed to stand for ten days, when the
wedges were driven back six inches further. After ten
days more the wedges were driven back sufficiently to
render the arch altogether clear of the centering. By
this means the mortar was firmly imbedded into all the
joints, and the arch came gradually to its ultimate bearing without any undue crushing. In order to ascertain
whether any change of form took place, three straight
lines were drawn in black chalk on the extreme face of
the arch previous to commencing the operation of striking
the centre,—one horizontally in the centre of the voussoirs forming the crown, and two from the haunches of
the arch, each intersecting the first line at about 25 feet
on each side of the keystone; so that if there had been
any derangement of the curve or irregular sinking, it
would at once have been clearly apparent. After the
centres had been removed, it was found that the sinking
of the arches varied from $2\frac{1}{2}$ to $3\frac{1}{4}$ inches, which was as
nearly as possible the allowance made by the engineer
in designing the work; the whole plan being worked
out with admirable precision and accuracy.

The method of fixing and removing the centres was
entirely new; being precisely the same as was afterwards followed by Mr. Robert Stephenson in fixing
the wrought iron ribs of the Conway and Britannia
bridges,—that is, by constructing them complete on a
platform adjacent to the river, and floating them between
the piers on barges expressly contrived for the purpose.
They were then raised into their proper places by four
strong screws, 8 inches in diameter and 4 feet long,

fixed in a strong cast iron box firmly bedded in the solid floor of the barge. The apparatus worked so well and smoothly, that the whole centre, consisting of eight ribs, each weighing about fifty tons, was usually placed within the week.

The means employed by Mr. Rennie for forming his road upon the bridge were identical with those adopted by Mr. Macadam at Bristol some six years later. But the arrangement constituted so small a part of our engineer's contrivances, that, as in many other cases, he made no merit of it. When the clay puddle placed along the intended roadway was sufficiently hard, he spread a stratum of fine screened gravel or hoggins, which was carefully levelled and pressed down upon the clay. This was then covered over with a layer of equally broken flints, about the size of an egg ; after which the whole was rolled close together, and in a short time formed an admirable "macadamized" road. Mr. Rennie had practised the same method of making roads over his bridges long before 1809 ; and he continued to adopt it in all his subsequent structures.

The whole of the stone required for the bridge (excepting the balustrades, which were brought ready worked from Aberdeen) was hewn in some fields adjacent to the erection on the Surrey side. It was transported on to the work upon trucks drawn along railways, in the first instance over temporary bridges of wood ; and it is a remarkable circumstance that nearly the whole of the material was drawn by one horse, called "Old Jack"—a most sensible animal, and a great favourite. His driver was, generally speaking, a steady and trustworthy man, though rather too fond of his dram before breakfast. As the railway along which the stone was drawn passed in front of the public-house door, the horse and truck were usually pulled up while Tom entered for his "morning." On one occasion the driver stayed so long that "Old Jack," becoming impatient, poked his head into the open door, and taking his master's coat-collar between his teeth, though in a gentle sort of manner, pulled him out from the midst of his companions, and thus forced him to resume the day's work.

WATERLOO BRIDGE.

[By Percival Skelton, after his original Drawing.]

The bridge was opened with great ceremony by His Royal Highness the Prince Regent, attended by the Duke of Wellington and many other distinguished personages, on the 18th of June, 1817. It was originally named the Strand Bridge; but after that date the name was altered to that of "Waterloo," in honour of the

Duke. At the opening, the Prince Regent offered to confer the honour of knighthood on the engineer, who respectfully declined it. Writing to his friend Whidbey, he said, "I had a hard business to escape knighthood at the opening." He was contented with the simple, unadorned name of John Rennie, engineer and architect of the magnificent structure which he had thus so successfully brought to completion. Waterloo Bridge is indeed a noble work, and probably has not its equal for magnitude, beauty, and solidity. Dupin characterised it as a colossal monument worthy of Sesostris or the Cæsars; and what most struck Canova during his visit to England was, that the trumpery Chinese bridge, then in St. James's Park, should be the production of the Government, whilst Waterloo Bridge was the enterprise of a private company. Like all Rennie's works, it was built for posterity. That it should not have settled more than a few inches—not five in any part—after the centres were struck, is an illustration of solidity and strength probably without a parallel. We believe that to this day not a crack is visible in the whole work.

The necessity for further bridges across the Thames increased with the growth of population on both sides of the river; and in the year 1813 a Company was formed to provide one at some point intermediate between Blackfriars and London Bridge, of which Mr. Rennie was appointed the engineer. The scheme was at first strongly opposed by the Corporation on the ground of the narrowness of the river at the point at which it was proposed to erect the new structure; but the public demands being urgent, they at length gave way and allowed the necessary Act to pass, insisting, however, on the provision of a very large waterway, so that the least obstruction should be offered to the navigation. Mr. Rennie prepared a design to meet the necessities of the case, and in order to secure the largest possible waterway, he projected his well-known Southwark Bridge, extending from Queen Street, Cannon Street, to Bridge Street, Southwark. It consists of three cast iron arches, with two stone piers and abutments. The arches are flat segments of circles, the centre one being not less than 240 feet span (or 4 feet larger than Sunderland Bridge, the largest cast iron arch that had until then been erected), rising 24 feet, and springing 6 feet above high water of spring tides. The two side arches are of 210 feet span, each rising 18 feet 10 inches, and springing from the same level. The two piers were 24 feet wide each at the springing, and 30 feet at the base.

The works commenced with the coffer-dam of the south pier on the Southwark side, and the first stone was laid by Admiral Lord Keith about the beginning of 1815. All the centering for the three arches was fixed by the autumn of 1817, and the main ribs were set by the end of April, 1818. The centres were struck by the end of the month of June following, and completely removed by the middle of October; and the bridge was opened for traffic in March, 1819.

In the course of this work great precautions were used in securing the foundations of the piers. The river was here at its narrowest and deepest point, the bed being 14 feet below low water of ordinary spring tides. The coffer-dams were therefore necessarily of great depth and strength to resist the pressure of the body of water, as well as the concussion of the barges passing up and down the river, which frequently drove against them. Hence the dams were constructed in the form most capable of resisting external pressure, and yet suitable to the dimensions of the foundations.[1] The masonry

---

[1] They were made elliptical, and consisted of two main rows of piles 14 inches square, placed 6 feet apart. These were cemented together on each side at the level of low water at half-tide, and 2 feet above high water of spring tides, with horizontal walings or braces, also 14 inches square, firmly secured to every tenth pile by wrought iron screw bolts 2¼ inches diameter passing through each row of piles, and fixed at each end by

and iron work of the bridge were erected with great care and completeness. The blocks of stone in the piers were accurately fitted to their places by moulds, and driven down by a heavy wooden ram. The least possible quantity of finely tempered mortar was used, so that

---

large bracket-pieces of timber and wrought iron plates and nuts, so that the whole could be firmly screwed up and braced together. The piles were well headed and shod with wrought iron, and driven from 15 to 20 feet into the solid bottom of the river. On the outside of the two main rows of piles there was another placed six feet from the inner, and driven to the same depth, but their heads only extended up to the level of half-tide. They were tied together with horizontal braces and wrought iron screw bolts, in the same manner as the two main rows. The joints of all were well caulked, and the spaces between the three rows were filled with well-puddled clay, so as to be completely impervious to water. On each dam there was a trunk three feet square, fitted with a valve, which could be lowered or raised at pleasure, so that the water could be let off to the level of low tide, or filled at any time, in the event of any accident occurring in the building of the piers. The abutment coffer-dams were constructed in a semi-circular form, and consisted of two rows of piles of like dimensions, similarly rammed between with clay to keep out the water. The enclosed spaces were pumped out by means of steam-engines of 20-horse power. The engines worked double pumps of 15 inches diameter by an arrangement of slide-rods. When necessary, as at the City abutment, where the soil was more porous, an additional engine with pumps was employed. By these complete methods the water was kept under until the foundations were got in at the great depth we have above stated. The piers and abutments rest on solid platforms of piles, cills, and planking about 2 feet 6 inches thick. The piles, of fir, elm, and beech,

20 feet long and 12 inches in diameter, shod with wrought iron, were driven in regular rows three feet apart, until a ram of 15 cwt. falling from a height of 28 feet moved them downward only half an inch at a blow. Their heads were then cut off level; the earth and clay removed from about them to the depth of a foot, and the space filled in with Kentish rag-stone, well rammed and grouted together with lime and gravel. Cills of fir, beech, and elm, 12 inches thick and 14 inches wide, were then accurately fitted and spiked to each pile-head, in the transverse direction of the piers and abutments, with wrought iron jagged spikes 18 inches long and six-eighths of an inch square. The spaces between the cills were solidly filled with brickwork, and another row of cills of the same dimensions, laid at right angles to those below, was fitted and spiked to them over each row of piles, in the same manner as above explained. The spaces were then filled with brick-work; the whole surface was covered with a solid flooring of elm-plank 6 inches thick, well bedded in mortar, and spiked down to each cill with wrought iron jagged spikes 10 inches long and five-eighths of an inch square. It may here be observed, as to the use of timber at this depth, that when it is exposed to an equable degree of moisture it is found almost imperishable—timber having been taken up, as fresh as when laid down, from the foundations of structures laid in water for more than a thousand years. Besides, timber is found, better than any other material, capable of distributing the pressure over the whole surface, as well as binding all the parts of the foundations together.

every part should have a perfectly true permanent bearing. Great care was also taken in the selection of the blocks. The exterior of the piers was constructed of hard silicious stone brought from Craigleith Quarry near Edinburgh, and Dundee; the interior, from the bottom

SOUTHWARK BRIDGE.

[By Percival Skelton, after his original Drawing.]

of the foundations to the springing of the arches, of hard Yorkshire grit; while that part of the piers and abutments from which the arches spring consisted of the hardest and closest blocks of Cornish and Aberdeen granite: in fine, it may be affirmed that a more

solid piece of masonry does not exist than Southwark
Bridge.

The iron work consists of eight arched ribs, the main
strength of the arches being embodied in their lower
parts, which are solid. The lower or main arch is
divided into thirteen pieces, with a rib 5½ inches thick
at the top and bottom, and 2½ inches in the centre.
The joints radiate outwards from the lower edge, and
form so many cast iron instead of stone voussoirs, from
6 to 8 feet deep and 13 feet long. At the junction of
each of these main rib pieces there are transverse
plates of the same depth, having flanges cast upon
them on both sides in a wedge form, so that the ends
of the main rib piers fit into them on one side, whilst
on the other there is a cast iron wedge, driven in
between the rib and flange piece, and enabling the whole
to be accurately adjusted and connected together. In
addition to this, each rib piece had a flange, cast at each
end with a certain number of holes three quarters of
an inch in diameter, into which wrought iron screw
bolts were introduced to connect the whole firmly together in the direction of the arch. These rib pieces
were also of great importance during construction, the
chief dependence being placed upon their lateral thrust
in holding the arches together. At each pier and
abutment there was a similar cast iron bed or abutting
plate, let accurately 1½ inch into the stone; but between
the end of each main rib which sprang from this plate
there was a groove cut out of the solid stone behind
the springing plates and main iron ribs of the arches,
18 inches wide, 3 inches thick at the top, and 2 inches
at bottom. This groove was accurately dressed and
polished. Three cast iron wedges, 9 feet long, 6 inches
wide, and 3¼ inches thick at top and 2 inches at bottom,
were then made and most accurately chipped and filed,
so as to fit exactly the groove above mentioned to within
12 inches of its bottom. When the whole of these wedges

at both ends of the arch had been put into their places, they were carefully driven home to the bottom of the grooves at the same time by heavy wooden rams, by which means the ribs of the arches were relieved from the centres and took their own bearing.   In other words, the arches were keyed from the abutments only, instead of from the centre, as is usual in bridges of stone.   This was an extremely delicate and nice process, as it required that the variations of the thermometer should be carefully observed, in order that each operation should be carried on at as nearly as possible the same degree of temperature, otherwise the form of the arch would have been distorted, the vertical and lateral pressure of the different parts would have been affected, and an undue strain thrown upon the abutments as well as the different parts of the arch.   But so nicely was the whole operation arranged and adjusted, that nothing of the kind occurred : the parts remained in perfect equilibrium ; not a bolt was broken, and not the smallest derangement was found in the structure after the process had been completed.

The spandrel pieces attached to the top of the main ribs were cast in the form of open diamonds or lozenges, connected together in the transverse direction by two tiers of solid crosses laid nearly horizontally—all closely wedged and firmly bolted together.   In addition to the transverse connecting plates cast in open squares, there were also diagonal braces of cast iron, commencing at the extremity of the outer rib of each arch and intersecting each other so as to form a diamond-shaped space in the centre.   These were also secured at their ends by wedges and bolts, like the main rib pieces.   After the main ribs of the arches were relieved from the centres, and had taken their bearing, before the centres were removed from beneath them, experiments were made how far they might be affected by expansion and contraction, in proportion to the different degrees of temperature to which the bridge might be exposed ; and for this purpose different gauges were made, of brass, iron, and wood. These gauges were firmly attached to the middle or crown of the wooden centres, and divided into sixteenths of an inch, and at each a Fahrenheit thermometer was placed; so that, the ends of the arch being fixed, the variation in the temperature would be indicated by the rise and fall in the centre. The observations were made daily—in the morning, at midday, and at sunset—for several months during summer and winter, when it was ascertained that the arches rose and fell about one-tenth of an inch for every 10 degrees of temperature, more or less.

The whole iron work is covered with solid plates, having flanges cast on their upper side. These plates are laid in the transverse direction and on the top of the spandrel walls, so that they form a solid and compact cast iron floor to support the roadway. The cornice, which is cast hollow, is of the plain Roman-Doric order, and is secured to the roadway-plates by strong stays and bolts at proper intervals. The parapet consists of a plinth, also cast hollow, with a groove at the top to receive the railing, which is cast in the form of open diamonds corresponding with the spandrels. The roadway is 42 feet wide from outside to outside, and formed in the same manner as that over Waterloo Bridge, already described. The total quantity of cast iron in the bridge is 3620 tons, and of wrought iron 112 tons. It has been said that an unnecessarily large quantity of material has been employed; and no doubt a lighter structure would have stood. But looking at the imperfections of workmanship and possible flaws in the castings, Mr. Rennie was probably justified in making the strengths such as he did, in order to ensure the greatest possible solidity and durability—qualities which eminently characterize his works, and perhaps most of all, his majestic metropolitan bridges. Although the Southwark Bridge was built before the Railway era,

which has given so great an impetus to the construction
of iron bridges, it still stands pre-eminent in its class,
and is a model of what a bridge should be.   Its design
was as bold as its execution was masterly.   Mr. Robert
Stephenson has well said of it that, " as an example of
arch-construction, it stands confessedly unrivalled as
regards its colossal proportions, its architectural effect,
and the general simplicity and massive character of its
details." [1]

[1] Article on Iron Bridges in ' Encyclopedia Britannica.'

WATERLOO BRIDGE.   [By R. P. Leitch.]

CHAPTER VII.

MR. RENNIE'S DOCKS AND HARBOURS.

THE growth of the shipping business, and the increase in our home and foreign commerce, led to numerous extensive improvements in the harbours of Britain about the beginning of the present century. The natural facilities of even the most favourably situated ports, though to some extent improved by art, no longer sufficed for the accommodation of their trade. Comparatively little had as yet been done to improve the port of London itself, the great focus of the maritime and commercial industry of Britain.[1] It is true its noble river the Thames provided a great amount of shipping room and a vast extent of shore convenience between Millwall and London Bridge; but the rise and fall of the tide twice in every day, and the great exposure of the vessels lying in the river to risks of collisions, and other drawbacks, were felt to be evils which the shipping interest found it necessary to remedy. Besides the crowding of the river by ships and lighters—the larger vessels having to anchor in the middle of the stream as low as Blackwall, from which their cargoes were lightered to the warehouses higher up the Thames—the warehouse

[1] The increase in the trade of London is exhibited by the following abstract of vessels entered at the port at different periods since the beginning of last century :—

YEARS.	BRITISH.		FOREIGN.		COLLIERS AND COASTERS.	
	Vessels.	Tonnage.	Vessels.	Tonnage.	Vessels.	Tonnage.
1702	839	80,040	496	76,995	No return.	No return.
1751	1498	198,023	184	36,346	Do.	Do.
1798	1649	397,096	1771	229,991	10,133	1,250,449
1860	6320	1,828,911	4857	1,152,499	18,346	3,152,853

accommodation was found very inadequate in extent as
well as difficult of access. There was also a regular
system of plunder carried on in the conveyance of the
merchandise from the ship's side to the warehouses, the
account of which, given in Mr. Colquhoun's work on
the 'Commerce and Police of the Thames,' affords a
curious contrast to the security and regularity with
which the shipping operations of the port are carried on
at the present day. Lightermen, watermen, labourers,
sailors, mates and captains occasionally, and even the
officers of the revenue, were leagued together in a
system of pilfering valuables from the open barges.
The lightermen claimed as their right the perquisites
of "wastage" and "leakage," and they took care that
these two items should include as much as possible.
There were regular establishments on shore for receiving
and disposing of the stolen merchandise.[1] The Thames
Police was established, in 1798, for the purpose of
checking this system of wholesale depredation; but, so
long as the goods were conveyed from the ship's side in
open lighters, and the open quays formed the principal
shore accommodation—sugar hogsheads, barrels, tubs,
baskets, boxes, bales, and other packages, being piled
up in confusion on every available foot of space—it was
clear that mere police regulations would be unequal to
meet the difficulty. It was also found that the confused
manner in which the imports were brought ashore led

---

[1] Mr. Colquhoun, the excellent
Police Magistrate, estimated that, in
1798, the depredations on the foreign
and coasting trade amounted to the
almost incredible sum of 506,000l.,
and on the West India trade to
232,000l.—together 738,000l.! He
stated the number of depredators—including mates, inferior officers, crews,
revenue-officers, watermen, lightermen, watchmen, &c.—to be 10,850;
and the number of opulent and inferior receivers, dealers in old iron,
small chandlers, publicans, &c., interested in the plunder, to be 550!
Colquhoun's book, and its descriptions of the lumpers, scuffle-hunters,
long-apron men, bumboat.men and
women, river-pirates, light-horsemen,
and other characters who worked at
the water-side, with their skilful ap-
pendages of jiggers, bladders with
nozzles, pouches, bags, sacks, pockets,
&c., form a picture of life on the Thames
sixty years since worthy to rank with
Mayhew's 'London Labour and London Poor' of the present day.

to a vast amount of smuggling, by which the honest merchant was placed at a disadvantage at the same time that the revenue was cheated. The Government, therefore, for the sake of its income, and the traders for the security of their merchandise, alike desired to provide an effectual remedy for these evils.

Mr. Rennie was consulted on the subject in 1798, and requested to devise a plan. Before that time various methods had been suggested, such as quays and warehouses, with jetties, along the river on both sides; but all these eventually gave place to that of floating docks or basins communicating with the river, surrounded with quays and warehouses, shut in by a lofty enclosure-wall, so that the whole of the contained vessels and their merchandise should be placed, as it were, under lock and key. By such a method it was believed the goods could be loaded and unloaded with the greatest economy and despatch, whilst the Customs duties would be levied with facility, at the same time that the property of the merchants was effectually protected against depredation. At the beginning of the century a small dock had existed on the Thames, called the Greenland Dock; but it was of very limited capacity, and only used by whaling vessels. Docks had existed at Liverpool for a considerable period, which had been greatly extended of recent years; so that there was no novelty in the idea of providing accommodation of a similar kind on the Thames, though it is certainly remarkable that, with the extraordinary trade of the metropolis, the expedient should not have been adopted at a much earlier period.

The first of the modern floating docks actually constructed on the Thames was the West India, occupying the isthmus that formerly connected the Isle of Dogs with Poplar, and of which Mr. Jessop [1] was the

---

[1] Mr. William Jessop, C.E., was among the most eminent engineers of his day. His father was engaged under Smeaton in the building of the Eddystone Lighthouse; and, dying in 1761, he left the guardianship of his

engineer.   At the same period, in 1800, a company was
formed by the London merchants for the purpose of
constructing docks at a point as near the Exchange as
might be practicable, for the accommodation of general
merchandise, and of this scheme Mr. Rennie was appointed the engineer.   He proposed several designs for
consideration on a scale more or less extensive, adopting
his usual course of submitting alternative plans, from
which practical men might make a selection of the one
most suitable for the purposes of their business; at the
same time inviting suggestions, which he afterwards
worked up into his more complete designs.   As the
future trade of London was an unknown quantity, he
wisely provided for the extension of the docks as circumstances might afterwards require.

In carrying out the London Docks it was deemed
advisable, in the first instance, to limit the access to the
present Middle River Entrance at Bell Dock, 150 feet

family to that engineer, who adopted
William as his pupil, and carefully
brought him up to the same profes-

WILLIAM JESSOP, C.E.

sion. With Smeaton Jessop continued
for ten years; and, after leaving
him, he was engaged successively on
the Aire and Calder, the Calder and
Hebble, and the Trent Navigations.

He also executed the Cromford and
the Nottingham Canals; the Lough-
borough and Leicester, and the Horncastle Navigations; but the most
extensive and important of his works
of this kind was the Grand Junction
Canal, by which the whole of the
north-western inland navigation of the
kingdom was brought into direct connection with the metropolis.   He was
also employed as engineer for the Caledonian Canal, in which he was succeeded by Telford, who carried out the
work. He was the engineer of the West
India Docks (1800-2) and of the Bristol
Docks (1803-8), both works of great
importance. He was the first engineer
who was employed to lay out and
construct railroads as a branch of his
profession; the Croydon and Merstham
Railroad, worked by donkeys and
mules, having been constructed by
him as early as 1803.   He also laid
down short railways in connection
with his canals in Derbyshire, Yorkshire, and Nottinghamshire.   During
the later years of his life he was
much afflicted by paralysis, and died
in 1814.

long and 40 feet wide, with the cill laid five feet below
low water of spring tides.   The entrance lock communicated with a capacious entrance basin, called the
Wapping Basin, covering a space of three acres, and this
again with the great basin called the Western Dock,
1260 feet long and 960 feet wide, covering a surface of
20 acres.   The bottom of the dock was laid 20 feet
below the level of high water of an 18 feet tide.   The
quays next to the river were five feet above high water,
increasing to nine feet at the Great Dock.   From the
east side of the latter, it was ultimately proposed to make
two or more docks, communicating with each other
and with a larger and deeper entrance lower down the
river at Shadwell; all of which works have since been
carried out.

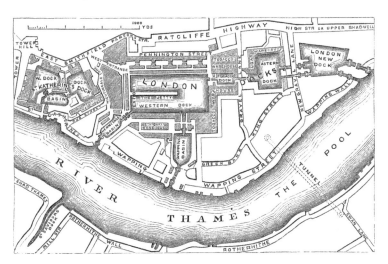

PLAN OF LONDON DOCKS.

As the site of the Docks was previously in a great
measure occupied by houses, considerable time necessarily elapsed before these could be purchased and
cleared away; so that the works were not commenced
until the spring of 1801, when two steam-engines were
erected, of 50 horse power each, for pumping the water,

and three minor engines for other purposes, such as grinding mortar, working the pile-engine, and landing materials from the jetty—an application of steam power as an economist of labour which Mr. Rennie was among the first to introduce in the execution of such works. The coffer-dam for the main entrance, and the excavation of the Docks, were begun in the spring of 1802;[1] after which time the works were carried forward with great vigour until their completion on the 30th of January, 1805, when they were opened with considerable ceremony.

At a subsequent period Mr. Rennie designed the present westernmost or Hermitage entrance lock and basin, the former of which is 150 feet long and 38 feet wide, with the cill laid two feet below low water of spring tides; the basin and main dock covering a surface of

---

[1] The locks were founded upon piles driven firmly into the soil, with rows of grooved and long-end sheeting piles in front of and behind the gates, in order to prevent the water from getting under them. The chamber between the lock-gates was formed by an inverted arch of masonry 2 feet 6 inches thick, strongly embedded in brickwork. The side-walls of the lock recesses and chamber were 7 feet thick, with strong counterfoots behind at the proper intervals. The whole of the locks and chambers were built of fine masonry, composed principally of hard blue sandstone from Dundee. All the retaining walls of the basins and docks were made curvilinear in the face, drawn from a radius of 80 feet, the centre being level with the top of the wall, and the bottom being inclined at the same angle as the radius. The wall was of a parallel thickness of 6 feet, except three or four footings at the back, where there were also counterfoots 3 feet 4 inches square, 15 feet asunder. The dock walls were founded generally upon a strong bed of gravel, which rendered piling unnecessary, and were built upon a flooring of beech and elm plank 6 inches thick. Under the front and back of this flooring ran a strong cill 12 inches square, to which the planks were firmly spiked. The walls and counterfoots were built of brickwork, the front, for 14 inches inwards, being formed of vitrified pavier bricks, and the remainder of good hard burned stock, the joints being a quarter of an inch thick in front and three-eighths thick at the back; the whole well bedded in excellent mortar made by a mill. There were two through or binding courses of stone, 14 inches thick in front, increasing in thickness backwards according to the radius of the front of the wall. The whole of the locks were furnished with cast iron turning or swivel-bridges erected across them. The works were generally done by contract, but the locks, which required greater care, accuracy, and completeness, were executed by daywork, under the engineer's immediate direction. [For further particulars as to these docks see Sir John Rennie's able work on 'British and Foreign Harbours;' Art. London Docks.]

one acre and a quarter. Another small dock of one acre was afterwards added on the north-east side of the Great Basin, exclusively devoted to the tobacco trade; and it was ultimately extended to the Thames at Shadwell, as contemplated in the original design.

After the Docks had been opened for trade, Mr. Rennie gave his careful attention to the working details, and he was accustomed from time to time to make suggestions with a view to increased despatch and economy in the conduct of the business. Thus, in 1808, he recommended that the whole of the lifting cranes in the Docks should be worked by the power of a steam-engine instead of by human or horse labour. He estimated that the saving thus effected, in the case of only twenty-six cranes, would amount to at least 1500l. a year, besides ensuring greater regularity and despatch of work; and, if applied to the whole of the cranes along the Docks and in the warehouses, a much greater annual saving might be anticipated. It was, however, regarded as too bold an innovation for the time; and we believe the suggestion has not been carried out to this day, the cranes in the London Docks being still worked by hand labour, at a great waste of time and money, as well as loss of business. Another of Mr. Rennie's valuable suggestions, with a view to greater economy, was the adoption of tramways all round the quays, provided with trucks, by means of which the transfer of goods from one part of the Dock to another might be effected with the greatest ease and in the least possible time. But this too was disregarded. Labour-saving processes were not then valued as they now are. The application and uses of machinery were as yet imperfectly understood, and there were in most quarters powerful prejudices to be overcome before it could be introduced. To this day the goods in the London Docks are hauled in trollies, waggons, or hand-barrows from ship to ship, or from the vessels to the respective bonded warehouses; and it still remains matter of surprise that a system so clumsy, so wasteful of time, so obstructive to rapid loading and unloading in dock, should be permitted to continue.

Shortly after these works were set on foot, and when the great importance and economy of floating docks began to be recognised by commercial men, another project of a similar character was started, to provide accommodation exclusively for vessels of the East India Company, of from 1000 to 1800 tons burden. A company was formed for the purpose, and an Act was obtained in 1803, the site selected being immediately to the west of the river Lea, at the point at which it enters the Thames, and where at that time there were two small floating basins or docks, provided with wooden locks, and surrounded with wooden walls, called the Brunswick and Perry's Docks. These it was determined to purchase and include in the proposed new docks, of which, however, they formed but a small part. Mr. Rennie and Mr. Ralph Walker were associated as engineers in carrying the works into execution, and they were finished and opened for business on the 4th of August, 1806. They consisted of an entrance lock into the Thames 210 feet long and 47 feet wide, with the cill laid 7 feet below low water of spring tides. This lock is connected with a triangular entrance basin, covering a space of $4\frac{1}{2}$ acres, on the west side of which it communicates by a lock with a dock expressly provided for vessels outward bound, called the Export Dock, 760 feet long and 463 feet wide, covering a surface of $8\frac{1}{3}$ acres. At the north end of the entrance basin is the Import Dock, 1410 feet long and 463 feet wide, covering a surface of $18\frac{2}{3}$ acres. The depth of these basins is 22 feet below high water of ordinary spring tides. The total surface of dock room, including quays, sheds, and warehouses, is about 55 acres. The original capital of the East India Dock Company was 660,000*l.*; but

Mr. Rennie constructed and completed the Docks for a sum considerably within that amount. Eventually they were united to the West India Docks, under the joint directorate of the East and West India Dock

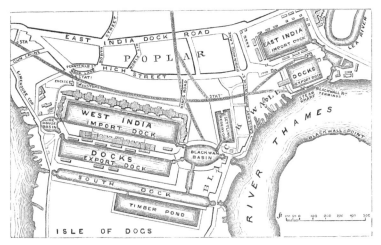

PLAN OF EAST AND WEST INDIA DOCKS.

Company.[1]   Mr. Rennie also introduced into these Docks many improved methods of working; his machinery, invented by him for transporting immense blocks of mahogany by a system of railways and locomotive cranes, having, in the first six months, effected a saving in men's wages more than sufficient to defray

---

[1] Among the improvements adopted by Mr. Rennie in these docks may be mentioned the employment of cast iron, then an altogether novel expedient, for the roofing of the sheds. One of these, erected by him in 1813, was 1300 feet long and 29 feet 6 inches in span, supported on cast iron columns 7½ inches in diameter at bottom and 5¾ at top. Another, still more capacious, of 54 feet clear span between the supports, was erected by him over the mahogany warehouses in 1817. He also introduced an entirely new description of iron cranes, first employing wheel-work in connection with them, by which they worked much more easily and at a great increase of power. He entirely re-arranged the working of the mahogany sheds, greatly to the despatch of business and the economy of labour. His quick observation enabled him to point out new and improved methods of despatching work, even to those who were daily occupied in the docks, but whose eyes had probably become familiar with their hurry-scurry and confusion.

their entire original cost, besides the increased expedition in the conduct of the whole Dock business.

Some of Mr. Rennie's harbour works at other places were of considerable magnitude and importance; the growing trade of the country leading to his frequent employment in constructing new harbours or extending and improving old ones. In almost every instance he had the greatest possible difficulty in inducing the persons locally interested to provide harbour space sufficiently extensive as well as secure. When asked to give his advice on such questions, he began with making numerous practical inquiries on the spot; he surveyed the adjacent coast, took soundings all round the proposed harbourage, noted the set of the currents, the direction of the prevailing winds, the force and action of the land streams, and the operations of the scour of the tides upon the shore. He also inquired into the trade to be accommodated, the probability of its expansion or otherwise, and prepared his plans accordingly. Writing to Mr. Foster, of Liverpool, in 1810, he said, "It seems to me that your merchants are much less liberal in their ideas than is generally supposed. The account you give me furnishes another strong proof of the necessity of enlarging your scale of docks." Adverting to another scheme on which he had been consulted, he added, "It is my intention to impress upon the minds of the promoters the necessity for a much larger scale of docks than is proposed; and though they may blame me now, they will thank me afterwards; as larger accommodation will not only afford great and immediate relief to the shipping now, but will save the expenditure of much money hereafter."

As early as 1793 he was employed by the Commissioners of British Fisheries to report as to the best means of improving the harbour of Wick,—the only haven capable of affording shelter for ships, in certain states of the wind, which was to be found along an extent of 120 miles of rock-bound coast. In his masterly report he boldly proposed to abandon the old system of jetties, and to make an entirely new harbour beyond the bar; thus at once getting rid of this great and dangerous obstacle to improvement, securing at the same time greater depth of water, better shelter, and the means of easier access and departure for vessels of all burdens. In order to accommodate the trade of Wick, he recommended that a canal should be made from the new harbour, having a basin at its termination in the town, where vessels would be enabled to float, and to load and unload at all times. He also proposed an effective plan of sluicing, with the view of scouring the outer harbour when necessary. It is much to be regretted that this plan was not carried out, and that so important a national work has been postponed almost until our own day; nor does the plan since adopted, though exceedingly costly, seem calculated to secure the objects which would have been obtained by executing Mr. Rennie's more comprehensive yet much more economical design. He was consulted about the same time respecting the improvement of the harbour of Aberdeen; but though want of means then prevented his recommendations from being acted on, his report[1] produced a salutary effect in pointing out the true mode of dealing with a difficult subject, and most of his suggestions have since been carried out by other engineers.

Of still greater importance was his report on the improvement of the navigation of the river Clyde, for the accommodation of the rapidly increasing trade of Glasgow. Perhaps in no river have the alterations executed after well-devised plans been more extraordinary than in this. Less than a century ago, the Clyde at Glasgow was accessible only to herring-boats, whereas

---

[1] See Sir John Rennie's ' British and Foreign Harbours;' Art. Aberdeen.

now it floats down with every tide vessels of thousands
of tons burden, capable of wrestling with the storms of
the Atlantic.   Watt, Smeaton, and Golborne had been
consulted at different times, and various improvements
were suggested by them.   Watt laid out a ship-canal
from Glasgow to the sea.   Smeaton proposed to construct a dam and lock at Marlin Ford, so as to allow
vessels drawing only four feet of water to pass up to
the quay at the Broomielaw.   The clearing out of the
channel by artificial means was, however, found the most
effectual method of opening up the navigation of the
river, and at length all other plans gave way to this.
Golborne had run out jetties at various points, by which
the scour of the tide had been so directed that considerably greater depth had been secured.   Mr. Rennie
examined the entire river below Glasgow in 1799, and
the result was his elaborate report of that year.   He
recommended numerous additions to the jetties, as well
as many improvements in their direction.   He also
advised that a system of dredging should be commenced, which was attended with the best possible
results ; and the same course having been followed by
succeeding engineers, the Clyde has now become one of
the busiest navigable thoroughfares in the world.   The
plan which he shortly after prepared and submitted of
a range of commodious docks along both banks of the
river at the Broomielaw, showed his sagacity and foresight in an eminent degree ; but unhappily it was considered too bold, and perhaps too costly, and was not
carried out.[1]

---

[1] Mr. Rennie proposed to form two
docks on the Broomielaw side of the
river—one 1350 feet long and 160
feet wide, with two entrances, and
another 900 feet long and 200 feet
wide; with a third dock upon the
Windmill Croft, on the south side of
the river, 300 feet long and 200 feet
wide; the whole presenting a total
length of quayage of 6120 feet, besides
a river quay wall 1150 feet long.
This magnificent plan, proposed more
than half a century since, viewed by
the experience of this day, shows how
clearly Rennie anticipated the commercial growth and manufacturing
prosperity of Glasgow, for which these
projected docks would have afforded

At the opposite end of the island he was consulted (in 1796) as to the best method of improving the harbour of Torbay, and submitted a series of able plans, only a small part of which were carried into effect. Shortly after (in 1797) we find him inspecting the sluicing arrangements of the harbour then under construction at Grimsby, when he furnished a plan of the great lock which it was necessary to place at the entrance of the canal leading to the dock, and which was in his opinion indispensable for scouring the harbour entrance and keeping it clear of silt. This lock was executed according to his plans by the local engineer; but it appeared that sufficient precautions had not been taken in founding and proportioning the dimensions of the retaining walls, on which Mr. Rennie had not been requested to give an opinion, the work appearing to be of so simple and ordinary a character. But shortly after the building was begun, a considerable portion gave way, and he was again sent for to inquire and report as to the cause of the failure. He found that the defect lay in the nature of the ground on which the foundation was built, which was so soft that it would not bear the weight of solid walls of the ordinary construction. Always ready with an expedient to meet a difficulty, he directed that, without diminishing the quantity of material employed, it should be distributed over a greater base, for the purpose of securing a larger bearing surface. With this object he prepared his plan of the requisite structure, adopting the expedient of *hollow walls*, which he afterwards employed so extensively in his pier and harbour works. They not only bore upon a larger base,

ample accommodation, at an estimated capital cost (at the time the plans were made) of only 130,000*l.* What would not Glasgow give now to have the benefit of Rennie's docks? Indeed it is remarkable that, to this day, so little has been done to realize his idea, and provide dock accommodation for the trade of the Clyde, which is now quite as much needed as the same kind of accommodation was in the Thames at the beginning of the present century.

but were found even stronger than solid walls containing an equal quantity of material, and those at Grimsby have stood firm until the present day. The contrivance was thought so valuable, that some years after Mr. Rennie had invented it, Sir Samuel Bentham (in 1811) took out a patent for the plan; but of this Mr. Rennie took no notice, having himself, as we have seen, been the original inventor of the process. Indeed, his attention had long before this time been directed to the best form of walls for resisting the pressure of water; and, as early as the year 1793, we find him recommending the adoption of curved walls, in place of the inclined straight-faced walls with perpendicular back formerly adopted.[1]

Another important harbour on which Mr. Rennie was early employed was that of Holyhead, situated at the point of the island of Anglesea nearest to the Irish capital. Although so conveniently placed for purposes of embarkation, everything had as yet been left to nature, which had only provided plenty of deep water and many bold rocks. But Holyhead had neither pier nor jetty, nor any convenience whatever adapting it for harbour uses. Besides, the place was almost inaccessible from inland by reason of narrow, rugged, and in many places almost precipitous roads. The dangerous ferries

---

[1] The occasion on which this plan was first recommended was in Mr. Rennie's report (1793) on the Hutchison Bridge across the Clyde. That bridge, erected by another engineer, fell down on the removal of the centres, on which Mr. Rennie was sent for, post haste, by the Lord Provost and magistrates of Glasgow to confer with them on the subject; and his advice as to the rebuilding of the bridge on another site was subsequently adopted. It appeared, from an inspection of the ruined piers, that a breast or quay wall had been built on the south side of the river, and to the west of the bridge, which had not been executed according to contract. The report stated : " The above walls should be enlarged in their dimensions and altered in their construction; they ought to be carried at least to the level of the river bed, and made five feet thick at the base next to the bridge, and four feet thick at the top, battering one-fifth of their height *in a curvilinear form*, the beds of the stones being *radiated to the centre of the curve* ; as the height lessens, the dimensions of the walls may be diminished in the same proportion, and, if built as above described, I have no doubt of the works being permanent."

at Conway and Menai also presented serious obstacles to travelling by that route; and hence the port of Liverpool, and Park Gate on the Dee near Chester, continued to be the principal places of embarkation for persons proceeding to Ireland, until the beginning of the present century. When the Act of Union was passed, the Government determined to bring the two countries into closer communication with each other; first by means of convenient roads through North Wales, and next by capacious harbours at Holyhead on the one coast and at Kingstown on the other. In the year 1802 Mr. Rennie was requested to report upon the subject, and he proceeded to Wales for the purpose of examining the Conway and Menai ferries, and the capabilities of Holyhead as a port. It was on that occasion that he recommended the construction of the permanent fixed bridges across both Straits after the plans which we have above referred to; but nothing was done towards carrying out his suggestions, and the whole question slept until the year 1809, when he was requested by the Government to prepare plans of a harbour at Holyhead, as the first step towards the desired improvement; and his design having been approved, the works were begun in the following year.

The form of the harbour of Holyhead has been determined, in a great measure, by the cliffs which there overhang the sea, and on the verge of which stand the ancient church and cemetery of the place. The works designed by Mr. Rennie consisted of a pier 1150 feet long, extending in a direction nearly due east from the inner side of Salt Island, which is separated by a narrow channel from the main island of Holyhead. The pier terminated at a depth of about 14 feet at low water of spring tides. At 80 feet distant from the extremity of the main pier there was a jetty 60 feet long, carried out at right angles to its inner face, to check any swell which might come round the

pier-head from entering the harbour, and to throw it
upon the opposite shore. The roadway was 50 feet
wide, and 8 feet above the level of high water of spring
tides, the parapet being 7 feet higher. The outer or sea
side of the pier was formed by a flat paved slope of
rough stone, laid at an inclination of 5 to 1. The quay-
wall was curved on the face one-fifth of the height; the
thickness of the masonry being 10 feet upon the average, strengthened at the back by strong counterfoots,

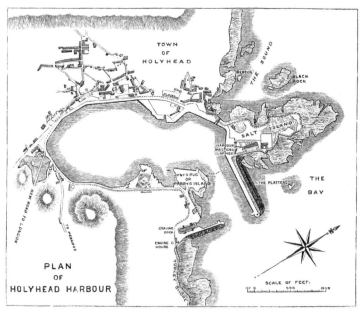

at the regular distance of 15 feet apart. The foundation of this wall was laid below low water by means of
long stones, inclined to each other, in the same manner
as at Howth Harbour, where the plan had been found
to answer remarkably well. The centre of the pier
was composed of loose rubble, taken from the adjacent
shore, packed solidly; the outside being paved with
large angular blocks varying from one to ten tons in
weight, well wedged together. The inside of the parapet
was built of solid masonry. The pier-head and jetty

were founded below low water by means of the diving-bell. The works were begun in 1810 and finished in 1824, and during their progress a small pier was run out from the Pibeo rock on the opposite shore, 550 feet long, leaving an opening between it and the main pier of 420 feet. This pier was provided with a jetty on the inside, similar to that on the main pier, composed of the same kind of materials, and finished in like manner. Within it was a small dry dock for merchant vessels.

HOLYHEAD HARBOUR.　[By R. P. Leitch.]

The total low-water space covered by these two piers was about six acres; but there was more than double that area at high water, besides a large shallow space of about thirty acres for timber.

In conformity with his usual practice, Mr. Rennie so laid out this harbour as to be capable of extension, on the same principles, according as the trade of the port might require. Part of his original design was to devote a large space in the inner portion of the bay, which is dry at low water, to a wet dock of 23 acres.

Had this been carried out, it would have proved of immense advantage to the numerous land-bound vessels which have occasion to put into the port. His original plan also contemplated a pier extending from the outer end of Salt Island, parallel to the one above described, 1500 feet long, terminating at a depth of 25 feet at low water; and as it would have been about 1400 feet distant from the other, and was to be provided with a jetty at right angles to its extremity, it would have provided an additional low-water harbour of 40 acres. The estimated cost of this work was 240,000*l.*; and if to this be added the probable outlay on the additional wet dock above mentioned of 124,000*l.*, it will be found that a total low-water space of 46 acres, and an additional tidal space of 25 acres, together with a wet dock of 23 acres, or a total floating area of about 94 acres, with an ample extent of quay accommodation, sufficient for any amount of packet or general commercial business, would have been provided at a comparatively moderate expenditure. Unhappily Mr. Rennie's plans were not carried out; and though his original design admirably answered the purpose intended, and the whole of the packet service was satisfactorily performed at the old port for many years, when an extension of Holyhead Harbour was determined upon, the Government (after Mr. Rennie's death) employed an engineer who proceeded upon an entirely new plan, the execution of which, when completed, will probably cost upwards of two millions sterling; and, after all, when the rocky and bad nature of the holding-ground within it is taken into account,[1] its security and convenience are still matters of considerable doubt amongst naval men.

During the period in which he was engaged in carrying out the works at Holyhead, Mr. Rennie was also

---

[1] It will be remembered that the *Great Eastern* was nearly wrecked in consequence of the bad holding-ground within the new harbour in the year 1859.

constructing harbours at Howth and Kingstown, with
the same object of facilitating the communication between the ports of England and Ireland. Howth Harbour was opened for packets in 1819, previous to which
time they had sailed from the Pigeon House, at the
mouth of the Liffey, in Dublin Bay. When the piers
at Kingstown Harbour were sufficiently advanced to be
available for the service, the packets were removed
to that port, the depth of water being greater, and the
situation on the whole more convenient.[1]

Among the other harbour works constructed by Mr.
Rennie in England, were the Hull Docks. These were
of great importance, and urgently needed for the accommodation of the large trade of that rising port. What
is called the Humber Dock was begun in 1803 and
finished in 1809.[2] The principal difficulty encountered
in the execution of these works was in getting in the
foundations of the dock walls—the bottom presenting
a great depth of soft mud. They were set on timber
piles and platforms well bound together, with truncated
arches of stone over them. A powerful steam-engine
was employed to draw the water from the coffer-dam in
front of the Humber entrance, to enable the foundations
of the cill to be got in, and the lock gates (which were

---

[1] Mr. Rennie's plan of Kingstown
Harbour consisted of two piers of four
arms each, carried out from the shore
3700 feet distant from each other,
their heads inclined inwards at an
angle of 122 degrees, and terminating
in a depth of 26 feet at low water of
spring tides. The width between the
outer angles of the two outer arms of
the pier was 1150 feet, the entrance
pointing N.E. ½ E. The total space
enclosed was 250 acres. The works
were commenced in 1817, the first
stone being laid by the Earl of Whitworth, the Lord Lieutenant; and the
works were still in progress at Mr.
Rennie's death in 1821. The harbour subsequently fell under the
jurisdiction of the Board of Irish
Works, and all sorts of new plans
were adopted at variance with the
original design of Mr. Rennie, in
carrying out which it is to be feared
that the harbour has been seriously
injured.

[2] This dock is 900 feet long by 370
wide. It covers a surface of 7¼ acres,
and is capable of holding about seventy
sail of square-rigged vessels. The
entrance lock communicating with
the tidal harbour opening into the
Humber is 42 feet wide and 158 feet
long between the gates, with the cill
land 6 feet below low water of spring
tides.

of stout oak) to be fixed.   It was in the course of executing the Hull Harbour works that Mr. Rennie invented
the dredging-machine, as it is now used, for the purpose
of clearing the basins of mud and silt.   Various unsuccessful attempts had before been made to contrive an
apparatus with this object.   A series of rollers, armed
with spikes to rake up the deposit, followed by buckets
and spoons to lift it from the bottom, worked by means
of a walking wheel between two barges, was the most
common practice; but it was clumsy, tedious, and
inefficient.   Other machines for a similar purpose were
driven by tread-wheels.   At length the idea was taken
up of fixing a series of buckets to an endless chain,
worked by horse power.   Mr. Rennie carefully investi-
gated all that had previously been attempted in this
direction, and then proceeded to plan and construct a
complete dredging-machine, with improved cast iron
machinery, to which he yoked the power of the steam-
engine.   He was thereby enabled to raise as much as
300 tons of mud and gravel in a day from a depth
of 22 feet; and the expedient proved completely successful.   The same kind of machine was extensively
used by Mr. Rennie in executing his various harbour
works.   One of these, constructed for the excavation of
the Perry Dock at Blackwall in 1802, was even furnished with a powerful apparatus for splintering rocks
and large stones which could not otherwise be removed;
and it answered the purpose most effectually.   In the
Clyde, the Thames, the Mersey, and the Witham, as
well as in various foreign rivers, the dredging-machine,
as contrived by Mr. Rennie, has been found invaluable;
and there is scarcely a port or harbour in the United
Kingdom in which it has not been most beneficially
employed.
    In addition to these docks and harbours, Mr. Rennie
furnished the plans of the new quays and docks at Greenock on the Clyde in 1802; and those at Leith, the port

of Edinburgh, in 1804. Neither of these schemes was carried out to the full extent, chiefly for want of funds; but the improvements effected at the former port were considerable, and at Leith two large docks, 1500 feet long, and two small ones, 750 feet long, constructed along the shore between the old tidal harbour and the village of Newhaven, provided a large amount of additional accommodation for the growing trade of that port. He was also consulted in 1805 respecting the improvement of Southampton; and the measures which he then recommended in his elaborate and able report formed the beginning of a series of works of almost national importance. Mr. Rennie's clear-sighted prognostications of the future prosperity of the port—arising from its great natural advantages in respect of security, capability of extension, the excellent anchorage of Southampton Water, the central situation of the place on the south coast, and its moderate distance from London—have been amply fulfilled; its subsequent connection with the capital by railway having given an impulse to its improvement and prosperity far beyond what even his sagacious mind could at that time have foreseen.[1]

---

[1] It would occupy much space to mention in detail the various harbours in the United Kingdom which Mr. Rennie was employed to examine, report upon, and improve; but the following summary may suffice:—In England, he examined and reported on Rye Harbour · (1801); Dover (1802); Hastings projected harbour (1806); Berwick, where he constructed the fine pier at the mouth of the Tweed, 2740 feet in length (1807); Margate Harbour (carried out 1808); Liverpool Docks, on which he made an elaborate report (1809); North Sunderland (1809); Shoreham (1810); Newhaven (1810); Harbour of Refuge in the Downs, north of Sandown Castle, on which he made a careful report (1812); Prince's Dock, Liverpool, of which he furnished the designs (1812); Bridlington (1812);

Sidmouth (1812); Rye, a second report (1813); Blyth (1814); Ramsey, Isle of Man (1814); Port Leven, Mount's Bay, Cornwall (1814); Bridgewater (1814); Whitehaven (1814); Scarborough (1816); the improvement of the navigation of the river Tyne (1816); Yarmouth (1818); Fishguard, Wales (1819); Kidwelly, Wales (1820); and Sunderland (1821). He also suggested various improvements, many of which were carried out, in the following harbours of Scotland, besides those above mentioned:—Loch Buy, Isle of Skye (1793); Port Mahomack, near Tarbet Ness (1793); Kirkcudbright and Saltcoats (1799); Craigmore, near Boroughstoness (1804); Montrose (1805); Ayr, where the improvements recommended by him were carried out (1805); Peterhead

In his harbours, as in all his engineering works, Mr.
Rennie proceeded upon certain definite principles, which
he sought to arrive at after a careful study of the whole
subject. He was averse to all makeshifts and temporary
expedients. When he was asked to give his advice as
to the *best* means of rendering a harbour efficient, he
stated his views fully and conscientiously, holding nothing in reserve. He set forth the whole cost which he
believed would be incurred, and no less. He abhorred
setting traps in the shape of low estimates to tempt men
to begin undertakings when he knew that they would be
exceeded. He spoke out the whole truth. "You want a
harbour," he would say, "of such strength as to be *safe*,
with piers able to resist the greatest possible force of the
sea. Well, here is the plan I recommend : it is the best
that I can suggest. But I tell you the whole cost which
I think will be incurred in its construction. Adopt the
plan or not, as you think proper." He would never
consent to reduce the strength of his piers and retaining
walls under the limits which he thought essential for
stability. He would not risk his reputation and character
upon slop-work ; he would rather lose his chance of employment altogether. Hence so many of his large but
effective designs for the improvement of our most important harbours remained unexecuted, or were only carried
out to a limited extent, sometimes by engineers who had
not mastered the fundamental principles on which his
plans were founded, and in such a manner as occasionally
to lead to vast inconvenience and almost endless expense.

In his report on the Earl of Elgin's proposed harbour at Charleston, on the north shore of the Frith

---

(1806); Frazerburgh, only partially
carried out (1806); Charleston (1807);
Alloa (1808); St. Andrew's (1808);
Portnessock, Galloway (1813); Ardrossan (1811 and 1815); and Portpatrick (1819). In like manner he
was consulted, and reported, as to the
following Irish harbours :—Westport
(1805); Ardinglass (1809); Dublin
(1811); Balbriggan (1818); Dona-
ghadee (1819); and Belfast (1821).
He was also consulted respecting dry
docks at Malta (1815), and a harbour
and docks at Bermuda (1815).

of Forth, in the year 1807, he very clearly laid down the broad principles on which he held that such works should be designed : " Every harbour," he said, " should be so constructed as to have its mouth as much exposed as possible to the direction from whence vessels can most conveniently enter 'in stormy weather when they are least manageable; but the Heads should be made of such a form as to admit of the least sea entering it, or so as to occasion as little swell within the haven as possible. This cannot by any practicable construction be *entirely* avoided ; but means should be provided within the harbour so as to reduce the recoil of. the waves to a minimum,—for it is the undertow or retiring sea, after the breaking of a wave, that renders vessels most unmanageable by making the helm lose its effect. At such a time the mariner is at a loss what to do, or how to manage his vessel ; and for the want of due attention to these particulars many of the most considerable artificial harbours in the kingdom are exceedingly difficult of access, and some of them are most unsafe even when entered." The great point, he held, was not only to make a harbour to keep out the sea, but to do so in such a manner as not to render its entrance from the most exposed or dangerous quarter difficult in stormy weather, when its shelter was most needed ; and while it must be so designed as to afford a safe shelter for shipping, it must also be easy to enter and easy to get out of.[1] That so many modern harbours, constructed

---

[1] From the following brief description it will be observed how skilfully he carried out these views in laying out the intended harbour at Charleston. He proposed to construct two great piers, one placed at the western extremity of the little inlet, to which a railway was being laid down—the straight part extending outwards about 154 yards, from which there were to be two kants of about 64 yards each, the last going 57 yards below low-water mark. From thence there was to be a return bend about 70 yards long, in a direction considerably to the north of east. At 50 yards from the extremity of this pier, another of the same length was proposed to be made, forming an angle with it of about 120 degrees, with two other kants similar to the former, and a larger one extending to the shore; the entrance being 50 yards wide, and the outer arm or kant of the east pier making

at great cost, are found comparatively inaccessible in
severe weather, is, we believe, to be accounted for
mainly by the circumstance that they have been laid
out after no definite rule or principle whatever. If they
succeed, it is often the result of a happy accident; and
if they prove failures, it is supposed that it could not
have been helped. Even within the last twenty years
several expensive havens have been constructed, which
have proved to be so dangerous that they can scarcely
be used. But by Mr. Rennie's forms of piers, vessels,
if they have only steerage-way, *must* enter the harbour in safety. They cannot strike on the pier-heads,
if the most ordinary care be used, the very recoil
of the waves forcing them forward into port; and,
as any swell which might enter would have ample
space to expend itself, the ship could either be brought
up, or take the beach without damage if necessary.
Again, a sailing vessel, on leaving the harbour, supposing the wind to be blowing right in, could lie out
upon either tack and make an offing, if it were prudent
to put to sea at all. And although the narrowed distance
between the two pier-heads might be termed the entrance,

---

an angle of 120 degrees with it, so
that both the outer arms made similar kants with each other. A large
space would thus be enclosed, which,
he believed, would make a very commodious and capacious harbour. "By
the above construction," he says in
his report, "though it may seem that
its exposure will admit of the swells
from the south and south-west getting
into the harbour, yet when it is considered that the angle at which a
wave will strike the Heads will occasion a rebound in a similar angle to
that in which it is struck, and as this
will be the case from each Head, it
follows that these reflected waves,
meeting each other, will occasion a
resistance which will have the effect
of preventing a considerable part of
the sea-wave from entering the harbour, and what does enter it will ex-

pend its fury on the flat beach within
and soon become quiet." This might,
he added, be in a great measure prevented by extending the pier-heads
further seaward, but which the large
additional expense precluded him
from recommending; and, indeed,
there would always be abundant shelter for the shipping under one or
other of the pier-heads. Besides, as
the Frith was only about two miles
wide at the place, the probability
was that there would be no such
heavy seas as to render so expensive a measure necessary. The
plan was, however, carried out to
only a limited extent, and we merely
quote the report for the valuable
principles to be observed in the construction of harbours, which are here
so clearly enunciated.

yet in effect it is not so; for the moment the vessel gets within the outer angles of the two return arms or kants, she may be said to be in or out of the harbour, as the case might be. In this way the fullest width of entrance and the smallest space for the admission of swell are ingeniously and effectually secured.

Whilst occupied on the works of the Ramsgate Harbour, of which he was appointed engineer in 1807, Mr. Rennie made use of the diving-bell in a manner at once novel and ingenious. It will be remembered that Smeaton had employed this machine in the operations connected with the building of the harbour;[1] but his apparatus being of wood, was exceedingly clumsy, and very limited in its uses. In that state Mr. Rennie found it when he was employed to carry on the extensive repairs of 1813. The east pier-head was gradually giving way and falling into the sea at its most advanced and important point. No time was to be lost in setting about its repair; but from the peculiarly exposed and difficult nature of the situation, this was no easy matter. The depth at the pier-head was from 10 to 16 feet at low water of spring tides; besides, there was a rise of 15 feet at spring and 10 feet at neap tides, with a strong current of from two to three knots an hour setting past it both on the flood and at the ebb. The work was also frequently exposed to a heavy sea, as well as to the risk of vessels striking against it on entering or leaving the harbour. Mr. Rennie's first intention was to surround the pier-head by a dam; but the water was too deep and the situation too exposed to admit of this expedient. He then bethought him of employing the diving-bell; but in its then state he found it of very little use. No other mode of action, however, presenting itself, he turned his attention to its improvement as the only means of getting down to the work, the necessity for repairing

---

[1] See Life of Smeaton, p. 69.

which had become more urgent than ever. Without loss
of time he proceeded to design and construct a bell of
cast iron, about 6 feet in height, 4½ feet wide, and 6 feet
long, having one end rather thicker and heavier than
the other, that it might sink lower, and thus enable the
exhausted or breathed air more readily to escape. At
the top of the bell eight solid bull's-eyes of cast glass
were fixed, well secured and made water-tight by means
of leathern and copper collars covered with white lead,
and firmly secured by copper screw bolts. To the top
of the inside were attached two strong chains for the
purpose of fastening to them any materials that might
be required for the work, and flanges were cast along
the sides of the bell, on which two seats were placed,
with footboards, for the use of the men while working.
In the centre of the top was a circular hole, to which
a brass-screwed lining was firmly fixed, and into this a
brass nozzle was screwed, having a leathern water-tight
hose fastened to it, 2½ inches in diameter. The hose was
in lengths of about 8 feet, with brass-screwed nozzles at
each end, so that it could be lengthened or shortened at
pleasure, according to the depth of water at which the
men in the bell were working. For the purpose of duly
supplying the machine with air, a double air-pump was
provided, which was worked by a sufficient number of
men. The air-pump was connected with the hose referred
to, and was either placed on the platform above or in a
boat which constantly attended the bell while under water.
Two stout wrought iron rings were fixed on the top of
the machine, to which ropes or chains were attached for
the purpose of lowering or raising it. The whole weighed
about five tons; and it was attached to a circular framework of timber, strengthened by iron, erected over where
the intended new circular pier-head was to be built, and
so fixed to a pivot near the centre of the work, that
it was enabled easily to traverse its outer limits. On
the top of the framework was a truck, made to move

backwards or forwards by means of a rack on the frame, and a corresponding wheel provided with teeth, worked by a handle and pinion. On the truck were placed two powerful double-purchase crabs or windlasses, one for working the diving-bell suspended from it, and the other for lowering stone blocks or other materials required for carrying on the operations at the bottom of the sea. By these ingenious expedients the building

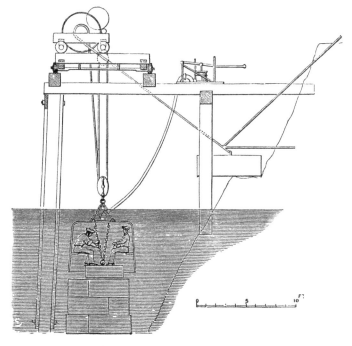

METHOD OF USING THE DIVING-BELL.

apparatus was so contrived as to move all round the new work, backwards and forwards, upwards and downwards, so that every part of the wall could be approached and handled by the workmen, no matter at what depth, whilst the engineer stationed on the pier-head above could at any time ascertain, without descending, whether the builders were proceeding in the right direction, as well as the precise place at which they were at work.

Everything being in readiness for commencing operations, the divers entered the bell and were cautiously
lowered to the place at which the building was to
proceed. A code of signals was established by which
the workmen could indicate, by striking the side of the
bell a certain number of strokes with a hammer, whether
they wished it to be moved upward, downward, or horizontally; and also to signal for the descent of materials
of any kind. By this means they were enabled, with
the assistance of the workmen above, to raise and lower,
and place in their proper bed, stones of the heaviest
description; and by repeating the process from day to
day, and from week to week, the work was accomplished
with as much exactness and almost as much expedition
under water as though it had been carried on above
ground.

Thus the entire repairs were completed by the 9th
of July, 1814; and to commemorate the ingenuity and
skill with which Mr. Rennie had overcome the extraordinary difficulties of the undertaking, the trustees of
the harbour caused a memorial stone to be fixed in the
centre of the new pier-head, bearing a bronze plate, on
which were briefly recorded the facts above referred to,
and acknowledging the obligation of the trustees to their
engineer. They also presented him at a public entertainment with a handsome piece of plate in commemora-
tion of the successful completion of the work. The
diving-bell, as thus improved by Mr. Rennie, has since
been extensively employed in similar works; and
although detached divers, with apparatus attached to
them, are made use of in deep sea works, the simplicity,
economy, and expeditiousness of the plan invented by
Mr. Rennie, and afterwards improved by himself, continue to recommend it for adoption in all undertakings
of a similar character.

CHAPTER VIII.

The Bell Rock Lighthouse.

About eleven miles eastward from the mainland of Scotland, near the entrances to the Friths of Forth and Tay, lies an extensive ledge of rocks, which for a long time was the terror of the seamen navigating that coast. It is nearly two miles in length, being the crest of a mountain rising from the sea bottom, only a small part of which is visible at high water. This sunken reef was a source of such peril, that, as early as the fourteenth century, the Abbot of Arbroath caused a bell to be placed upon the principal rock, the swinging of which by the motion of the waves warned seamen of its dangers ; and from this circumstance it came to be called the Bell Rock. It is affirmed that a notorious pirate, in order to plague the Abbot, cut the bell from the rock, but was himself afterwards wrecked on the very spot ; and on this tradition Southey founded his beautiful ballad of 'Ralph the Rover.' [1]

Nothing was done to replace the bell, or to set a beacon upon the reef ; and it remained in its dangerous state—the Eddystone of the northern seas—until the beginning of the present century, when the increasing

---

[1] The following is the tradition as given by an old writer :—"By the east of the Isle of May, twelve miles from all land in the German Sea, lyes a great hidden rock called Inchcape, very dangerous to the navigators, because it is overflowed every tide. It is reported that, in old times, there was upon the said rock a bell, fixed upon a tree or timber, which rang continually, being moved by the sea, giving notice to the saylors of the danger. This bell or clocke was put there by the Abbot of Aberbrothock, and, being taken down by a sea-pirate, a yeare thereafter he perished upon the same rock, with ship and goodes, by the righteous judgment of God."—Stoddart's 'Remarks on Scotland.'

commerce of Scotland, and the large number of vessels wrecked there, had the effect of directing public attention to the subject. As in the case of the Eddystone reef, the sailors' fear of it was such, that in order to avoid its dangers, they hugged the land so close as very frequently to run ashore.[1]

A Board of Commissioners had been appointed, under the powers of an Act passed in 1786, for the purpose of erecting lighthouses at the most dangerous parts of the coast of Scotland; and by the end of the century several had been built,—one on the Isle of May at the entrance to the Frith of Forth, another on the Cumbraes at the mouth of the Frith of Clyde, and others on rocky promontories on the eastern and western coasts, including the Orkneys. The lights exhibited were of a rude kind, and consisted of coal fires in chauffers; though, all that was needed being a light, they probably answered their purpose, but in a clumsy way. The most dangerous reef of all, however, was still left without any protection; and doubtless the delay in providing a light upon the Bell Rock arose from the great difficulty and expense of erecting a suitable structure on such a site.

In the winter of 1799, a tempest, memorable for its violence and fatal effects, ravaged the coasts, and drove from their anchors all the ships lying in Yarmouth

---

[1] Captain Basil Hall relates that, when a boy, he was constantly hearing of vessels getting wrecked through fear of that terrible Bell Rock, which lay about ten leagues due north of the house in which he was born—at Dunglass, on the borders of East Lothian, not far from the bold promontory on which Fast-Castle stands, overlooking the German Ocean; and he relates that "ships bound for the Forth, in their constant terror of the dangerous reef, were not content with giving it ten or even twenty miles of elbow-room, but must needs edge off a little more to the south, so as to hug the shore, in such a way that, when the wind chopped round to the northward, as it often did, these overcautious navigators were apt to get embayed in a deep bight to the westward of Fast-Castle. If the breeze freshened before they could work out, they paid dearly for their apprehensions of the Bell Rock, by driving upon ledges fully as sharp, and far more extensive and inevitable. Thus," he says, "at that time, from three to four, and sometimes half-a-dozen vessels used to be wrecked every winter, within a mile or two of our very door."—'Fragments of Voyages and Travels,' vol. i., p. 15, 16. Edinburgh, 1831.

Roads. The greater part were wrecked on the northern coast; and it was believed that many of them might have been saved, had a light been fixed on the Bell Rock to point out the entrances to the Friths of Forth and Tay. Among the other lamentable shipwrecks which took place on the Inchcape about the same time, was that of the *York*, a seventy-four-gun ship, which went down with all her crew. The reef was also a constant source of danger to the shipping of Dundee, then rising in importance, as it lay right in the main track of vessels making the mouth of the Tay from the German Ocean.

Many were the plans suggested for a lighthouse on the Bell Rock. In 1799 Captain Brodie submitted to the Commissioners of Northern Lights his design of a cast iron tower, to be supported on four pillars; but it was not adopted. In the mean time temporary beacons of timber were employed; but these rarely stood the storms of a single winter; and three successive structures of this kind were completely swept away. Mr. Robert Stevenson and Mr. Downie also proposed plans for the consideration of the Board between 1800 and 1804; but neither of them was adopted. Considerable diversity of opinion continuing to exist, the Commissioners determined to employ Mr. Rennie to examine the site and report as to the best course to be pursued. He accordingly proceeded to Scotland, and visited the Inchcape on the 17th of August, 1805, in company with Mr. Hamilton, one of the Commissioners, and Mr. Stevenson, their surveyor.

After mature deliberation, he sent in his report on the 30th of December following. The purport of it was, a recommendation to erect a substantial lighthouse of stone, similar to that on the Eddystone, as being, in his opinion, the only structure calculated to meet the necessities of the case. He regarded a wooden building as objectionable, because of the perish-

able character of the material, and its liability to be
destroyed by fire.   Although it would be possible to
erect a lighthouse of cast iron, its cost at that time would
have been equal to one of stone, with which, in point of
durability, it was not to be compared.   " I have therefore," he concluded, " no hesitation in giving a decided
opinion in favour of a stone lighthouse."   With such examples as the Tour de Cordouan near the entrance of the
Garonne, and the Eddystone off the coast of Cornwall, he
held that there could be no doubt as to the superiority of
this plan to any other that could be proposed.   Although
the Inchcape was not so long uncovered by the tide as
the Eddystone rock, and there might be greater delay in
getting in the first four or five courses of the foundation,
this was only a question of time ; and he had no doubt
that this difficulty could be overcome, and the whole structure completed in the space of about four years.   In his
report he further says : " Mr. Stevenson, to whose merit
I am happy to bear testimony, has been indefatigable in
obtaining information respecting this rock, and he has
made a model of a stone lighthouse nearly resembling
that of the Eddystone, in which he has proposed various
ingenious methods of constructing the work by way of
facilitating the operations.   I own, however, after fully
considering them, and comparing them with the construction of Mr. Smeaton—I mean in the process of
building—and also reflecting that there are undoubted
proofs of the stability of the Eddystone, that I am inclined to give the latter the preference; its general
construction, in my opinion, rendering it as strong
as can well be conceived."   But, taking into account
that the foundation of the proposed building lay so
much lower in the sea, he suggested that the column
should be somewhat higher, so that the eave of the cupola
should be about 100 feet above the surface of the rock,
the Eddystone being only 84 feet 6 inches,—though this
alteration would involve a somewhat greater diameter

of the base. He further pointed out that the pillar should be surrounded by such an extent of rock as to diminish the force of the waves breaking at its foot. He also proposed that the floor of the lower room of the lighthouse should be 50 feet above the level of the rock, and from thence to the top of the platform 35 feet; making a total height of 85 feet to the platform or gallery. He recommended that argand lights should be employed, with parabolic reflectors; and he suggested for consideration the employment of carburetted hydrogen gas, then coming into extensive use for lighting purposes. The cost of the lighthouse, so constructed, he estimated at about 42,000*l*.

The Commissioners adopted Mr. Rennie's report, and proceeded to Parliament for the requisite powers, which were obtained in the session of 1806; after which (on the 3rd of December following) they unanimously appointed him the chief engineer for conducting the work. At his recommendation Mr. Stevenson was appointed the assistant-engineer to superintend the operations on the spot, and under him were placed two able foremen superintendents, Mr. Peter Logan over the masons, and Mr. Francis Watt over the joiners employed upon the lighthouse. Mr. Rennie was then requested to report further in detail, with plans and specifications of the various work, which he prepared and duly submitted to the Board. In his second report of the 26th December following, he entered at great length into the description of stone to be used in the building, based upon a personal inspection of the quarries at Mylnefield near Dundee, at Arbroath, and at Aberdeen; and on his recommendation it was determined to use blocks from the Rubieslaw quarry at Aberdeen for the outer, and Dundee stone for the inner masonry. He also repeated his advice, that in carrying out the work, the plan of construction adopted by Smeaton in the building on the Eddystone should be mainly followed; one of the few deviations consisting in the substitution of dovetailed pieces of stone for chain bars in the joints both of the walls and the floors. These recommendations having been adopted, Mr. Rennie was authorised to proceed with the requisite preparations for the work; and, after making all due arrangements, and giving his representative suitable instructions, he left the practical operations to be carried out by him accordingly.[1]

The whole of the year 1807 was occupied in constructing the necessary vessels and in erecting the requisite machinery and building-shops at the working yard at Arbroath, which was fixed upon as the most convenient point on the coast for carrying on the land operations. Some progress had been made at the rock itself, where a smith's forge was fixed and a temporary beacon erected, whilst a floating light, fitted up in an old fishing dogger, was anchored near the reef until the lighthouse could be erected. Preparations had also been made for proceeding with the foundations, the necessary excavation being conducted at considerable peril, in consequence of the violence of the waves and the short period during which the reef was uncovered during each day. The dangerous nature of the employment may be illustrated by the following brief account of an accident which happened to the workmen on the 2nd of September, before the excavation for the first course had been completed. An additional number of masons had that morning come off from Arbroath in the tender of forty tons, named *The Smeaton*, and having landed them on the rock, the vessel rode at salvagee, with a crosshead made fast to the floating buoy. The wind rising, the men began to be uneasy as to the security of the *Smeaton's* riding-ropes, and a party went off in a boat to

---

[1] A detailed account of the operations was afterwards published by the assistant-engineer, in his interesting work entitled 'An Account of the Bell Rock Lighthouse.' By Robert Stevenson, Civil Engineer. Edinburgh, 1824.

examine whether she was all secure; but before they could reach the vessel's side they found she had already gone adrift, leaving the greater part of the men upon the reef in the face of a rising tide. By the time the *Smeaton's* crew had got her mainsail set and made a tack towards their companions, she had drifted about three miles to leeward, with both wind and tide against her, and it was clear that she could not possibly make the rock until long after it had been completely covered. There were thirty-two men in all upon the Inchcape, provided with but two boats, capable of carrying only twenty-four persons in fine weather. Mr. Stevenson seems to have behaved with great coolness and presence of mind on the occasion, though he confessed that of the two feelings of hope and despair, the latter considerably predominated. Fully persuaded of the peril of the situation, he kept his fears to himself, and allowed the men to continue engrossed with their occupations of boring and excavating. After working for three hours, the water began to rise upon the lower parts of the foundations, and the men were compelled to desist. The forge fire became extinguished; the smith ceased from hammering at his anvil, and the masons from their hewing and boring; and when they took up their tools to depart, and looked around them, their vessel was not to be seen, and the third of their boats had gone after the *Smeaton* and was drifting away in the distance! Not a word was uttered; but the danger of their position was instantly comprehended by all. They looked towards their master in silence; but the anxiety which had been growing on his mind for some time had now become so intense, that he was speechless. When he attempted to speak, his mouth was so parched that his tongue refused utterance. Turning to one of the pools on the rock, he lapped a little water, which gave him relief, though it was salt; but what was his happiness when, on raising his head, some one called out "A boat! a boat!" and

sure enough, a large boat was seen through the haze
making for them. She proved to be the Bell Rock
pilot-boat, which had come off from Arbroath with letters,
and her timely arrival doubtless saved the lives of the
greater part of the workmen. They were all taken off
and landed in safety, though completely drenched and
exhausted.

Mr. Rennie, accompanied by his son George, visited
the rock on the 5th of October, 1807, the day before the
works were suspended for the winter. They came off
from Arbroath and stayed on board the lighthouse yacht
all night, where Mr. Stevenson states that he " enjoyed
much of Mr. Rennie's interesting conversation, both on
general topics and professionally upon the progress of
the Bell Rock works, on which he was consulted as chief
engineer." On the following morning Mr. Rennie
landed to inspect the progress made in the excavation,
being received with a display of colours from the beacon
and three cheers from the workmen. They continued
at work for only about three hours that day, after which
the whole working party, accompanied by the chief engineer and his son, the resident engineer, and the foremen of the works, returned to land for the winter.

The preparation of the stone blocks for next summer's operations then proceeded on shore; and by the
spring large numbers were dressed, and waiting to be
floated off. In May, 1808, the excavations on the rock
were proceeded with, and on the 10th of July the
first stone was laid with considerable ceremony. Mr.
Rennie paid his next visit on the 25th of November following, for the purpose of inspecting the work done and
reporting progress to the Commissioners. From his
report it appears that three courses of masonry had by
that time been laid in a very complete manner. At
his suggestion, a modification was adopted in the cement
used for the building, and also in the use of the granite
blocks delivered from the Aberdeen quarry, some of

which had been found defective. By the end of 1809,
at our engineer's next visit of inspection, the tower
had been built to a height of 30 feet, and was comparatively secure against the effects of the most violent
seas. In his report to the Commissioners he stated that
he found that the form of slope which he had adopted for
the base of the tower, as well as the curve of the building,
fully answered his expectations—that they presented
comparatively small obstruction to the roll of the waves,
which played round the column with ease—and he
expressed the opinion that the lighthouse, when finished,
" would be found to be the most perfect work of its kind."
In this report he recommended a modification in the
details of the upper part of the building. In dovetailing
the stones together, the method employed at the Eddystone had up to this point been followed; but from the
top of the staircase he proposed that a somewhat different
plan should be adopted. "The stone floors in the Eddystone," he said, "were formed by an arch in the shape
of a dome springing from the surrounding walls, to
strengthen which chain-bars were laid in the walls. I
propose that these should be done with large stones
radiating from a circular block in the middle, to which
their interior ends are to be dovetailed as well as the
radiated joints, and these connected to the surrounding
walls by means of a circular dowel. By this means the
lateral pressure on the walls will be removed, the whole
connected together as one mass, and no chain-bars will
be wanted except under the cornice. Thus the whole
will be like a solid block of stone excavated for the
residence of the lightkeepers, stores, &c." He concluded
with some practical advice as to the construction of the
lantern after an improved method which he proposed,
in order that it might be in readiness in the course of
the ensuing summer, by which time he anticipated that
the building would be ready for use if the weather proved
favourable. These recommendations were adopted, and

the work having proceeded satisfactorily, the whole was completed by the end of 1810, and the light was regularly exhibited after the 1st of February, 1811. When finished, the tower was 10 feet higher than the original design, being 95½ feet to the top of the cornice, and 127 feet to the top of the lantern. The additional height to which Mr. Rennie had thought it necessary to carry the lighthouse during its construction had the effect of raising the total cost to 61,331*l.*; but he believed the increased outlay would be fully justified by the greater security given to the building, and its increased efficiency for the purpose for which it was intended.

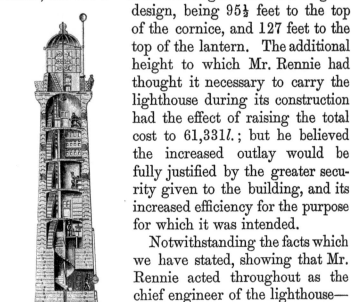

Notwithstanding the facts which we have stated, showing that Mr. Rennie acted throughout as the chief engineer of the lighthouse—that he furnished the design, arranged the details of the building, settled the kind of materials to be used down even to the mode of mixing the mortar, and from time to time made va-

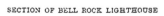

SECTION OF BELL ROCK LIGHTHOUSE

rious alterations and modifications in the plans of the work during its progress, with the sanction of the Commissioners—his name has not usually been identified with the erection of this structure; the credit having been almost exclusively given to Mr. Robert Stevenson, the resident engineer, arising, no doubt, from the circumstance of Mr. Rennie being in a great measure ignored in the ' Account of the Bell Rock Lighthouse,' afterwards pub-

lished by that gentleman. Mr. Stevenson was unquestionably entitled to great merit for the able manner in which
he performed his duty, which Mr. Rennie was always
prompt to acknowledge; but had any failure occurred,
in consequence of a defect in the plans, Mr. Rennie, and
not Mr. Stevenson, would have been held responsible. As,

THE BELL ROCK LIGHTHOUSE. [By Percival Skelton.]

however, it proved a success, it is but fair that the former
should not be deprived of the merit which unquestionably
belonged to him as chief engineer, which office he continued to hold until the completion of the undertaking.
It is a matter of impossibility that engineers in extensive practice should personally superintend the various

structures designed by them, which are proceeding at the same time in different parts of the country. Hence the appointment, at their recommendation, of resident assistants, whose business it is to see that the details of the design are faithfully carried out, and that the work is executed in all respects according to the determined plans. To take two instances—Telford's Menai Bridge and Stephenson's Britannia Bridge—in the former of which cases Mr. Provis was appointed resident, and in the latter Mr. Edwin Clarke. Both of these gentlemen afterwards published detailed histories of their respective works; but neither of them ignored the chief engineer, nor claimed the merit which attached to the successful erection of these great structures.

During Mr. Rennie's lifetime various paragraphs were published in the Edinburgh papers, claiming for Mr. Stevenson the sole credit of having designed and erected the lighthouse. At this he was naturally annoyed; and the more so when he learnt that Mr. Stevenson was about to "write a book" on the subject, but without communicating with him thereon. "I have no wish," he says, in a letter to a friend, "to prevent his writing a book. If he details the truth fairly and impartially, I am satisfied. I do not wish to arrogate to myself any more than is justly my due, and I do not want to degrade him. If he writes what is not true, he will only expose himself. I bethink me of what Job said, 'Oh that mine enemy would write a book!'" The volume, however, was not published until three years after Rennie's death; and it was not before the publication of Sir John's work on Breakwaters, that his father's claims as chief and responsible engineer of the lighthouse were fairly asserted and afterwards fully and clearly established.[1]

---

[1] The correspondence which took place on the subject will be found recorded in the 'Civil Engineer and Architect's Journal,' vol. xii., 1849.

CHAPTER IX.

MR. RENNIE'S WAR DOCKS AND OTHER GOVERNMENT WORKS.

FROM an early period Mr. Rennie's eminent practical abilities pointed him out for employment in the public service; and he was consulted by the Ministry of the day more particularly as to the machinery used in the Government establishments, the formation of royal docks, and the construction of breakwaters. At the recommendation of Mr. Smeaton he was called upon by the Victualling Department of the Navy to advise them as to the improvement of their flour-mills at Rotherhithe, which were worked by the rise and fall of the tide; and the manner in which he performed that service was so satisfactory, that it shortly led to his advice being taken on other subjects, some of which might at first sight be supposed to lie beyond the range of his engineering experience.

Great alarm prevailed in 1803 as to the threatened intentions of our warlike neighbours across the Channel. The 'Moniteur' and the 'Brussels Gazette' were openly speculating as to the time it would take the French army to reach London, and it therefore behoved England to be upon her guard. All the possible lines of approach from the coast to the capital were carefully examined; and it appeared to military men that the eastern side of London was the most accessible to the advance of an enemy landing near the mouth of the Thames or on the Essex coast. Mr. Rennie was employed to examine the valley of the Lea, to ascertain whether means could be devised for suddenly laying it under water, if necessary, to check the approach of a hostile army from that quarter. After careful consideration, he laid before the Government a plan with that object, which included a series of dams furnished with sluices,—one at the junction of the Lea with the Thames, another across the valley at Bow, a third at Temple Mills, and others higher up the river as far as Amwell. These were contrived so that the waters should be penned up, and the valley flooded at will. The works were, however, only partially carried out, and some five dams were formed above Bromley, with jetties at different points to increase the current; but happily these defensive works were never required, Napoleon's warlike ambition having shortly after become diverted in another direction.

About the same time Mr. Rennie was consulted as to the best means of improving the defences of the coast of Kent; and he laid out and constructed for the Government the Hythe Military Canal, defended by earthworks, and protected by a breastwork on the land side. It extends for some twenty miles westward across Romney Marsh to the river Rother, near Rye.[1] He was also consulted as to the best site for a low-water harbour on the south-east coast for accommodating frigates to watch the opposite shore,—Napoleon's legions being then assembled on the heights above Boulogne, and his flat-bottomed boats crowding its port. Mr. Rennie's opinion was in favour of Folkestone as the best site for such a port, where a ridge of rocks outside its then small tidal basin offered unusual facilities for the formation of a haven capable of accommodating vessels of considerable burden. This work was not, however, carried out, and Folkestone Harbour is happily now devoted to more pacific purposes.[2]

---

[1] See Descriptive View of Romney Marsh, Vol. I., p. 8.

[2] Mr. Rennie seems to have been frequently in communication with

When Fulton proposed the scheme of his famous torpedo for blowing up ships at sea, by stealthily approaching them under water, Earl Stanhope made so much noise
about it in the House of Lords, that a commission was
appointed to investigate its merits, of which Mr. Rennie
was a member.   Little importance was attached to
Fulton's pretended "invention;" nevertheless it was
determined to afford him an opportunity of exhibiting
the powers of his infernal machine, and an old Danish
brig riding in Walmer Roads was placed at his disposal.
He succeeded, after an unresisted attack of two days
—during which he had also the assistance of Sir Home
Popham—in blowing up the wretched carcass, and with
it his own pretensions as an inventor.[1]

Among other subjects on which we find Mr. Rennie
consulted by the Government authorities, were the im-

---

the military authorities of the day on
warlike matters.   Thus, in 1809, he
was applied to for a gang of workmen
to proceed to Flushing, during the
unfortunate Walcheren expedition, to
assist in destroying the piers, flood-
gates, and basins of that port; after
effecting which they returned home.
The contractors were Messrs. Brough
and Macintosh.   The sum expended
on artificers' and workmen's wages,
amounting to 2961*l.* 0*s.* 7*d.*, was dis-
bursed by Mr. Rennie, who obtained
the requisite vouchers from the Treasury.

[1] Mr. Rennie had a very mean
opinion of Fulton, regarding him as a
quack who traded upon the inventions of others.   He considered that
little merit belonged to him in regard to the invention of the steamboat.   Thus, Jonathan Hulls, Miller
of Dalswinton, and Symington had
been at work upon the invention long
before Fulton; Miller having actually
made a voyage to Sweden and back
with his steamboat as early as 1789,
eighteen years before Fulton made his
first successful experiment on the
Hudson.   Fulton's alleged invention

of cast iron bridges was not more original.   Writing to Mr. Barrow of
the Admiralty, in 1817, Mr. Rennie
says: "I send you Mr. Fulton's book
on Canals, published in 1796, when
he was in England, and previous to
his application of the steam-engine to
the working of wheels in boats.   On
the designs (as to bridges, &c.)
contained in that book, his fame, I
believe, principally rests; although
he acknowledges that Earl Stanhope
had previously proposed similar
plans, and that Mr. Reynolds of
Coalbrookdale, in Shropshire, had
actually carried them into execution;
so that all the merit he has—if merit
it may be called—is a proposal for
extending the principle previously applied in this country.   The first iron
bridge was erected at Coalbrookdale
in 1779, and between that and the
publication of Fulton's book in 1796
many others were erected; so that,
in this department, he has little to
boast of.   I consider Fulton, with
whom I was personally acquainted, a
man of very slender abilities, though
possessing much self-confidence and
consummate impudence."

provement of the machinery of the Waltham Powder Mills, especially as to the more economical application of the water power—the fixing of moorings in the tideways of the royal harbours—the clearing of the Thames in front of Woolwich Dockyard of its immense accumulation of mud—the erection of a quarantine establishment in the Medway [1]—the provision of wet docks for the Royal Navy—and the introduction of improved machinery at the various dockyards; on which last subject Mr. Rennie was especially competent to give advice.

A Commission of Civil Officers of the Navy was appointed, in 1806, to consider the best means of turning out work from the dockyards with the greatest despatch and economy. Private manufacturing establishments were then a long way ahead of the Government yards, where methods of working, long abandoned everywhere else, still continued in practice; and to call a mechanic or labourer on the Thames "a regular dockyarder" was to apply to him the lowest term of reproach that could be used. Foreign governments were introducing steam-engines and the most improved kinds of machinery, whilst our Admiralty were standing still, notwithstanding

---

[1] The quarantine establishment of the port of London was then situated at Stangate Creek, which joins the Medway about two miles above Sheerness. It consisted of several old two and three-decker hulks, into which goods were placed. Passengers while performing quarantine might well fret and fume at their detention, having before them a most uninteresting prospect—a wide extent of flat marsh-land, with a fringe of mud at low water. A small vessel of war was stationed at the entrance of the creek to prevent infringement of the regulations. The annoyance caused by this establishment was very great, and it was complained of more and more as our foreign commerce extended. On several occasions, vessels filled with passengers, having accidentally run foul of the ships performing quarantine, were compelled at once to heave-to, and undergo two or three days' detention before they could be released. To diminish this evil, the Government determined to erect a permanent quarantine establishment about three miles up Stangate Creek, at a place called Chetney Hill, a small rising ground situated in the marshes. It was proposed to isolate this hill by a canal, provided with a lock; and Mr. Rennie was requested to prepare the requisite plans, which he did (in 1806), and the works were executed at a heavy expense; but we believe they were never used, and the old hulks continued to be employed until the final abandonment of the quarantine system.

the war with France, which called for more than ordinary despatch in the building and repairs of ships of war. In Mr. Rennie's report to the Commission, he incidentally mentions mechanical appliances which he was engaged at the time in manufacturing for exportation. " I am erecting," he said, " a steam-engine for the royal dockyard at Copenhagen, for the purpose of blowing all the bellows in the smithies, and another for pumping water out of the docks. I also understand they mean to construct machinery there for forging anchor-palms and other large iron work. Rolling-mills for bars, bolts, hoops, &c., might also be employed with advantage. Saw-mills, such as I have constructed at Calicut, on the coast of Malabar, for sawing plank, beams, and other articles, would be very serviceable. Block-machinery and rope-works might likewise be worked by steam-engines, as well as mills for rolling copper, machinery for working cranes, and other purposes." He pointed out, that dockyards ought to be so laid out as to enable work of the same kind to be carried on by continuous operations as in a well-ordered manufactory. He showed that the water at the entrance of all the dockyards, excepting Plymouth, was too shallow to enable large ships to be docked for repairs, without dismantling them and taking out their guns and stores, which was a cause of much delay, damage, and expense; and he urged the provision of a dockyard in which the largest ships might lie afloat at low water, and be docked and undocked in all states of the tide. He would also have powerful steam-engines provided, by which any dock might be pumped dry in a few hours, so as to enable repairs to be at once proceeded with. He had no doubt that the cost of constructing such a harbour and dock would be saved to the nation in the course of a very few years. He also urged, as of still greater importance, the necessity for concentrating all dockyard work as much as possible. Himself the head of a large manufacturing

establishment, he was well aware that the more the
several branches are kept apart from each other, the
less is the efficiency secured, and the greater is the waste
of material as well as loss of time. He therefore urged
above all things concentration, and he broadly held that
without it economy was impossible.

Portsmouth, Plymouth, Sheerness, Chatham, Woolwich, and Deptford were mostly far apart, some of them
very badly adapted for the purposes of royal dockyards,
and at nearly all of them the same costly process of
patching, cobbling, and waste was going forward. Indeed, he held that it would be much cheaper, viewed
as a money question only—not to mention the increased
despatch of business and the improved quality of the
work done—to construct an entirely new dockyard, where
every department could be laid out in the most complete and scientific manner. These views looked so
reasonable, and they pointed to results so important, that
the Board of Naval Revision determined to pursue the
investigation ; and they requested Mr. Rennie to examine
all the royal dockyards, and report as to the improvements that might be made in them with the above
object; and also on his plan of a new and complete
naval arsenal suitable to the requirements of the nation.

The result of his inquiries was set forth in the elaborate report delivered by him on the 14th May, 1807.
He had found most of the royal harbours in a state
of decay, silted up with mud or sand, and in a generally discreditable condition. Of all the naval arsenals,
he found Plymouth had suffered the least, in consequence of less alluvial matter flowing into the harbour
from the rivers discharging themselves into the Sound
—the principal objection to that port being that it was
exposed to the violence of south-westerly and south-
easterly winds. Portsmouth he found to be in a very
defective state, much silted up with mud, the depth on the
bar having become reduced within a century from 18 to

14 feet, whilst the works generally were in a condition of great decay. The docks were also, in his opinion, quite inadequate to the accommodation of ships requiring repairs, whilst the storehouses, workshops, and building slips were ill laid out, having been run up in haste after no well-digested plan, involving bad work and waste both of money and of time.[1] Deptford Royal Dockyard, the oldest on the Thames, was objectionable because of the decreasing depth of water, which rendered it less and less available for ships of large burden; and hence it was gradually being abandoned for ship-building purposes; Mr. Rennie recommending that it should for the future be exclusively used as a victualling yard. Woolwich also he considered ill adapted for the purposes of a royal harbour and arsenal: it was situated too high up the river, where the water was shallow, and the place was incapable of enlargement to the required extent, except at an enormous cost. He held that large ships, even if built there, must go down the stream into deeper water before they could take in their guns, stores, and provisions, thus involving the risk of damage and the certainty of delay and increased expense. With reference to the naval arsenals on the Medway, although considerable improvements had been made in the dry and wet docks of Chatham, yet he held that the place was, from its situation, incapable of being adapted to the important purposes of a naval establishment of an extent and accommodation commensurate with the national requirements. Besides, the navigation of the Medway from the Nore was very

---

[1] By way of illustrating his views, Mr. Rennie used to say: "Let any stranger visit Portsmouth Dockyard, the head establishment of the British navy, he will be astonished at the vastness and number of buildings, and perhaps say, 'What a wonderful place it is!' knowing nothing about the subject. But I can compare the place to nothing else than to a pack of cards, with the names of different buildings, docks, &c., marked upon them, and then tossed up into the air, so that each, in falling, might find its place by chance,—so completely are they devoid of all arrangement and order."

intricate, and the upper part too shallow for ships of large burden. Then, as for Sheerness, his opinion was that the yard there, besides being on too small a scale to adapt it for the purposes of a national harbour and arsenal, would be exposed to great risk in event of a war, being almost incapable of effectual defence. "On the whole," he observed, "it appears to me, on consideration of all the facts I have been able to collect respecting the principal naval arsenals of the empire, that they are far from possessing such properties, either in point of situation, extent, arrangement, or depth of water in the harbours, as the large and growing naval power of this country demands."

In reviewing the sites of the different arsenals, Mr. Rennie says he was struck with surprise to find them mostly placed on the *lee shore* of their respective harbours—the worst of all positions. Plymouth, Portsmouth, Chatham, and Sheerness were thus situated; Deptford was on the weather shore, and Woolwich had the prevailing wind blowing across it. He also pointed out that, at all the royal dockyards, vessels had to take in their stores and be rigged and fitted out in open harbours; and, with the exception of Plymouth, the materials were clumsily and expensively conveyed from the shore to the ship's side in lighters. There being no wet docks at any of the royal yards, vessels in ordinary lay moored out in the open tideway; and the expense of moving them, of placing men on board to watch them, and the various accidents to which they were thus liable, were the constant cause of heavy loss to the nation. Viewing the subject in all its bearings, and with a view to despatch, efficiency, and economy, he strongly recommended the construction of capacious wet and dry docks, in some convenient situation on a weather shore, provided with sufficient storehouses and workshops, fitted with engines and machinery on the most improved plans for the building and repair of ships; and he expressed

the strong opinion that the political importance of adopting such a measure would far outweigh the expense,
however large, which it might be necessary in the first
place to incur.

The subject was felt by the Admiralty authorities to
be of so much importance, that Mr. Rennie was again
requested to report as to the most advisable site for
a great naval harbour and arsenal such as he had proposed; and he was requested by Lord Howick (then
First Lord) more particularly to state his views as
to the eligibility of Northfleet on the Thames, which
had been suggested as the most desirable situation.
In his visit of inspection to the place, Mr. Rennie
was accompanied by his venerable friend James Watt,
and his confidential assistant Mr. Southern, together
with Mr. Whidbey, acting master-attendant at Woolwich Dockyard; and he had the benefit of their great
experience in assisting him to mature the design which
he shortly laid before the Admiralty for their consideration. The site seemed to him most convenient for
the purpose. "Northfleet," he said, "possesses every
advantage that can possibly be wished for in a naval
station, and it is capable of being rendered as complete
and perfect an establishment as can be made for building,
repairing, and fitting out vessels of war of all classes,
on the largest scale." The Thames at that point presented abundant depth of water; there was a large space
of flat land available for the harbour and docks, which
might be so laid out as to be almost indefinitely extended;
the situation was on the weather shore, well protected,
and capable of being strongly defended; it would be in
direct connection by water with Woolwich, Deptford,
and London, as well as with Chatham and Sheerness;
and as a great national harbour and arsenal he regarded
its situation on the Thames, at the entrance to the greatest
port in the world, as in all respects the most suitable and
appropriate. Accompanying his report was a carefully

devised and most elaborate plan,[1] which unhappily was never carried out.

---

[1] The site of the proposed arsenal was the flat portion of land near Northfleet, about eight hundred acres in extent, lying in the angular space formed by Fidlers' and Northfleet Reaches. Its depth close to the shore was about seven fathoms at low water, or sufficient for vessels of the largest burden. The main entrance-lock was to be at the Northfleet end of the docks, within which was to be an entrance-basin 1815 feet long and 600 feet wide, covering about twenty-five acres. Dry docks were to be placed conveniently near, from which the water was to be pumped by powerful steam-engines, so that vessels might be docked directly from the basin, and have their bottoms examined with the least loss of time. Part of the entrance-basin was to be appropriated for an anchor-wharf, another for a gun-wharf; next the stores and victualling wharves, with their appropriate buildings; the whole arranged on a system, so that the materials required on shipboard might be passed forward to their respective wharves from one stage of preparation to another, with the greatest despatch and economy. At the north end of the basin were to be the mast and boat pounds, with their adjoining workshops, connected also with the Thames and the main western basin by separate entrances. The main western basin was to be at right angles to the entrance-basin, 4000 feet long and 950 feet wide, covering a surface of about eighty acres. Alongside were to be six dry docks and eight building slips, all fitted in the most complete manner with the requisite saw-pits, seasoning-sheds, mould-lofts, timber-stores, and smitheries, conveniently situated in their rear. The whole of the heavy work, such as bellows-blowing, tilt-hammering, forging of anchors, and iron-rolling, was to be performed by the aid of steam-engines and machinery of the most perfect kind. Seventy sail of the line, with a proportionate number of smaller vessels, might conveniently lie in this basin, and yet afford abundant space for the launching of new vessels. Another basin, 980 feet long and 500 feet wide, was proposed for timber ships, on the south-west extremity of the great basin, with a separate entrance into the Thames a little below Greenhithe. The whole of the arsenal was to be connected together by a system of railways extending to every part and all round the wharves. The plan was most complete, some

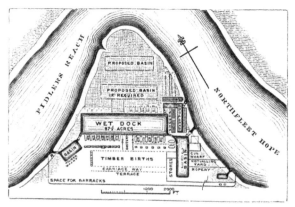

PROPOSED DOCKS AT NORTHFLEET

Mr. Rennie was also consulted respecting the improvement of the old royal dockyards, and submitted

---

of the details being highly ingenious. Had it been adopted, the following would have been amongst its certain advantages. We use Mr. Rennie's own words: "All materials," said he, "used for building ships will be brought to the ship with a prodigious saving of labour and time, compared to what can be done in any of the present yards. When launched from the slip, the vessel will immediately go into the dry dock to be coppered; after which she will proceed to the masthouse-quay to take in her masts, which will be placed in her with the assistance of the sheers, and in a state of perfect dryness. She will then proceed to the entrance-basin, and successively take in her ballast, stores, rigging, sails, cables and cordage, victuals and water, guns and shot, anchors, and, finally, her crew and boats, and so proceed to sea, almost without the possibility of any delay occurring in the successive stages of her equipment. Then, in case of a ship coming in for repairs: where they are slight, she will go into either of the dry docks in the entrance or eastern basin and there have her copper repaired, her bottom scrubbed, or any other slight repairs effected, and return immediately to sea, without experiencing more than a few days', perhaps only a few hours', detention. If she requires a more thorough repair, she will deliver her anchors at the wharf next to the pound in the entrance-basin, then her guns on the gun-wharf, then turn out her crew into one of the receiving-vessels moored in the entrance-basin for the purpose, with what provisions they may require during the process of repair, delivering the remainder into the victualling yard. She will next deliver her cables into the cable storehouse, and, successively, her cordage, her sails and rigging, and her small stores, as she passes along the wharves with their storehouses specially appropriated for their reception. She will

then deposit her ballast on the ballast wharf, and pass through the communication lock into the great wet dock, where she will go alongside the sheers to have her masts taken out; and, finally, go into one of the dry docks to be repaired. In coming out of the dock, she will move by the same course reversed, and, having taken her men on board in the entrance-basin, she will then be ready to proceed to sea again." In concluding, he observes: "I should hope that the benefits accruing to the service from the superior quality of the stores manufactured under the immediate superintendence of the agents of Government, as well as the economy produced by the substitution of machinery for hand-labour, would prove an ample indemnity for the expense incurred in the establishment of the respective manufactories." But the cost was the real difficulty; Mr. Rennie's estimate of the total outlay requisite to complete the works amounting to four millions and a-half sterling. Yet the plan was so masterly and comprehensive, and so obviously the right thing to be done, that the Portland Administration determined to carry it out, and the necessary land was bought for the purpose. Frequent changes of Ministry, however, took place at the time; the resources of the country were heavily taxed in carrying on the war against Napoleon in Spain; the public attention was diverted in other quarters; and no further steps were taken to carry out Mr. Rennie's design. He knocked at the door of one Administration after another without effect. In 1810 we find him writing to Lord Mulgrave, the First Lord; to the Right Hon. George Rose; to the Earl St. Vincent, and others; but though the more the plans were scrutinised the more indisputable did their merits appear, he could find no Ministry strong enough to carry them out. When peace came, Government and people were alike sick of wars,

various plans with this object, the most important of which were only carried out after the Northfleet project had been finally abandoned. Of the new works executed after our engineer's designs, those at Sheerness were the most extensive. The royal dockyard there was felt to be a public disgrace. It consisted of a number of old wooden docks and basins, formed by ships' timbers roughly knocked together from time to time as necessity required. The ground, in its original state, had been merely an accumulation of mud and bog, surrounded by a wide extent of flat wet land. It is easy to conceive how, in ancient times, Sheerness must have been regarded as of importance, occupying as it does the extreme north-western point of the Isle of Sheppey, and commanding the entrance both of the Thames and the Medway. But, with the improvements in modern artillery, the place has in a great measure ceased to be of value as a public arsenal, being incapable of efficient defence against a hostile fleet. Mr. Rennie entertained a strong opinion as to the inferiority of the site compared with others which he pointed out; but from the time when he was directed to prepare plans for the reconstruction of the dockyards, his business was confined

---

naval armaments, and glorious victories; and believing that all danger from France was at an end,—the French fleet having been destroyed or captured, and Napoleon banished to St. Helena,—it was supposed that the old royal harbours, patched and cobbled up, might answer every purpose. So the land at Northfleet was sold, and the whole subject dismissed from the public mind. But, after the lapse of half a century, the wisdom of Mr. Rennie's advice has become more clearly apparent even than before. For years past, the waste in our dockyards, which it was the chief object of his Northfleet design to prevent, has become one of the principal topics of public discussion, and it has been the standing opprobrium of every successive Naval Administration. What Mr. Rennie urged fifty years since still holds as true as ever—that *without concentration economy is impossible*. So long as Government goes on tinkering at the old dockyards, spending enormous sums of money in the vain attempt to render them severally efficient, and maintaining separate expensive staffs in so many different places,—building a ship in one yard and sending it round the island to another, perhaps more than a hundred miles distant, to be finished and fitted, and then to another to take in its guns, stores, &c.,—so long shall we have increasing reason to complain of the frightful waste of public money in our royal dockyards.

to carrying out his instructions in the best possible manner.

The plan finally decided upon was that of a river-front, extending from the Garrison Point to near the Old Town Pier of Sheerness, of the length of 3150 feet, including the entrances, and enclosing within it three basins: one to the north, 480 feet long and from 90 to 200 feet wide, containing a surface of about two acres, 4 feet below low water of spring tides, with two frigate-docks and a building-slip and boat-slips; a central tidal basin 220 feet square, of the depth

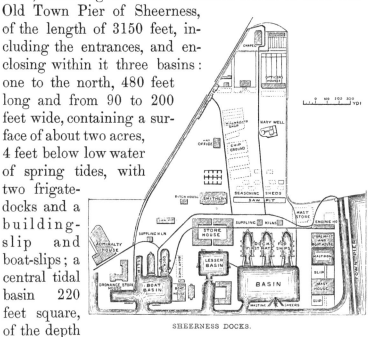

SHEERNESS DOCKS.

of 2 feet below low water, with storehouses around it for the reception and delivery of victualling and other stores; and on the west end of the dockyard a basin 520 feet long and 300 feet wide, covering a surface of nearly four acres, provided with dry docks for ships of the line on the south side, with their cills and the bottom of the basin laid 9 feet below low water of spring tides, westward of which were the mast-ponds, mast-locks, and workshops. In the rear, on the south of these works, were placed sawpits, timber-berths, and the officers' houses. The total surface of the dockyard was 64¾ acres. The foundation-stone of the docks was laid by the late Lord Viscount Melville in 1815, and the works then commenced and continued without interruption until the year 1826, when the whole were completed.

Their execution was attended with many difficulties, and necessarily required a great deal of Mr. Rennie's care and attention. The foundations were a soft running sand, extending to an almost fathomless depth. The strong currents flowing past the place rendered it necessary to adopt an entirely new system of operations, which were carried out to an extent never before attempted in so exposed a situation. That form of sea-wall was devised which should most effectually resist the strong pressure of the current, and the heavy swell beating upon its outer side, without yielding to the lateral pressure or thrust of the water of the basins and the earth by which it was backed. At the same time, the weight necessary to ensure stability must not be such as to sink vertically. Mr. Rennie adopted the means to secure these objects which he had employed with such success at Grimsby Docks in 1797, namely, to take the like quantity of materials as would have been necessary for an ordinary wall, and dispose of them in such a form, that the same weight should be distributed over a greater surface, thus diminishing the vertical pressure. In the foundations of the walls he also adopted the method employed by him in similar works, of driving the piles and cutting off their heads at an angle inclining inwards, or towards the land side, laying the courses of stone at the same angle; by which a greater resistance was offered to the pressure of the earth, and the building was prevented from being pushed outwards, as was more or less the case in most of the walls built on the old construction. The entrance gates to the great basin were also planned and executed with great skill, Mr. Rennie carrying into effect the same simple but correct principles laid down by him in his report on the Northfleet docks, making the direction of the entrance suitable to the current in the Medway, from which the ships entered the port.

The coffer-dams in which these Cyclopean walls were built demand a passing notice because of their gigantic dimensions and the great depth at which they were placed, not less than from 27 to 28 feet below low water of spring tides. They were composed of four rows of piles about 12 to 14 inches square. The two centre rows were carried about four feet above the level of high water of the highest spring tides, and were driven from 20 to 30 feet into the sea-bottom. The outer and inner rows were about seven feet from the two centre main rows, but only extended up to the level of the half tide, and were driven from 15 to 20 feet into the ground. All the piles were hooped and shod with wrought iron, firmly bound together at the bottom, middle, and top, with timber braces and wrought iron bolts and ties in every direction. The joints were caulked with oakum covered with pitch, and the spaces between the piles were filled with well-tempered clay, so that they were as nearly water-tight as possible, notwithstanding the tremendous weight of the sea outside at the top of every tide. Though the dams somewhat changed form by this pressure and inclined inwards, the piles were prevented being forced in that direction by powerful counteracting braces, and the whole stood fast until the completion of the work, contrary to the expectations of many, who regarded it as an altogether impracticable thing to construct coffer-dams of such magnitude in so exposed a situation, where the pressure to be resisted was so enormous. The only accident which occurred to the coffer-dams was in 1820, when the old *Bellerophon* line of battle ship, which had been anchored outside to break the swell, was forced from her moorings by a fierce storm from the north-east, and driven against the piling, a large extent of which gave way, and the waters rushed in upon the works. Fortunately, the building of the wall was by this time considerably advanced, so that no great damage was done. On the fall of the tide the dams were repaired without difficulty, and the works proceeded to completion.

Among Mr. Rennie's other dock works may be mentioned the new river-wall, with a ship-basin and two building-slips, at Woolwich; the new dockyard, building-slips, and dry dock at Pembroke; the new entrance to Deptford Basin; and various improvements at Chatham,[1] Portsmouth, and Plymouth. It was in front of

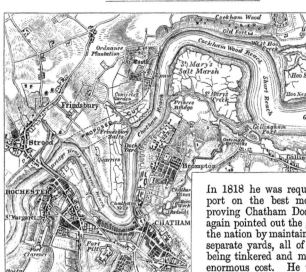

PLAN OF MEDWAY IMPROVEMENT.

In 1818 he was requested to report on the best means of improving Chatham Dockyard, and again pointed out the great loss to the nation by maintaining so many separate yards, all of which were being tinkered and mended at an enormous cost. He then boldly recommended that the Government should sell Deptford Dockyard, which he held to be comparatively useless, and devote the proceeds to the improvement of Chatham, after the plan which he submitted. It consisted in cutting a new channel from a point in the river Medway, a little below Rochester Bridge, to another point lower down the river at Upnor Castle, thus straightening its channel, increasing its current, and consequently improving its depth. By this simple means the whole of the bend in the present river would be converted into a capacious wet dock, extending along the whole front of the dockyard, shut in by

[1] One of Mr. Rennie's most ingenious plans was that proposed by him for the improvement of Chatham Dockyard. He saw large sums of money expended from year to year in the ineffectual patching-up of the old yards, with little good result; and he foresaw that eventually the Government, if it would secure efficiency and avoid waste, must fall back upon some such plan as his Northfleet Docks to obtain the requisite economy by concentration of dockyard work.

this last port that Mr. Rennie constructed his celebrated Breakwater, of which we next proceed to give
an account.

---

gates at either end. He also proposed to make another cut from its lower end to join the Medway at Gilling-ham, where there is ample depth of water at all times for ships of the largest burden. Lord Melville was much struck with the simplicity and at the same time the comprehensive character of the plan, and desired to have an estimate of the cost, which Mr. Rennie furnished. The whole amount—including land, labour, and materials—according to the engineer, would not exceed 685,000*l.*, against which there was to be set the heavy annual cost for moorings in the tide-way (which would be saved) or equal to a capital sum of 200,000*l.*; the expense of watching vessels lying at moorings, amounting to about 15,000*l.* a year, or equal to a capital sum of 300,000*l.*; and the amount

realised by the sale of the disused dockyard at Deptford, which of itself would have been almost sufficient to defray the entire cost of this magni-ficent new arsenal, not to mention the saving in the steam and other vessels employed in carrying stores to the men-of-war lying in ordinary along the course of the Medway, and the great despatch and economy which would have been secured in all the opera-tions connected with the building, fitting, and repairs of ships. The plan, however, was too comprehensive for the time, and was not adopted. Patching is still going on at Chatham, at a cost far exceeding that required to carry out Mr. Rennie's design; but it still remains to be seen whether anything like the same amount of concentration and efficiency will be secured.

CHAPTER X.

PLYMOUTH BREAKWATER.

THE harbour of Plymouth being situated at the entrance
of the English Channel, nearly opposite two of the
principal French coast arsenals, has always been regarded as a naval station of national importance. The
great area of the Sound, its depth of water, and its connection with the two spacious and secure inner harbours
of Hamoaze and Catwater, admirably fit it for such a
purpose. The Sound is more than three miles wide in
all directions, and includes an area of about four thousand acres, with a depth of water varying from four to
twenty fathoms at low water of spring tides. Its shores
are bold and picturesque, rising in some places in almost
perpendicular cliffs ; in others, as at Mount Edgcumbe,
the land, clothed with the richest verdure, slopes gently
down to the high-water line.

The only natural defect of the haven—and it was felt
to be a serious one—was that the Sound lay open to
the south, and was consequently exposed to the fury
of the gales blowing from that quarter during the
equinoxes. The advantages presented by its excellent
anchorage and great depth of water were thus in a considerable degree neutralized, and the cases were not
unfrequent of large vessels being forced from their
moorings during storms, and driven on shore.

From an early period, therefore, the better protection
of the outer harbour of Plymouth was deemed to be
a matter of much importance, and various plans were
proposed with that object. One of these was, to carry
out a pier from near Penlee Point, at the south-western

entrance to the Sound, for a distance of above 3000
feet, into deep water; another was, to construct a pier
several thousand feet long, from Staddon Point to the
Panther rock; and a third contemplated a still more

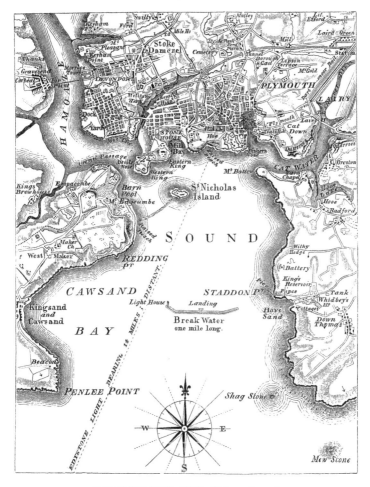

MAP OF PLYMOUTH SOUND. [Ordnance Survey.]

extensive pier, of above 8000 feet in length, extending
westward into deep water, from the south-eastern entrance to the Sound. While the question was under
discussion, the Lords of the Admiralty called upon

Mr. Rennie to report upon these several plans, and to advise them what, in his opinion, was the best course to be pursued. In his preliminary inquiry, which was of the most thorough kind, he was assisted by Mr. Joseph Whidbey and Mr. James Hemmans. Mr. Whidbey had been a sailing master in the Royal Navy, and was a most able and meritorious officer. He had sailed round the world with Vancouver, and raised himself from the station of a man before the mast to the highest position a non-commissioned officer could reach. His varied experience had produced rich fruits in a mind naturally robust and vigorous. As might be expected, he was an excellent seaman. He was also a person of considerable acquaintance with practical science, and had acquired from experience a large knowledge of human nature, of a kind that is not to be derived from books. He was afterwards raised to the office of Master-Attendant at Woolwich Dockyard, and was greatly beloved and respected by all who knew him. Mr. Hemmans was Master-Attendant of Plymouth Dockyard, and possessed an intimate practical knowledge of the locality, which proved of much value in the course of the investigation.

The result of Mr. Rennie's careful study of the object to be accomplished, and the best mode of fulfilling the requirements of the Admiralty, was embodied in the report presented by him on the 22nd of April, 1806. He there expressed himself of opinion that, of the three plans which had been proposed, that of a pier extending westward from Andurn Point, near the Shagstone, at the south-eastern extremity of the Sound, was the best; but even that was objectionable, as calculated to produce shoaling of the harbour by favouring the deposit of silt,—a process which was then going on, and which it was most desirable to prevent. Looking at the main object of the proposed work, which was to render the Sound a safe haven for vessels riding at anchor there, as well as to increase the security of Plymouth inner harbour, he considered that another method might be adopted, with greater certainty of success, at much less cost, and without any risk of silting up the entrance of the harbour.

It may here be explained that, in its original state, there were three channels of entrance into the Sound: one on the west side, 3000 feet wide, between the Scottish Ground rocks and the Knapp and Panther rocks; another, on the eastern side, 1800 feet wide, between the shore and the Tinker, St. Carlos, and Shovel rocks; and a third in mid channel, also about 1800 feet wide, which was rarely used, being through dangerous rocks on either side. Any works which might be constructed across this middle channel, it appeared to Mr. Rennie, would be of no detriment to the navigation of the Sound. On the other hand, by narrowing the passage through which the tidal waters flowed in and out, the tendency would be to increase the scour, and consequently to deepen the two remaining channels—an object only second in importance to the protection of the harbour itself. His plan, accordingly, was to form a breakwater across this middle channel by throwing into the sea large angular blocks of rubble, of from two to twelve tons each, leaving them to find their own base, and to continue this process until a sufficient mass had been raised to the level of half-tide, so as to make a ridge, about 70 yards broad at the base and 10 yards at the top, these dimensions to be afterwards varied and increased according to circumstances. He proposed that the Breakwater should be of a total length of 5100 feet; of which 3000 feet, forming the central part, was to be in a straight line across the Sound, and 1050 feet at either end inclined inwards at an angle of 160 degrees. It was also proposed to run out a pier of 2400 feet, in two arms, from Andurn Point, at the south-eastern entrance to the Sound, but this part of the

design was not carried into effect. The total estimate of cost, including the Point Andurn pier, was
1,102,440*l.*

The formation of breakwaters by a similar process
had been practised from an early period; and by such
means the moles of Venice, Genoa, Rochelle, Cherbourg,
and other ports, had been formed. Mr. Rennie had
himself adopted the same method in forming the harbours of Howth, Kingstown, and Holyhead; and the
success which had followed his operations at those
places left no doubt in his mind as to the practicability
of constructing an efficient breakwater across Plymouth
Sound, though its situation was admitted to be much
more difficult. Mr. Rennie considered that a ridge
of rough, heavy stones would be the simplest and least
costly, as it would probably be found the most efficient,
plan of protection that could be devised. By throwing
in the blocks in the given line, and allowing the waves
by their own action to form the slope or angle of repose
at which the materials would lie, the expenditure of an
infinity of labour and money would be avoided. In this
kind of work he held that success was to be secured by
following the laws of nature; and, with reference to the
proper slope, his own expression was that "the waves
were the best workmen."

The report in which Mr. Rennie embodied these recommendations excited great interest amongst naval
men, and led to much discussion. Many pronounced the
scheme to be visionary and impracticable; and it was
alleged that, even if it could be executed, such a work
would prove altogether useless. Others held that it
would destroy the Sound for purposes of navigation, and
lead to its complete silting up; whilst a greater number
criticised and condemned it in its details. Amongst
other critics, one of the most severe was the late General
Sir Samuel Bentham, a brother of Jeremy the philosopher,
who strongly recommended a plan of his own, con-

sisting of one hundred and forty wooden towers, with stones sunk between them in a double line. Five years' controversy thus passed, and numerous designs were prepared, submitted, and considered, all of which were referred to Mr. Rennie, who still remained as firmly satisfied as ever of the expediency of his design, and vindicated its superiority to all others which had been proposed. At length the Lords of the Admiralty were won over to his views, as well as Earl St. Vincent, Lord Keith, and many other leading naval officers. Lord Melville was then at the head of the Admiralty, and when he had become fully persuaded as to the merits of Mr. Rennie's plan, and ascertained that there was a growing unanimity of opinion in its favour, he exerted himself strenuously to have it carried into effect.

The result was, that an Order of Council was issued on the 22nd June, 1811, by which the requisite powers were given to proceed with the works. Mr. Rennie was appointed engineer in chief, and Mr. Whidbey resident engineer, with Mr. William Stewart as assistant. It was very difficult to find contractors willing to undertake the execution of a work of so novel and extensive a ·character, except at prices which the engineer could not sanction; and it was therefore determined only to contract for the labour of the several operations in detail, so that there might be an opportunity for revising the scale of prices from time to time,[1] the Government undertaking to find the requisite plant and materials. A piece of ground was purchased from the Duke of Bedford at Oreston, up the Catwater, containing twenty-five acres of limestone, well adapted for the purposes of the work; and steps were taken to open out the quarry,

---

[1] The propriety of this arrangement was proved by the fact, that whereas the price paid in 1812 for taking and depositing rubble in the Breakwater was 2s. 9d. per ton, it was afterwards reduced to 1s. per ton, as the contractors and workmen became better acquainted with the nature of the work.

to lay down railways to the wharves, to erect cranes, to build vessels suitable for conveyance of the stone, and to provide the various appliances required for carrying out the undertaking.

On the 12th of August, 1811, the birthday of the Prince Regent, the first stone of the main ridge was deposited on the Shovelbank rock, nearly in the centre of the work.   Lord Keith, commander of the Channel Fleet, was present, attended by the chief naval, military, and civil authorities of the port, together with the staff and men of the building establishment.   From this time forward the operations were carried on with every despatch when weather would permit, beginning at the centre and working towards the extremities.   The lines of the Breakwater were carefully marked out by buoys, to which the barges laden with the stone blocks were attached whilst they were being emptied into the sea ; after which they returned to the quarry wharves, about five miles distant, for fresh cargoes.   For nearly two years this process of emptying in the rubble proceeded, until in March, 1813, portions of the work began to be visible at low water, and by the end of July there appeared a continuous line of about 720 yards.   By the month of March in the following year the ridge had been so raised, that its effect in tranquillizing the waters of the Sound during violent south-westerly gales was very considerable, and vessels of all classes sought its protection, and came to anchor behind it with perfect confidence. The Admiral's ship on the station, the *Queen Charlotte*, of 120 guns, had been accustomed heretofore to ride in Cawsand Bay ; but it was now brought to an anchor under the lee of the Breakwater.   Among other vessels which took shelter there in 1814 was a large French three-decker, which rode out a severe gale in safety. The year after, when Napoleon entered Plymouth Sound on board the *Bellerophon*, he expressed himself in such terms of the work, as gave Mr. Rennie much gratifica-

tion on the ex-Emperor's expressions being communicated to him.

By the 11th of August, 1815, not less than 615,057
tons of stone had been deposited, and a length of 1100
yards was raised above low water of spring tides. The
complete success of the work was now beyond dispute,
and exceeded the most sanguine anticipations. Even
the most sceptical became convinced of its great practical
utility, and many who had before offered vehement
opposition to its being begun, became clamorous for its
completion on even a larger scale than Mr. Rennie had
originally intended. And it was at length determined
by the Admiralty, after advising with the engineer, to
carry the whole Breakwater twenty feet, instead of ten
feet, above the level of low water of spring tides.
Whilst the original plan of Mr. Rennie was calculated
to afford complete security to the larger class of vessels,
this addition to its height doubtless gave more adequate
protection to the smaller craft. The finishing of the
work above the low water line, however, involved a
more expensive kind of workmanship; for the greatest
force of the waves is exercised between the lines of high
and low water. Hence it became necessary to render
this upper part of the Breakwater so strong as to present
the greatest possible amount of resistance. Mr. Rennie
suggested the propriety of increasing the seaward slope
to about 5 to 1, so as to give greater strength, and
present an increased resistance to the force of the waves.
But this recommendation was overruled for the time,
and the work proceeded. No estimate of the additional
cost was asked for, and the operations were continued
without intermission, the stone blocks being deposited
at the rate of 1030 tons a-day during the year 1816—
a greater average than has since been accomplished in
carrying out any similar work, notwithstanding the
improved modern appliances of stages provided with

s 2

steam power and railways worked by powerful locomotives.[1] Towards the end of the year about 300 yards of the west end of the Breakwater had been raised to the full height of two feet above high water, and 20 feet above low water of spring tides.

Success, as usual, produced over confidence; and the authorities on the spot, believing that if the sea slope of the rubble were roughly put together at an inclination of 3 to 1, it would present sufficient resistance, as well as in order to reduce the expense, directed it to be so executed. But about this time a succession of severe gales set in from the south-west and sent a tremendous sea upon the Breakwater, especially one of great violence which occurred on the 17th of January, 1817. On examining the work, after the return of moderate weather, it was ascertained that a length of about 200 yards of the rubble of the upper part had been displaced or deranged,—that several blocks of stone, varying from two to five tons, had by the force of the waves been thrown over from the south or sea slope to the north or land slope,—and that their further effect had been to increase the inclination of the former to $5\frac{1}{3}$ to 1 instead of 3 to 1, as it had originally stood. Nevertheless, the great mass of the Breakwater remained unmoved, and large numbers of vessels, availing themselves of the secure protection which it provided, had been enabled to ride out the storm in safety. Unfortunately, however, the *Jasper* sloop of war and the *Telegraph* schooner, anchored without the line of protection, were driven on shore and wrecked under the citadel, when a melancholy loss of life took place.

Mr. Rennie was of opinion that this storm had been

---

[1] The largest quantity of stone deposited in one year was in 1821, when not less than 373,773 tons were quarried, lightered, and emptied into the work.

of service by forcing the stones together and con-
solidating the work more firmly. His recommendation
as to the necessity of increasing the seaward slope having
been so singularly confirmed by the action of the waves,
he now advised that it should be allowed to remain as
left by the storm of the 17th, and that the rest of that
face of the Breakwater should be made uniform with
it. It would, however, appear that Mr. Whidbey, the
resident engineer, contrived to finish most of the exterior face at a slope of only 3 to 1, as before; and
thus it stood without any material interruption until
several years after Mr. Rennie's death. By that time
nearly the whole of the intended rubble, amounting
to 2,381,321 tons, had been deposited, and the main
arm, with 200 yards of the west arm, making 1241
yards in length, had been raised to the required level.
The work had arrived at that stage when it had to
experience the full force of another terrific storm,
which took place on the 23rd of November, 1824. It
blew at first from the south-south-east, and then veered
round to the south-west; and the effect of this concur-
rence of winds was to heap together the waters of the
Channel between Bolt Head and Lizard Point, and drive
them with terrible force into the narrow inlet of Plymouth Sound. This storm was not only greatly more
violent, but of much longer duration, than that of 1817.
When the Breakwater could be examined, it was found
that, out of the 1241 yards of the upper part which had
been completed with a slope of 3 to 1, 796 yards had
been altered as in the previous storm, and the immense
blocks of stone which formed the sea face of the work
had by the force of the waves been rolled over to the
landward side, thus reducing the sea slope as before to
about 5 to 1. The accuracy of Mr. Rennie's view as to
the proper slope—which was indicated by the action of
the sea itself—was thus a second time confirmed; and
the same view having been taken by the eminent engi-

neers [1] who were called upon to make an inspection of
the work, and to report as to the best means of rendering

PLYMOUTH BREAKWATER.

[By Percival Skelton, after his original Drawing.]

it permanently secure, it was determined to make the
permanent slope of the same inclination, and the works
were so carried out accordingly.[2]   The total quantity of

---

[1] These were Mr. Telford, Mr.
Josias Jessop, Sir J. Rennie, and Mr.
G. Rennie.   For more full particulars
as to the history and construction of the
Breakwater, we refer the reader to Sir
John Rennie's elaborate work entitled
'An Historical, Practical, and Theo-
retical Account of the Breakwater in
Plymouth Sound.'   London, 1848.

[2] The slopes were paved with
blocks of the largest stone, firmly
wedged together; the centre line was
removed 36 feet further seawards;
the top width was reduced 5 feet;
a strong binding course of dovetailed
granite masonry was built at the bottom of the sea slope, which was laid
one foot convex from the bottom to
the top; whilst the land slope was

laid with close-fitting rubble at the
inclination of 2 to 1.   It was, however, found, in the course of the work,
that the rough paving of the rubble
alone was scarcely strong enough to
withstand the violence of the waves
without a certain degree of yielding;
and Sir J. Rennie, having been consulted by the Admiralty, recommended that, in addition to the granite basement binding course, there
should be another similar course both
in the centre and at the top of the
sea slope; and that the remainder
should be paved with rough-dressed
limestone ashlar, set in courses at
right angles to the slope, about three
feet deep on the average—each course
binding bond well with the one ad-

rubble deposited to the end of 1848—when the work may be said to have been completed—was 3,670,444 tons, besides 22,149 cubic yards of masonry, or an amount of material at least equal to that contained in the Great Pyramid. The whole cost of this magnificent work was about a million and a half sterling, including a convenient and spacious watering place at Bovesand Bay.

As forming a convenient and secure haven of refuge for merchant ships passing up and down Channel, along the great highway between England, America, and India—as a capacious harbour for vessels of war, wherein fifty ships of the line, besides frigates and smaller vessels, can at all times find safe anchorage—Plymouth Breakwater may in all respects be regarded as a magnificent work, worthy of a great maritime nation.

---

jacent,—the lower parts of the granite bonding courses being laid level, but the upper parts forming part of the slope. It was still found that there was a difficulty in preventing the outer edge or base of the sea slope, where the main lower granite bonding courses were placed, from being undermined by the waves; and it was determined to place a trenching or foreshore on the outside of the sea slope, 40 feet wide in the centre of the Breakwater, increasing to 50 feet wide at the commencement of the western arm, and diminishing towards the eastern arm to the width of only 30 feet. This foreshore was about 2 feet above the level of low water of spring tides next to the toe or base; and the surface was roughly paved with rubble well wedged together. The whole of the slope was paved with well-dressed courses of ashlar masonry without mortar, 3 feet 6 inches deep, well bedded down upon the rubble below. The extremity of the western arm was furnished with a solid head of circular masonry, 75 feet diameter at the top, with slopes of 5 to 1 all round. At the point at which the lighthouse has since been placed, an inverted arch of solid blocks was formed, the whole well bonded, dovetailed, and dowelled together, and firmly united with the other parts of the solid rock. These works answered admirably, and Plymouth Breakwater now rests as firm as a rock upon the bottom of the sea.

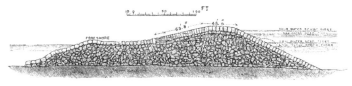

SECTION OF THE BREAKWATER AS FINISHED

CHAPTER XI.

MR. RENNIE'S LAST WORKS—HIS DEATH AND CHARACTER.

ON undertakings such as these, of great magnitude and importance, was Mr. Rennie engaged until the close of his useful and laborious life. There was scarcely a project of any large public work on which he was not consulted; sometimes furnishing the plans, and at other times revising the designs of others which were submitted to him. Numerous works of minor importance also occupied much of his attention, as is shown by the extent of his correspondence and the number of his reports, which contain an almost complete repository of engineering practice. Whilst he was engaged in designing and superintending the construction of his great London Bridges, the formation of Plymouth Breakwater, the building of the docks at Sheerness, the cutting of the Crinan Canal, and the drainage of the Fens by the completion of the Eau Brink Cut, he was at the same time consulted as to many important schemes for the supply of large towns with water. His report on the distribution of the water supplied by the York Buildings Company in the Strand—in which he proposed for the first time to appropriate a distinct service to the several quarters of the district supplied—was a masterpiece in its way; and the principles he then laid down have been generally followed by subsequent engineers. He also reported on the improved water-supply of Manchester, Edinburgh, Bristol, Leeds, Doncaster, Greenwich and Deptford, and many other large towns in England

and Scotland, as well as in the colonies and in foreign countries.[1]

In addition to the various mills and manufactories fitted up by him with new and improved machinery, we may mention that he advised the Bank of England on the subject of the more rapid manufacture of bank notes by the employment of the steam-engine; and he entirely re-arranged the Government machinery at Waltham for the better manufacture of gunpowder. He erected the anchor-forge at Woolwich Dockyard, considered to be the most splendid piece of machinery in its day; he supplied Baron Fagel (then Dutch minister in this country) with designs of dredging-engines for clearing the mud out of the rivers and canals of Holland; and he designed and constructed the celebrated machinery for making ropes according to Captain Huddart's patent.[2]

---

[1] In 1817, his fame having gone abroad as the most skilled water engineer of the day, Captain Dufour, of Geneva, came to England for the purpose of consulting him as to the extension and improvement of the waterworks of that city. Captain Dufour was introduced to Mr. Rennie by the mutual friend of both, the eminent Dr. Wollaston. Mr. Rennie made a careful and detailed report on the surveys and plans submitted to him, especially on the engine and pumping machinery of the proposed works; and his advice was followed, very much to the advantage of the citizens of Geneva.

[2] Captain Joseph Huddart, F.R.S., was a singularly estimable character. He was a man of great nautical experience, sound judgment, and excellent skill as a mechanic and engineer, and was often consulted by Mr. Rennie in reference to marine works of more than ordinary importance. His origin was humble, like that of so many of the early engineers; and, like them also, he was drawn to the pursuit by the force of his genius, rather than by the peculiar direction of his education. He was born at Allonby, in Cumber-

land, in 1740, the son of a shoemaker and small farmer. From his mother he inherited a determined spirit and a vigorous constitution, combined with sound moral principles, which he

CAPTAIN JOSEPH HUDDART, F.R.S.

nobly illustrated in his life. He received an ordinary share of education at the common school of his village, to which he added a knowledge of

In his capacity of advising engineer to the Admiralty,
Mr. Rennie embraced every opportunity which his posi-

---

mathematics and astronomy, obtained
from the son of his teacher, who had
studied these branches at Glasgow
University. He seems to have been
an indefatigable learner, for he also
acquired some knowledge of music
from an itinerant music-teacher, whom
he very shortly outstripped.   His
mechanical tastes early displayed
themselves.   Watching some millwrights employed in constructing a
flour-mill, he copied the machinery
which they erected, in a model which
he finished as they completed their
mill. He also made a model of a
74-gun ship, after the drawings given
in Mungo Murray's 'Treatise on Navigation and Shipbuilding,' which he
was so fortunate as to fall in with.
At an early age he was employed in
herding his father's cows on a hill-side
overlooking the Solway Frith, and
commanding a view of the coast of
Scotland. There he took his books,
with a desk of his own making, and
while not forgetting the cattle, employed himself in reading, drawing, and
mathematical studies.   When a little
older, his father set him on the cob-
bler's stool, and taught him shoemaking, though the boy's strong inclination was to be a sailor.   But
large shoals of herrings making their
appearance about this time in the
Solway Frith, a small fishing company was started by the Allonby
people, in which his father had a share,
and young Huddart was sent out
with the boats, very much to his
delight. He now began to study
navigation, carrying on shoemaking
in the winter and herring-fishing at
the time the shoals were on the
coast. On the death of his father, he
succeeded to his share in the fishery,
and took the command of a sloop
employed in carrying the herrings to
Ireland for sale. During his voyages
he applied himself to chart-making,
and his chart of St. George's Channel,
which he afterwards published, is
still one of the best. The herrings
having left the frith, Huddart got the

command of a brig, his excellent character securing him the post, and he
made a successful voyage in her to
North America and back. His progress was steady and certain. A few
years later we find him in command
of an East Indiaman. After many
successful voyages, in which he happily brought all his ships to port, and
never met with any serious disaster,
he retired from the service; having
been in command of a ship of greater
or less burden for a period of twenty-
five years. He now published many
of his charts, the results of the observations he had made during his numerous voyages. His eminent character, not less than his known scientific
knowledge, secured his introduction to
the Trinity House as an Elder Brother,
and to the direction of the London
and East India Docks, in which situations he was eminently useful. The
lighting of the coast proceeded chiefly
under his direction, and many new
lighthouses were erected and floating-
lights placed at various points at his
recommendation. Among others, he
superintended the construction of the
lighthouse at Hurst Point. He also
surveyed the harbours of Whitehaven, Boston, Hull, Swansea, St.
Agnes, Leith, Holyhead, Woolwich
Dockyard, and Sheerness; several of
these in conjunction with his friend
Mr. Rennie, who was always glad to
have the benefit of his excellent judgment. He made many improvements
in ship-building; but the invention
for which the nation is principally
indebted to him is his celebrated rope-
making machinery, by which every
part of a cable is made to bear an equal
strain, greatly to the improvement of
its strength and wearing qualities.
This machinery, constructed for him
by Mr. Rennie at Limehouse, was
among the most perfect things of the
kind ever put together. Captain
Huddart died at his house in High-
bury Terrace, London, in 1816, closing
a life of unblemished integrity in the
seventy-fifth year of his age.

tion afforded him of recommending the employment of steam power in the Royal Navy. His advice met with the usual reception from the inert official mind: first indifference; next passive resistance; then active opposition when he pressed the matter further. Naval officers, who had grown old in sailing tactics, could ill brook the idea of navigating ships of war by a mechanical invention like the steam-engine, by which skill in seamanship, of which the old salts were so proud, would be entirely superseded. The navy had done well enough heretofore without steam; why introduce it now? It was a smoky innovation, and if permitted, would only render ships liable to the constant risk of being blown up by boiler explosions. Lord Melville, however, listened to Mr. Rennie's suggestions, and at length consented to the employment of a small steam-vessel as a tug for a ship of war, by way of experiment. Mr. Rennie accordingly hired the Margate steamboat, *Eclipse*, to tow the *Hastings*, 74, from Woolwich to two miles below Gravesend, against a rising tide. The experiment was made on the 4th of June, 1819, and proved so successful that the Admiralty were induced to authorise a steamboat to be specially built at Woolwich for similar service. This vessel was named the *Comet*; it was built after the designs and under the direction of the late Mr. Oliver Lang, assisted to a considerable extent by Mr. Rennie, who attended more particularly to the designing and fitting of the engines, which were made by Boulton and Watt. The *Comet*, though a small vessel, was the parent of other royal ships of vastly greater dimensions. She was only 120 feet long between the perpendiculars, and 22 feet 6 inches in extreme breadth; the draught of water was about $6\frac{1}{2}$ feet, and the power of her engines about 40 horses. The Admiralty had great doubts as to the width of the paddle-boxes; but Mr. Rennie encouraged them to make the experiment after his design. " Steam-vessels," he observed, " are as yet only in their infancy, and can scarcely be expected to have arrived at anything approaching perfection. Much, however, will be learnt by experience; but unless some risk is run in the early application of the new power, no improvements are likely to be made." [1] For her size, the *Comet* proved a most efficient vessel—the best that had up to that time been constructed. She fully answered the purpose for which she was intended; and the result was so satisfactory, that vessels of increased size and power were from time to time built, until the prejudice amongst naval men against the employment of steam power having been got over, it was at length generally introduced in the Royal Navy. [2]

The last of Mr. Rennie's great designs was that of New London Bridge, which, however, he did not live to complete. The old bridge had been gradually falling into decay, and was felt to be an increasing obstacle to the navigation of the river. The starlings which protected the piers had of late years been seriously battered by the passing of hoys, barges, and lighters, on which they had inflicted equal injury in return; for vessels were constantly foundering on them, and many were sunk and their cargoes damaged or destroyed. Emptying stones into the river, to protect the decayed pile-work, had only

---

[1] Letter to the Admiralty, 22nd May, 1820.

[2] Mr. Rennie was engaged for many years in urging the introduction of steam power in the Royal Navy. In 1817 we find him writing to Lord Melville, Sir J. Yorke, Sir D. Milne, and others on the subject. It would appear that Lord Melville had declared that he was determined to employ steam-vessels as tugs, so soon as he could convince the Sea Lords of their advantages; on which Mr. Rennie compliments Sir D. Milne, saying that he is "glad to find that there is one admiral in the navy favourable to steamboats." In July, 1818, he laments that he cannot convince Sir G. Hope or Mr. Secretary Yorke of their utility, but that he is persuaded their adoption *must* come at last. On the 30th May, 1820, he writes James Watt, of Birmingham, informing him that the Admiralty had at last decided upon having a steamboat, notwithstanding the strong resistance of the Navy Board. "My reasons," he says, "I understand, were satisfactory; but unless the Admiralty cram it down the throats of the Navy Board, nothing will be done; for of all the ignorant, obstinate, and stupid boards under the Crown, the Navy Board is the worst. I am so disgusted with them that, could I at the present moment with decency relinquish the works under them which I have in hand, I would do so at once."

had the effect of further obstructing the navigation, the scour of the current having formed two great banks of stone across the whole bed of the river: one about 100 feet below the bridge, and the other about the same distance above it, with two deep hollows between them and the piers, from 25 to 33 feet deep at low water. The piers and arches were also becoming decrepit. Though the top-hamper of houses had long been removed, and the piers patched and strengthened at various times, the bridge was every year becoming less and less adapted for accommodating the increasing traffic to and from the City. At last it was regarded as a standing nuisance, and generally condemned as a disgrace to the capital. To maintain the structure, inefficient and unsafe though it was, cost the City not less, on an average, than 3500*l*. a year, and the expense was likely to increase. The Corporation felt that they could no longer avoid dealing decisively with the subject. They then resolved to take the opinion of the best engineers and architects; and Mr. Daw, the architect of the Corporation, Mr. Chapman, the engineer, and Messrs. Alexander and Montague, two eminent City architects, were consulted as to the best steps to be taken under the circumstances. The result of their deliberations was a recommendation to the Corporation to remove eight of the arches and to substitute four larger ones, as well as to make extensive repairs in the remaining arches, piers, and superstructure. Their plan was referred to Mr. Rennie, Mr. Chapman, and Messrs. Montague, for further consideration; and, as was Mr. Rennie's custom before making his report, he proceeded to master the whole of the facts, on which alone a sound opinion could be formed. He had the tides and currents watched and noted, and the river carefully sounded above and below bridge, from Teddington Lock to the Hermitage entrance of the London Docks. He examined the piers down to their foundations, and explored the bottom of

the river, making borings at various points between the
one shore and the other.   In the report which he made
to the Corporation on the 12th of March, 1821, a great
deal of new and accurate information was first brought
to light respecting the flow of the tide through the
arches, and the additional depth of water likely to be
secured by their removal.   Although it was pronounced
quite practicable to carry out the alterations which had
been recommended, and the erection of four new arches
in lieu of the eight old ones, he was of opinion that
the cost would be very considerable ; and, after all, the
old foundations would still present great defects, which
could never be wholly cured.   Mr. Rennie therefore
suggested the propriety of building an entirely new
bridge of five arches, with a lineal waterway of not less
than 690 feet, in lieu of the then waterway of 231 feet,
below the top level of the starlings, and 524 feet above
them.   Besides the greatly increased accommodation
which would be provided by such a structure for the
large traffic passing between London and Southwark, Mr.
Rennie held that not the least advantage which it promised was the much greater facility which it would
afford for the navigation of the river to and from the
wharves above bridge ; for coasters and even colliers,
with striking masts, might then be enabled to navigate
the whole extent of the City westwards.   The increased
waterway would also enable the waters descending from
the interior to flow more readily away, floods often
inflicting great damage along the shore, especially in the
winter months, when the arches of the old bridge became
choked up with ice.

The report was felt to be almost conclusive on the
subject ; and the more it was discussed the deeper grew
the conviction in the minds of all that its recommendations ought to be adopted.   The Corporation accordingly
applied to Parliament, in the year 1821, for an Act
enabling them to purchase the waterworks under the

arches of the old bridge, and to erect an entirely new structure. The bill, of course, had its opponents; some arguing that there was no necessity for a new bridge, and that its erection would be only a useless waste of money, whilst the old one could be repaired and made fit for traffic at so much less outlay. The case in favour of the new bridge was, however, too strong to be resisted, and Mr. Rennie's evidence was considered so clear and conclusive that committees of both Houses unanimously approved the bill, and it duly received the sanction of Parliament. Power was conferred by the Act enabling the Treasury to advance from the Consolidated Fund such sums as might be necessary· to supply any deficiency in the funds at the disposal of the Corporation applicable to the erection of the bridge; the Government regarding the work as one of national importance, and consequently entitled to public assistance.

During the progress of the bill through Parliament, Mr. Rennie prepared the general outlines of a design of the new structure. It consisted of five semi-elliptical arches, the centre one 150 feet span, the two side arches 140 feet, and the two land arches 130 feet, making a total lineal waterway of 690 feet; the height of the soffit or under-side of the centre arch being 29 feet 6 inches above the level of Trinity high-watermark. The general principle of this design was approved and embodied in the bill. Very shortly after the Act had passed, Mr. Rennie was seized by the illness which carried him off, and it was accordingly left to others to execute the great work which he had thus planned. The Corporation of London then appealed to the whole engineering and architectural world for competitive designs, and at least thirty were prepared in answer to their call. These were submitted to a Committee of the House of Commons in the year 1823, and after long consideration the plan originally proposed by Mr. Rennie was finally adopted; on which the Corporation of London selected his son, the present Sir John Rennie, engineer in chief, to carry it into effect; and the nomination having been approved by the Lords of the Treasury, in conformity with the pro-

NEW LONDON BRIDGE. [By Percival Skelton.]

vision of the Act, steps were forthwith taken to proceed with the work. It is scarcely necessary to say how admirably Mr. Rennie's noble design has been executed. New London Bridge, in severe simplicity and unadorned elegance of design—in massive solidity, strength, and perfection of workmanship in all its parts—not less than as regards the capacious size of its arches and the breadth and width of its roadway and approaches—is perhaps the finest work of its kind in the world.[1]

It will be observed from the preceding chapters, that Mr. Rennie's life was one of constant employment, and that, apart from his great engineering works, his career contains but few elements of biographic interest. Indeed his works constitute his biography, overlaying, as they did, almost his entire life, and occupying nearly the whole of his available time. His personal wants were few; his habits regular; and his pleasures of the most moderate sort, consisting chiefly in reading and in the enjoyment of domestic life. At the age of twenty-nine he married Miss Mackintosh, an Inverness lady, who made his home happy; and he became the father of nine children, six of whom survived him. In the early part of his career in London he lived in the Great Surrey

---

[1] The new bridge was erected about thirty yards higher up the river than the old one, and involved the construction of new approaches on both sides. The first coffer-dam was put in on the Southwark side, and the first pile was driven on the 15th of March, 1824; the foundation stone was laid with great ceremony by H. R. H. the Duke of York, on the 15th of June, 1825, assisted by the Lord Mayor (Garrett), the Aldermen, and Common Council. The bridge was finally completed and formally opened by His Majesty King William the Fourth on the 1st of August, 1831—the time occupied in its construction having thus been seven years and three months. The total cost of the bridge and approaches was about two millions sterling. All the masonry below low water is composed of hard sandstone grit, from Bramley Fall, near Leeds; and the whole of the exterior masonry above low water is of the finest hard gray granite, from Aberdeen, Devonshire, and Cornwall. The actual width of the arches as executed is as follows: the centre arch is 152 feet 6 inches span; the two arches next the centre are 140 feet; and the two land arches 130 feet. The details of construction of the coffer-dams, piers, and floating and fixing the centres, were similar to those adopted by Mr. Rennie in building Waterloo and Southwark bridges. The total length of the bridge is 1005 feet; width from outside to outside, 56 feet; width of the footpaths, 18 feet; and of the carriageway, 35 feet. The total quantity of stone built into the bridge is 120,000 tons. The builders were Messrs. Joliffe and Banks, the greatest contractors of their day.

SECTION OF NEW LONDON BRIDGE.

Road, from which he afterwards removed to Stamford
Street, not far from his works.

His close and often unremitting application early
began to tell upon his health. In 1812, when arrived
at the age of fifty-one, he was occasionally laid up
by illness. While occupied one day in inspecting the
works of Waterloo Bridge, he accidentally set his foot
upon a loose plank, which tilted up, and he fell into
the water, but happily escaped with only a damaged
knee. Though unable for some time to stir abroad, he
seized the opportunity of proceeding with the preparation
of numerous reports, and of working up a long lee-way
of correspondence. In the following year he was frequently confined to the house by a supposed liver-
complaint; but his correspondence never flagged. He
tried the effects of change of air at Cheltenham; but
he had no time for repose, and after the lapse of only a
week he was again in harness, giving evidence before
a Committee of the House of Commons on Lough Erne
drainage. He made another hurried visit to Cheltenham, but evidently took no rest; his absence from
active business only affording him an opportunity for
writing numerous letters to influential persons at the
Admiralty on the subject of his grand scheme of the
Northfleet Docks. To one of his correspondents we find
him saying he was "better, though only half a man
yet." In course of time, however, he partially recovered, and was forthwith immersed in business—engaged
upon his docks, bridges, and breakwaters.

He very rarely "took play." In 1815 his venerable
friend James Watt, of Birmingham, urged him to pay a
visit with him to Paris, shortly after the battle of Waterloo.
But Mr. Rennie was too full of work at the time to
accept the invitation, and the visit was postponed until
the following year, when he was accompanied by James
Watt, jun., then of Aston Hall, near Birmingham. This
journey was the first relaxation he had taken for a period

of thirty years; yet it was not a mere holiday trip, but partly one of business, for it was his object to inspect with his own eyes the great dock and harbour works executed by Napoleon during the Continental war, of which he had heard so much, and to gather from the inspection such experience as might be of use to him in the improvement of the English dockyards on which he was then engaged. The two set out in September, 1816, passing by Dover to Calais and thence to Dunkirk, where Mr. Rennie carefully examined the jetties, arsenal, docks, and building-slips at that port. From thence they proceeded to Ostend, and afterwards to Antwerp, where our engineer admired the great skill and judgment with which the dock works there, still incomplete, had been laid out. From Antwerp they went to Paris, where they stayed only two days, and then to L'Orient and Brest, accompanied by Mr. Joliffe, the contractor. At both these ports Mr. Rennie took careful note of the depths, dimensions, and arrangements of the harbours in detail, receiving every attention from the authorities. At Cherbourg, in like manner, he examined the building-yards and docks, as well as the progress made with the famous Digue [1]—

---

[1] The Digue is of considerably greater extent than the breakwater at Plymouth, being above 2¼ English miles long. Up to the time of Mr. Rennie's visit, the work had been a series of attempts and failures, which, however, eventually produced experience, and led to success. Wooden cones filled with small stones were first tried; they were sunk so as to form a sea-rampart; but the cones were shattered to pieces by the force of the waves, and the stones were scattered about in the bottom of the sea. Then loose rubble-stones were tried; but the blocks were too small, and these, too, were driven asunder. Larger blocks were then used; but, for a time, the smaller stones beneath acted as rollers to the larger ones. At length, however, these found their bearing, and when Mr. Rennie visited the place, the slope formed by the sea-ridge of rubble was as much as 11 to 1. This greatly increased the contents of the breakwater, while its stability was not much to be depended on. Many accidents occurred to the work, and several extensive breaches were made through it by the force of the sea. At low water the height of the Digue was at some parts only three feet; at others, considerably more; whereas, in some places, the top of the work was from seven to eight feet below low water of spring tides. At length, after many years' labour and vast expense, the work has been brought to completion; and it now forms a very excellent defence for the fine war roadstead and arsenal of Cherbourg, greatly exceeding the

T 2

a rival to his own Breakwater at Plymouth. At Cherbourg he was joined by Mr. Whidbey, who had come over in a vessel of war to meet him. Mr. Rennie returned to England after less than a month's tour; and though he had made a labour of his pleasure-trip, the change of air and scene did him good, and he entered with zest upon the business of clearing off the formidable arrears of work which had accumulated during his absence. We may add, as an illustration of his habit of turning everything, even his pleasure, to account, that one of the first things he did on his return home was to make an elaborate report to Lord Melville, then First Lord of the Admiralty, of the results of his investigations of those foreign harbours which he had visited in the course of his journey.

After a few years' more devoted application, his health again began to give way When consulted by Mr. Foljambe of Wakefield, in June, 1820, respecting a railway proposed to be laid down in that neighbourhood, he excused himself from entering upon the business because his hands were so full of work and his health was so delicate. Shortly after, we find him writing to a friend that the new works executed by him during the past year had cost about half a million, besides those in progress at Sheerness, which would cost a million. He was then busy with his investigations relative to New London Bridge, the report on which he prepared whilst laid up with gout. He persisted in going abroad as long as he could, and went to his doctor in a carriage for advice, instead of letting the latter come to him. But resistance, however brave, was useless against disease, and at length he was compelled to succumb. To the last he went on issuing instructions to his inspectors in different parts of the country relative to the works

humble dimensions which it presented when Mr. Rennie visited the place. The whole cost of the works amounted to upwards of seven millions sterling.

then in progress—the docks at Chatham and Sheerness, the harbours at Howth and Kingstown, the bridge at New Galloway, the Eau Brink Cut, the Aire and Calder Navigation, and the pumping-engines for Bottesham and Swaffham, in the Fens. He was especially anxious about the Eau Brink Cut, nearly ready for opening, urging his assistants to report from time to time, giving him full particulars of the progress made. In the midst of all this, he writes a letter to a harbour-master at Bridlington, giving him detailed instructions as to the arrangement of tide tables! His last business letter was written to the Navy Board respecting the proper kind of gates to be used for the dry dock at Pembroke : it was dated the 28th September, 1821. A little before this he had written to his friend Mr. Jerdan, the Edinburgh engineer, that he had completed all business connected with his preparation for the next session of Parliament, when he had many bills to carry through. But how often are the intentions of the bravest defeated! Day by day he grew weaker, struggling with the whole force of his will against the disease that was slowly mastering him. Although extremely ill, he insisted on rising from his bed, and tottered about, even taking an occasional airing in a carriage. In this state he continued until the 4th of October. On that day he did not get up. His mind had until then been as clear and vigorous as ever ; but now it began to wander. There was no resisting the hand of death, which was already upon him. He took no further heed of what passed around him, and about five in the evening a violent fit occurred, from which he never rallied. About an hour later he expired, in the sixty-first year of his age.

The portrait prefixed to this memoir expresses, so far as an accurate delineation of his features can do, the actual character of the man. It is grave and thoughtful, yet has an expression of mildness perfectly in unison with his gentle yet cheerful disposition. Raeburn painted

his portrait, and Chantrey chiselled his bust; but the
chalk drawing by Archibald Skirving,[1] after which our

[1] Archibald Skirving, like John Rennie, was the son of an East Lothian farmer. He was born in 1749, at Garleton, a farm belonging to the Earl of Wemyss. His father, Adam Skirving, was a well-known humorist and ballad-maker—one of his songs, 'Hey, Johnny Cope,' a description of the rout of the royal army at the battle of Prestonpans, being still popular in Scotland. Its publication gave great offence to some of Cope's officers, and one of them, Lieutenant Smith, went so far as to send Skirving a challenge, dated the George Inn, Haddington. When the messenger arrived with the missive, the farmer was in his yard, turning over manure. After reading the letter, he said, "Ye may gang back to Lieutenant Smith, and say to him, if he likes to come up-by here, I'll tak' a look at him; if I've a mind to fecht him, I'll fecht him; and if no, I'll do as he did—I'll rin awa'!" Many similar stories are told of the farmer's wit and humour, a considerable share of which was inherited by his son Archibald. In early life the latter went to Rome to study art, and remained in Italy nine years. He walked the whole way from Rome, but, while passing through France, the revolutionary war broke out, and he was apprehended and thrown into prison, where he lay for nine months. He subsequently studied painting under David. Returned to Scotland, he pursued his art in a somewhat desultory manner, not being under the necessity of applying himself to it with that patient and continuous devotion which is essential to attaining high eminence in any profession. He painted when, where, and whom he pleased; and sometimes pursued a very eccentric course with his sitters. One gentleman's portrait he painted in such a manner as to give special prominence to a large wart upon his face. A lady who insisted on sitting to him, he put off with the ungallant remark that she "would ruin him for yellow."

Notwithstanding his eccentricity, Skirving was an extremely clever artist, and his crayon drawings have rarely been surpassed for vigour and brilliancy. He executed probably the best head of Burns, with whom he was intimate; and the portrait of John Rennie, which Mr. Holl has rendered with great skill, will give a good idea of Skirving's power as a delineator of character. Skirving and Rennie were intimate friends, although in most respects so unlike each other. Yet Skirving had as true a genius in him, and might have secured as great a reputation in his own walk, as his friend Rennie, had he worked as patiently and industriously. As he grew older, he became more eccentric and sarcastic. He dressed oddly, in a broad-brimmed white hat, without any neckcloth. He was at the Earl of Wemyss's house at Gosford one day, when the Countess was conversing with him as to the acquirements of her daughters in art. The young ladies were meanwhile occupied in making grimaces at the odd man behind his back, forgetting that they were standing opposite a mirror, in which he could see all their movements. "The young ladies," observed the painter, "may have studied art, but I never saw such ugly faces as those they make," pointing to the glass before him. Allan Cunningham relates the story of Skirving's calling on Chantrey while he was finishing the bust of Bird, the artist. "Well!—and who is that?" asked Skirving. "Bird, the eminent painter." "Painter!—and what does he paint?" "Ludicrous subjects, Sir." "Ludicrous subjects!—Have you sat?" "Yes—he has had one sitting; but when he heard that a gentleman with a white hat, who wore no neckcloth, had arrived from the North, he said, 'Go—go; I know of a subject more ludicrous still: Mr. Skirving is come!'" This odd, but clever artist died at Inveresk, near Edinburgh, in 1819, at the advanced age of seventy.

engraving is made, is on the whole the most lifelike representation of the man as he lived. In person he was large, tall, and commanding; and strength was one of the attributes belonging to his family. But physical endurance has its limits, and we fear that Mr. Rennie taxed his powers beyond what they would fairly bear. He may be said to have died in harness, in the height of his fame, after threescore years, forty of which had been spent in hard work; still his death was premature, and in the case of a man of such useful gifts, was much to be lamented. But he himself held that life was made for work, and he could never bear to be idle. Work was with him not only a pleasure,—it was almost a passion. He sometimes made business appointments at as early an hour as five in the morning, and would continue incessantly occupied until late at night. It is clear that the most vigorous constitution could not long have borne up under such a tear and wear of vital energy as this.

He was very orderly, punctual, and systematic, and hence was enabled to get through a very large amount of business. No matter how numerous were the claims upon his time, nothing was neglected nor hurried. His reports were models of what such documents should be. They set forth all the facts bearing upon the topic under consideration in great detail; but with much plainness, force, and clearness. His harbour reports were especially masterly; in them he elaborately stated all the known facts as to the prevailing winds, currents, and tides, usually drawing very logical and conclusive inferences as to the particular plan which, under the circumstances, he considered it the most desirable to adopt. In his estimates he was careful to conceal nothing, stating the full sum which in his judgment the work under consideration would cost; nor would he understate the amount by one farthing in order to tempt projectors to begin any undertaking on which he was consulted.

He took the highest ground in his dealings with contractors. He held that the engineer was precluded by
his position from mixing himself up with their business,
and that if he dabbled in shares or contracts, either
openly or underhand, half his moral influence was gone,
and his character liable to be seriously compromised.
Writing to Playfair at Edinburgh, in 1816, he said—
"Engineers should be entirely independent of these
connexions—not dabblers in shares—and free alike of
contractors and contracts." By holding scrupulously
to this course, Mr. Rennie established a reputation for
truthfulness, honesty, and uprightness, not less honourable and exalted than his genius as an architect and
engineer was illustrious.

He was a man of powerful and equally balanced mind—
not so clever, as profound; not brilliant, but calm, serene,
and solid, like one of his own structures. While he lay
on his deathbed, his last letters to his assistants urged
upon them attention, punctuality, and despatch—qualities
which he himself had illustrated so well in his own life.
In his self-education he had overlooked no branch of
science cultivated in his day; and in those which bore
more especially upon his own calling, his knowledge was
well-arranged, complete, and accurate.

Withal he was an exceedingly modest, unpretending,
and retiring man. His great aim was to do the thing he
was appointed to do in the best possible manner. He
thought little of fame, but a great deal of character
and duty. If his time was so entirely pre-occupied that
he could not personally devote the requisite attention to
any new undertaking brought before him, he would decline to enter upon it, and recommended the employment
of some other leading engineer. He considered it his duty
himself to go into the minutest details of every business
on which he was consulted. He left as little as possible
to subordinates, making his calculations and estimates
himself; and he wrote and even copied his own reports;

deeming no point, however apparently unimportant, beneath his careful attention and study.

Hence great reliance was placed upon his judgment by those who consulted him; and the accurate though comparatively reserved manner in which he expressed himself before Committees of Parliament gave all the greater weight to his evidence. "What I liked about Rennie," says one who knew him well, "was his severe truthfulness." When under examination on such occasions, he could always give a strong, clear reason, in support of any scheme he recommended, based upon his own careful preliminary study of the whole subject. But when asked any question outside the line of his actual knowledge, he had the honesty to say at once, "I do not know." He would not guess nor attempt to give ingenious answers to show his cleverness, nor act the special pleader in the witness-box, but confine himself solely to what he positively knew.

In the course of his professional career, Mr. Rennie experienced the great advantage which he had derived from his early training as a millwright. His practical knowledge enabled him to select the best men to carry out his designs, and he took pride and pleasure in directing them how to do their work in the most efficient manner. His manufactory was indeed a school, in which some of the best mechanics of the day received a thorough training in machine work; and many of his workmen, like himself, eventually raised themselves to the rank of large employers of skilled labour. Mr. Rennie was never ashamed to put his hand to any work where he could teach a lesson or facilitate despatch, and to the end of his career he continued as "handy" as he had been at the beginning.

A curious illustration of his expertness at smith-work occurred during a journey into Scotland, when on his way to visit the Earl of Eglinton at Eglinton Castle. He went by the stage-coach, in company

with some Ayrshire farmers and one or two rather important "Paisley boddies."[1]   When travelling over a
very bad piece of the road, the jolting was such as to
break the axletree of the coach, and it came to a stand
on a solitary moor, with not a house in sight. Mr. Rennie
asked the coachman if there was any blacksmith near at
hand, and was told there was one a mile or two off.
"Well, then, help me to carry the parts of the axle there,
and I'll see to its being mended."   The blacksmith, however, was not at home; but Mr. Rennie forthwith lit the
forge fire, blew the bellows, and with the rather clumsy
assistance of one of his fellow-passengers, he very soon
welded the axle in a workmanlike manner, helped to
carry it back to the coach, and after the lapse of a few
hours the vehicle was again wheeling along the road
towards its journey's end.   Mr. Rennie's fellow-passengers, who had been communicative and friendly during
the earlier part of the journey, now became very reserved,
and the "boddies" especially held themselves aloof from
"the blacksmith," who had so clearly revealed his calling
by the manner in which he had mended the broken axle.
Arrived at their journey's end for the day, the travellers
separated ; Mr. Rennie proceeding onwards to Eglinton
Castle.   Next morning, when sitting at breakfast with his
noble host, a servant entered to say that a person outside
desired to have a word with the Earl.   "Show him in."
The person entered. He proved to be one of Mr. Rennie's
fellow-travellers ; and great indeed were his surprise and
confusion at finding the identical "blacksmith" of the
preceding day breakfasting with my Lord!   The Earl
was much amused when Mr. Rennie afterwards described
to him the incident of the mending of the broken axle.

One of his few hobbies was for old books; and if he
could secure a few minutes' leisure at any time, he would
wander amongst the old book-stalls in search of rare

---

[1] *Paisley Boddie*—a name applied in the West of Scotland to a person
belonging to Paisley.

volumes. Froissart's and Monstrelet's 'Chronicles' were amongst his favourites, and we find him on one occasion sending a present of duplicate copies to his friend Whidbey, accompanied with the wish that he might derive as much pleasure from their perusal as he himself always did from reading "honest John Froissart." He also commissioned his friends, when travelling abroad, to pick up old books for him; and in 1820 we find him indulging his "extravagance," as he termed it, so far as to request Sir William Jolliffe to bring 300*l*. worth of old books for him from Paris.

Although Mr. Rennie realized a competency by the practice of his profession, he did not accumulate a large fortune. The engineer was then satisfied with a comparatively moderate rate of pay,[1] and Mr. Rennie's charge of seven guineas for an entire day's work was even objected to by General Brownrigg, the head of the Ordnance Department at the time. "Why, this will never do," said the General, looking over the bill; "seven guineas a-day! Why, it is equal to the pay of a Field Marshal!" "Well," replied Mr. Rennie, "I am a Field Marshal in my profession; and if a Field Marshal in your line had answered your purpose, I suppose you would not have sent for me!" "Then you refuse to make any abatement?" "Not a penny," replied the engineer; and the bill was paid.

Mr. Rennie was blamed in his time for the costliness of his designs, and it was even alleged of him that he carried his love of durability to a fault. But there is no doubt that the solidity of his structures proved the best economy in the long run. Elevated by his genius and his conscientiousness above the thoughts of imme-

---

[1] We do not wonder to find Mr. Rennie complaining of the small remuneration of 350*l*. awarded to him by the Kennet and Avon Canal Company for constructing their canal works; and we are surprised to find his bill against the Manchester Waterworks Company, for his year and a-half's advice and service, amounting to only 159*l*. 7*s*., his charge to them for a whole day's labour being only 6*l*. 6*s*.

diate personal gain, no consideration would induce him
to recommend or countenance in any way the construction of cheap or slight work. He held that the engineer
had not merely to consider the present but the future
in laying down and carrying out his plans. Hence his
designs of docks and harbours were usually framed so as
to be capable of future extension; and his bridges were
built not only for his own time, but with a view to the uses
of generations to come. In fine, Mr. Rennie was a great
and massive, yet a perfectly simple and modest man; and
though his engineering achievements may in some measure have been forgotten in the eulogies bestowed upon
more recent works, they have not yet been eclipsed, nor
indeed equalled; and his London bridges—not to mention his docks, harbours, breakwater, and drainage of
the Lincoln Fens—will long serve as the best exponents
of his genius.

The death of this eminently useful man was felt
to be a national loss, and his obsequies were honoured
by a public funeral. His remains were laid near those
of Sir Christopher Wren in St. Paul's Cathedral, the
dome of which overlooks his finest works. The same
motto might apply to him as to the great architect near
whose remains his lie—"Si monumentum quaeris, cir-
cumspice."

Thomas Telford, F.R.S.

Engraved by W. Holl after the portrait, by Samuel Lane..

Published by John Murray Albemarle Street 1861.

LIFE OF THOMAS TELFORD.

GLASGOW BRIDGE. [By R P Leitch.]

LIFE OF THOMAS TELFORD.

CHAPTER I.

ESKDALE.

THOMAS TELFORD was born in one of the most solitary nooks of the narrow valley of the Esk, in the eastern part of the county of Dumfries, in Scotland. Eskdale runs north and south, its lower end having been in former times the western march of the Scottish border. Near the entrance to the dale is a pillar set upon a high hill, some eight miles to the eastward of the Gretna Green station of the Caledonian Railway,—which many travellers to and from Scotland may have observed,—a monument to the late Sir John Malcolm, Governor of Bombay, one of the distinguished natives of the district. It looks far over the English border-lands which stretch away towards the south, and marks the entrance to the mountainous parts of the valley which lie to the north. From that point upwards it gradually becomes narrower, the road winding along the river's banks, in some places high above the stream, dark-brown with peat water, which swiftly rushes over the rocky bed below. A few miles from the lower end of Eskdale lies the little capital of the district, the town of Langholm; and there, in the market-place, stands another monument to the virtues of the Malcolm family in the statue erected to the memory of Admiral Sir Pulteney Malcolm, a distinguished naval officer. Above Langholm the country becomes more hilly and moorland. In many places only

TELFORD'S NATIVE DISTRICT

a narrow strip of haugh land by the river's side is left available for cultivation; until at length the dale contracts so much that the hills descend to the very road, and there are only to be seen their steep heathery sides sloping up towards the sky on either hand, and a narrow stream plashing and winding along the bottom of the valley among the rocks at their feet.

From this brief description of the character of Eskdale scenery, it may readily be supposed that the district is very thinly peopled, and that it never could have been capable of supporting a large number of inhabitants. Indeed, previous to the union of the crowns of England and Scotland, the principal branch of industry that existed in the Dale was of a lawless kind. The people living on the two sides of the border looked upon each other's cattle as their own, provided only they had the strength to "lift" them. They were, in truth, even during the time of peace, a kind of outcasts, against whom the united powers of England and Scotland were often employed. On the Scotch side of the Esk were the Johnstones and Armstrongs, and on the English the Graemes of Netherby; but both clans were alike wild and lawless. It was a popular border saying that "Elliots and Armstrongs ride thieves all;" and an old historian says of the Graemes that "they were all stark moss-troopers and arrant thieves; to England as well as Scotland outlawed." The neighbouring chiefs were no better: Scott of Buccleugh, from whom the modern Duke is descended, and Scott of Harden, the ancestor of the novelist, were both renowned freebooters.

There stand at this day on the banks of the Esk, only a few miles from the English border, the ruins of an old fortalice, called Gilnockie Tower, in a situation which in point of natural beauty is scarcely equalled even in Scotland. It was the stronghold of a chief popularly known in his day as Johnnie Armstrong.[1] He was a mighty freebooter in the time of James V., and the terror of his name is said to have extended as far as Newcastle-upon-Tyne, between which town and his castle on the Esk he was accustomed to levy black-mail, or "protection and forbearance money," as it was called. The King, however, determining to put down by the strong hand the depredations of the march men, made a sudden expedition along the borders; and Johnnie Armstrong having been so ill-advised as to make his appearance with his followers at a place called Carlenrig, in Etterick Forest, between Hawick and Langholm, James ordered him to instant execution. Had Johnnie Armstrong, like the Scotts and Kers and Johnstones of like calling, been imprisoned

---

[1] Sir Walter Scott, in his notes to the 'Minstrelsy of the Scottish Border,' says that the common people of the high parts of Liddlesdale and the country adjacent to this day hold the memory of Johnnie Armstrong in very high respect.

beforehand, he might possibly have survived to found a British peerage; but as it was, the genius of the Armstrong dynasty was for a time extinguished, only, however, to reappear, after the lapse of a few centuries, in the person of the eminent engineer of Newcastle-upon-Tyne, the inventor of the Armstrong gun.

The two centuries and a half which have elapsed since then have indeed effected extraordinary changes.[1] The energy which the old borderers threw into their feuds has not become extinct, but survives under more benignant aspects, exhibiting itself in efforts to enlighten, fertilize, and enrich the country which their wasteful ardour before did so much to disturb and impoverish. The heads of the Buccleugh and Elliot family now sit in the British House of Lords. The descendant of Scott of Harden has achieved a world-wide reputation as a poet and novelist; and the representative of the Graemes of Netherby—whose country seat now sits so peacefully amidst its woods upon the English side of the border, overlooking Lower Esk—is one of the most venerable and respected of British statesmen. The border men,

[1] It was long before the Reformation flowed into the secluded valley of the Esk; but when it did, the energy of the Borderers displayed itself in the extreme form of their opposition to the old religion. The Eskdale people became as resolute in their covenanting as they had before been in their freebooting; and the moorland fastnesses of the moss-troopers became the haunts of the persecuted ministers in the reign of the second James. A little above Langholm is a hill known as "Peden's View," and the well in the green hollow at its foot is still called "Peden's Well"—that place having been the haunt of Alexander Peden, the "prophet." His hiding-place was among the alder-bushes in the hollow, while from the hill-top he could look up the valley, and see whether the Johnstones of Wester Hall were coming. Quite at the head of the same valley, at a place called Craighaugh, on Eskdale Muir, one Hislop, a young covenanter, was shot by Johnstone's men, and buried where he fell; a gray slabstone still marking the place of his rest. Since that time, however, quiet has reigned in Eskdale, and its small population have gone about their daily industry from one generation to another in peace. Yet, though secluded and apparently shut out by the surrounding hills from the outer world, there is not a throb of the nation's heart but pulsates along the valley; and when the author visited it, some two years since, he found that a wave of the great Volunteer movement had flowed into Eskdale; and the "lads of Langholm" were drilling and marching under their chief, young Mr. Malcolm of the Burnfoot, with even more zeal than in the populous and far more exposed towns and cities of the south.

who used to make such furious raids and forays—
have now come to regard each other, across the ima-
ginary line which divides them, as friends and neighbours ; and they meet as competitors for victory only at
agricultural meetings, where they strive for prizes for
the biggest turnips or the most effective reaping-ma-
chines ; whilst the men who followed their Johnstone
or Armstrong chiefs as prickers or hobilers to the fray
have, like Telford, crossed the border with powers
of road-making and bridge-building which have proved
a source of increased civilization and well-being to the
population of the entire United Kingdom.

The hamlet of Westerkirk, with its parish church and
school, lies in a narrow part of the valley, a few miles
above Langholm. Westerkirk parish is long and narrow,
its boundaries being the hill-tops on either side of the
dale. It is about seven miles long and two broad, with
a population of about 600 persons of all ages. Yet
this number is quite as much as the district is enabled
to support, as is proved by its remaining as nearly as
possible stationary from one generation to another.[1]
But what becomes of the natural increase of families?
" Oh, they swarm off! " was the explanation given to us

---

[1] The names of the families in the
valley remain very nearly the same
as they were three hundred years ago
—the Johnstones, Littles, Scotts, and
Beatties prevailing above Langholm ;
and the Armstrongs, Bells, Irwins,
and Graemes lower down towards
Canobie and Netherby. It is interesting to find that Sir David Lindesay, in
his curious drama published in ' Pin-
kerton's Scotish Poems ' (vol. ii., p.
156), gives these as among the names
of the borderers some three hundred
years since. One *Common Thift*,
when sentenced to condign punish-
ment, thus remembers his Border
friends in his dying speech :—

" Adew! my bruther Annan theives,
   That holpit me in my mischeivis ;

Adew! Grossars, Niksonis, and Bells,
Oft have we fairne owrthreuch the fells :
Adew! Robsons, Howis, and Pylis,
That in our craft hes mony wilis :
Littlis, Trumbells, and Armestranges ;
Baileowes, Erewynis, and Elwandis,
Speedy of flicht, and slicht of handis ;
The Scotts of Eisdale, and the Gramis,
I haf na time to tell your nameis."

Telford, or Telfer, is an old name
in the same neighbourhood, com-
memorated in the well known border
ballad of ' Jamie Telfer of the fair
Dodhead.' Sir W. Scott says, in the
' Minstrelsy,' that "there is still a
family of Telfers, residing near Langholm, who pretend to derive their
descent from the Telfers of the Dodhead."

U 2

by a native of the valley. "If they remained at home," said he, "we should all be sunk in poverty, scrambling with each other amongst these hills for a bare living. But our peasantry have a spirit above that: they will not consent to sink; they look up; and our parish schools give them a power of making their way in the world, each man for himself. So they swarm off—some to America, some to Australia, some to India, and some, like Telford, work their way across the border and up to London, though he is the only one from this valley who has yet reached Westminster Abbey."

One would scarcely have expected to find the birth-place of the builder of the Menai Bridge and other great national works in so obscure a corner of the kingdom. Possibly it may already have struck the reader with surprise, that not only were all the engineers described in the preceding pages self-taught in their professions, but they were brought up mostly in remote country places, far from the active life of great towns and cities. But genius is of no locality, and springs alike from the farm-house, the peasant's hut, or the herd's shieling. Strange indeed it is that the men who have built our bridges, docks, lighthouses, canals, and railways, should nearly all have been country-bred boys: Edwards and Brindley the sons of small farmers; Smeaton, brought up in his father's country house at Austhorpe; Rennie, the son of a farmer and freeholder; and Stephenson, brought up in a village, an engine-tenter's son. But Telford, even more than any of these, was a purely country-bred boy, and was born and brought up in a valley so secluded that it could not even boast of a cluster of houses of the dimensions of a village.

Telford's father was a herd on the sheep-farm of Glendinning. The farm consists of green hills, lying along the valley of the Meggat, a little burn, which descends from the moorlands on the east, and falls into

TELFORD'S BIRTHPLACE. [1]

[By R. P. Leitch.]

the Esk near the hamlet of
Westerkirk. John Telford's
cottage was little better
than a shieling, consisting
of four mud walls, spanned
by a thatched roof. It stood upon a knoll near the
lower end of a deep gully worn in the hillside by the
torrents of many winters. The ground stretches away
from it in a long sweeping slope up to the sky, and is
green to the top, except where the bare grey rocks in
some places crop out to the day. From the knoll may
be seen miles on miles of hills up and down the valley,
winding in and out, sometimes branching off into smaller
glens, each with its gurgling rivulet of peaty-brown water
flowing down from the mosses above. Only a narrow
strip of arable land is here and there visible along the

---

[1] The engraving represents the valley of the Meggat, showing the cottages of
Glendinning in the distance.

bottom of the dale, all above being sheep-pasture, moors, and rocks. At Glendinning you seem to have got almost to the world's end. There the road ceases, and above it stretch trackless moors, the solitude of which is only broken by the wimpling sound of the burns on their way to the valley below, the hum of bees gathering honey among the heather, the whirr of a blackcock on the wing, the plaintive cry of the ewes at lambing-time, or the sharp bark of the shepherd's dog gathering the flock together for the fauld.

In this cottage on the knoll Thomas Telford was born on the 9th of August, 1757, and before the year was out he was already an orphan. The shepherd, his father, died in the month of November, and was buried in Westerkirk churchyard, leaving behind him his widow and her only child altogether unprovided for. We may here mention that one of the first things which that child did, when he had grown up to manhood and could " cut a headstone," was to erect one with the following inscription, hewn and lettered by himself, over his father's grave :—

" IN MEMORY OF JOHN TELFORD, WHO, AFTER LIVING 33 YEARS AN UNBLAMEABLE SHEPHERD, DIED AT GLENDINNING, NOVEMBER, 1757",—

a simple but poetical epitaph, which Wordsworth himself might have written.

The widow had a long and hard struggle with the world before her ; but she encountered it bravely. She had her boy to work for, and, destitute though she was, she had him to educate. She was helped, as the poor so often are, by those of her own condition, and there is no sense of degradation in receiving such help. One of the risks of benevolence is its tendency to lower the recipient to the condition of an alms-taker. Doles from poors'-boxes have this enfeebling effect; but a poor neighbour giving a destitute widow a help in her time of need is felt to be a friendly act, and is alike elevating to the character of both. Though misery such as is witnessed in large towns was quite unknown in the valley, there was poverty; but it was honest as well as hopeful, and no one felt ashamed of it. The farmers of the dale were very primitive [1] in their manners and habits, and being a warm-hearted, though by no means a demonstrative race, they were kind to the widow and her fatherless boy. They took him by turns to live with them at their houses, and gave his mother occasional employment. In spring-time she milked the ewes, in summer she made hay, and in harvest she went a-shearing; so that she not only contrived to live, but to be cheerful.

The house to which the widow and her son removed, at the Whitsuntide following the death of her husband, was at a place called The Crooks, about midway between

COTTAGE AT THE CROOKS. [By Percival Skelton.]

Glendinning and Westerkirk. It was a thatched cot-house, with two ends; in one of which lived Janet Telford (though more commonly known by her own name of Janet Jackson) and her son Tom, and in the other her neighbour Elliot; one door being common to both.

Young Telford grew up a healthy boy, and he was so full of fun and humour that he became known in the

---

[1] It may be mentioned as a curious fact that about the time of Telford's birth there were only two tea-kettles in the whole parish of Westerkirk, one of which was in the house of Sir James Johnstone of Wester Hall, and the other in that of Mr. Malcolm of the Burnfoot.

valley by the name of "Laughing Tam." When he
was old enough to herd sheep he went to live with a
relative, a shepherd like his father, and he spent most
of his time with him in summer on the hill-side amidst
the silence of nature. In winter he lived with one or
other of the neighbouring farmers. He herded their
cows or ran errands, receiving for recompense his meat,
a pair of stockings, and five shillings a year for clogs.
These were his first wages, and as he grew older they
were gradually increased.

But Tom must now be put to school, and, happily,
small though the parish of Westerkirk was, it possessed
the advantage of that admirable institution the parish
school. To the orphan boy the merely elementary
teaching there provided was an immense boon. To
master this was the first step of the ladder he was afterwards to mount; his own industry, energy, and ability
must do the rest. To school accordingly he went, still
working a-field or herding cattle during the summer
months. Perhaps his own "penny fee" helped to pay
the teacher's hire; but it is supposed that his uncle
Jackson defrayed the principal part of the expense of
his instruction. It was not much that he learnt; but
in acquiring the arts of reading, writing, and figures, he
learnt the beginnings of a great deal.

Apart from the question of learning, there was another
manifest advantage to the poor boy in mixing freely at
the parish school with the sons of the neighbouring
farmers and proprietors. Such intercourse has an influence upon a youth's temper, manners, and tastes, which
is quite as important in the education of character as
the lessons of the master himself; and Telford often,
in after-life, referred with pleasure to the benefits
which he thus derived from his early school friendships.
Amongst those to whom he was accustomed to look
back with most pride, were the two elder brothers of the
Malcolm family, both of whom rose to high rank in the
service of their country; William Telford, a youth of

great promise, a naval surgeon, who died young ; and
the brothers William and Andrew Little, the former of
whom settled down as a farmer in Eskdale, and the
latter, a surgeon, lost his eyesight when on service on the

WESTERKIRK CHURCH AND SCHOOL.
[By Percival Skelton, after his original Drawing.]

coast of Africa. Andrew Little afterwards established
himself as a teacher at Langholm, where he educated,
amongst others, General Sir Charles Pasley, Dr. Irving,
the Custodier of the Advocates' Library at Edinburgh,
and others known to fame beyond the bounds of their
native valley. Well might Telford say, when an old
man, full of years and honours, on sitting down to
write his autobiography, " I still recollect with pride
and pleasure my native parish of Westerkirk, on the
banks of the Esk, where I was born."

CHAPTER II.

LANGHOLM—TELFORD LEARNS THE TRADE OF A STONEMASON.

THE time arrived when young Telford must be put to
some regular calling. Was he to be a shepherd like his
father and his uncle, or was he to be a farm-labourer,
or put apprentice to a trade ? There was not much
choice; but at length it was determined to bind him to
a stonemason. In Eskdale that trade was for the most
part confined to the building of drystone walls, and there
was very little more art employed in it than an ordinarily
neat-handed labourer could manage. It was eventually
determined to send the youth—and he was now a strong
lad of about fifteen—to a mason at Lochmaben, a small
town across the hills to the eastward, where a little more
building and of a better sort—such as farm-houses, barns,
and road-bridges—was carried on, than in his own immediate neighbourhood. There he remained only a
few months; for his master using him badly, the high-
spirited youth would not brook it, and ran away, taking
refuge with his mother at The Crooks, very much to her
dismay.

What was now to be done with Tam ? He was willing
to do anything or go anywhere rather than back to
his Lochmaben master. In this emergency his cousin
Thomas Jackson, the factor or land-steward at Wester
Hall, offered to do what he could to induce Andrew
Thomson, a small mason at Langholm, to take Telford for
the remainder of his apprenticeship; and to him he went
accordingly. The business carried on by his new master
was of a very humble sort. Telford, in his autobiography,

states that most of the farmers' houses in the district then
consisted of " one storey of mud walls, or rubble stones
bedded in clay, and thatched with straw, rushes, or
heather; the floors being of earth, and the fire in the
middle, having a plastered creel chimney for the escape
of the smoke; and, instead of windows, small openings in the thick mud walls admitted a scanty light."
The farm-buildings were of a similarly wretched description.

The principal owner of the landed property in the
neighbourhood was the Duke of Buccleugh; and shortly
after the young Duke Henry succeeded to the title and
estates in 1767, he introduced considerable improvements
in the farmers' houses and farm-steadings, and the dwell-
ings of the peasantry, as well as of the roads through
Eskdale. In this way a demand sprang up for masons'
labour, and Telford's master had no want of regular
employment for his hands. Telford had the benefit of this
increase in the building operations of the neighbourhood ;
not only in raising rough walls and farm enclosures, but
in erecting bridges across rivers wherever regular roads
for wheel carriages were substituted for the horse-tracks
formerly in use.

During the greater part of his apprenticeship Telford
lived in the little town of Langholm, taking frequent
opportunities of visiting his mother at The Crooks on
Saturday evenings, and accompanying her to the parish-
church of Westerkirk on Sundays. Langholm was then
a very poor town, being no better in that respect than
the district that surrounded it. It consisted chiefly of
mud hovels, covered with thatch—the principal building
in it being the Tolbooth, a stone and lime structure,
the upper part of which was used as a justice-hall and
the lower part as a gaol. There were, however, a few
good houses in the little town occupied by people
of the better class, and in one of these lived an

elderly lady, Miss Pasley,[1] one of the family of the Pasleys of Craig. As the town was so small that everybody in it knew everybody else, the ruddy-cheeked, laughing mason's apprentice soon became generally known to all the townspeople, and amongst others to Miss Pasley. When she heard that he was the poor orphan boy from up the valley, the son of the hard-working widow woman, Janet Jackson, so "eident" and so industrious, her heart warmed to the mason's apprentice, and she sent for him to her house. That was a proud day for Tam; and when he called upon her, he was not more pleased with Miss Pasley's kindness than delighted at the sight of her little library of books, which contained more volumes than he had ever before seen. He had by this time acquired a strong taste for reading, and indeed exhausted all the little book stores of his friends. His joy may therefore be imagined when Miss Pasley volunteered to lend him some books from her own library! Of course the young mason eagerly and thankfully availed himself of the privilege; and thus, while working as an apprentice and afterwards as a journeyman, he gathered his first stores of information in British literature, in which he was accustomed to the close of his life to take such pleasure. He almost always had some book with him, which he would snatch a few minutes to read during the intervals of his work; and in the winter nights he occupied his spare time in poring over the volumes that came in his way, usually with no better light than what was afforded by the cottage fire. On one occasion Miss Pasley lent him 'Paradise Lost,' and he took the book with him to the hill-side to read. His delight was such that it fairly taxed his powers of expression. He could only say "I read and read,

---

[1] Aunt of Sir Charles Pasley, lately deceased.

and glowred; then read, and read again." He was also a great admirer of Burns, whose writings so inflamed his mind that at the age of twenty-two, when barely out of his apprenticeship, we find him breaking out in verse.[1]

By diligently reading all such books as he could borrow from friends and neighbours, Telford made considerable progress in his learning; and, what with his scribbling of "poetry" and various attempts at composition, he had become so good and legible a writer that he was often called upon by his less-educated fellows to pen letters for them to their distant friends. He was always willing to help them in this way; and, the other working people of the town making use of his services in the same manner, all the little domestic and family histories of the place soon became familiar to him. One evening a Langholm man asked Tom to write a letter for him to his son in England; and when the young scribe read over what had been written to the old man's dictation, the latter, at the end of almost every sentence, exclaimed, " Capital! capital!" and at the close he said, " Well! I say, Tam! Werricht himsel' couldna ha' written a better!"—the said Wright being a well-known lawyer or " writer " in Langholm.

His apprenticeship over, Telford went on working as a journeyman at Langholm, his wages at the time being only eighteenpence a-day. What was called the New Town was then in course of erection, and there are houses still pointed out in it, the walls of which Telford helped to put together. In the town are three arched door-heads of a more ornamental character than

---

[1] In his 'Epistle to Mr. Walter Ruddiman,' first published in ' Ruddiman's Weekly Magazine,' in 1779, occur the following lines addressed to Burns, in which Telford incidentally sketches himself at the time, and hints at his own subsequent meritorious career:—

" Nor pass the tentie curious lad,
    Who o'er the ingle hangs his head,
    And begs of neighbours books to read;
        For hence arise
    Thy country's sons, who far are spread,
        Baith bold and wise."

the rest, of Telford's hewing; for he was already beginning to set up his pretensions as a craftsman, and took pride in pointing to the superior handiwork which proceeded from his chisel. About the same time the bridge connecting the Old with the New Town was built across the Esk at Langholm, and upon that structure he was also employed. Many of the stones in it were hewn by his hand, and on several of the blocks forming the land-breast his tool mark is still to be seen.

Not long after the bridge was finished, an unusually high flood or spate swept down the valley. The Esk was "roaring red frae bank to brae," and it was generally feared that the new brig would be carried away. Andrew Thomson, the master mason, was from home at the time, and his wife, Tibby, knowing that he was bound by his contract to maintain the fabric for a period of seven years, was in a state of great alarm. She ran from one person to another, wringing her hands and sobbing, "Oh! we'll be ruined—we'll all be ruined!" In her distress she thought of Telford, in whom she had great confidence, and called out, "Oh! where's Tammy Telfer—where's Tammy?" He was immediately sent for. It was evening, and he was soon found at the house of Miss Pasley. When he came running up, Tibby exclaimed, "Oh, Tammy! they've been on the brig, and they say it's shakin'! It'll be doon!" "Never you heed them, Tibby," said Telford, clapping her on the shoulder, "there's nae fear o' the brig. I like it a' the better that it shakes—it proves it's weel put thegither." Tibby's fears, however, were not so easily allayed; and insisting that she heard the brig "rumlin," she ran up—so the neighbours afterwards used to say of her—and set her back against the parapet as if to hold it together. At this, it is said, "Tam hodged and leuch;" and Tibby, observing how easily he took it, at length grew more calm. It soon became clear enough

that the bridge was sufficiently strong; for the flood subsided without doing it any harm, and it has stood the
furious spates of nearly a century uninjured.

Telford acquired considerable general experience
about the same time as a house-builder, though the
structures on which he was engaged were of a humble
order, being chiefly small farm-houses on the Duke
of Buccleugh's estate, with the usual out-buildings.
Perhaps the most important of the jobs on which
he was employed was the manse of Westerkirk, where

VALLEY OF ESKDALE, WESTERKIRK IN THE DISTANCE.
[By Percival Skelton, after his original Drawing.]

he was comparatively at home.   The hamlet stands
on a green hill-side at the entrance to the valley of
the Meggat.   It consists of the kirk, the minister's
manse, the parish-school, and a few cottages, every
occupant of which was known to Telford.   It is backed
by the purple moors, up which he loved to wander
in his leisure hours and read the poems of Fergusson
and Burns.   The river Esk gurgles along its rocky
bed in the bottom of the dale, separated from the
kirkyard by a steep green field; whilst near at hand,

behind the manse, stretch the fine woods of Wester Hall, where Telford was often wont to roam. We can scarcely therefore wonder that, amidst such pastoral scenery, the descriptive poetic faculty of the country mason should have become so decidedly and strongly developed. It was while working at Westerkirk manse that Telford sketched the first draft of his descriptive poem, entitled 'Eskdale,' which was published in the 'Poetical Museum' [1] in 1784.

These early poetical efforts were at least useful as stimulating his self-education. For the practice of poetical composition, while it cultivates the sentiment of beauty in thought and feeling, is probably the best of all exercises in the art of writing correctly, grammatically, and expressively. By drawing a man out of his ordinary calling, too, it often furnishes him with a power of happy thinking which may in after life be a fountain of the purest pleasure; and this, we believe, proved to be the case with Telford, even though he ceased in later years to pursue the special cultivation of the art.

---

[1] The 'Poetical Museum,' Hawick, p. 267. 'Eskdale' was afterwards reprinted by Telford when living at Shrewsbury, when he added a few lines by way of conclusion. The poem describes very pleasantly the fine pastoral scenery of the district:—

" Deep 'mid the green sequester'd glens below,
Where murmuring streams among the alders flow,
Where flowery meadows down their margins spread,
And the brown hamlet lifts its humble head—
There, round his little fields, the peasant strays,
And sees his flock along the mountain graze;
And, while the gale breathes o'er his ripening grain,
And soft repeats his upland shepherd's strain,

And western suns with mellow radiance play,
And gild his straw-roof'd cottage with their ray,
Feels Nature's love his throbbing heart employ,
Nor envies towns their artificial joy."

The features of the valley are very fairly described. Its early history is then rapidly sketched; next its period of border strife, at length happily allayed by the union of the kingdoms, under which the Johnstones, Pasleys, and others, men of Eskdale, achieve honour and fame. Nor did he forget to mention Armstrong, the author of the 'Art of Preserving Health,' who seems to have been educated in the valley; and Mickle, the translator of the 'Lusiad,' whose father was minister of the parish of Langholm; both of whom Telford took a natural pride in as native poets of Eskdale.

Shortly after, when work became slack in the district, Telford undertook to do small jobs on his own account —such as the hewing of gravestones and ornamental doorheads. He prided himself especially upon his hewing, and from the specimens of his workmanship which are still to be seen in the churchyards of Langholm and Westerkirk, he had evidently attained considerable skill. On some of these pieces of masonry the year is carved—1779, or 1780. One of the most ornamental is that set into the wall of Westerkirk church, being a monumental slab, with an inscription and moulding, surmounted by a coat of arms, to the memory of James Pasley of Craig.

He had now learnt all that his native valley could teach him of the art of masonry; and, bent upon self-improvement and gaining a larger experience of life as well as knowledge of his trade, he determined to seek employment elsewhere. He accordingly left Eskdale for the first time, in 1780, and sought work in Edinburgh, where the New Town was then in course of erection on the elevated land, formerly green fields, extending along the north bank of the " Nor' Loch." A bridge had been thrown across the Loch in 1769, the stagnant pond or marsh in the hollow had been filled up, and Princes Street was rising as if by magic. Skilled masons were in great demand for the purpose of carrying out these and the numerous other architectural improvements which were in progress, and Telford had no difficulty in obtaining abundant employment. He remained at Edinburgh for about two years, during which he had the advantage of taking part in first-rate work and maintaining himself comfortably, whilst he devoted much of his spare time to drawing, in its application to architecture. He took the opportunity of visiting and carefully studying the fine specimens of ancient work at Holyrood House and Chapel, the Castle, Heriot's Hospital, and the numerous curious illustrations of middle age domestic architecture with which the Old Town abounds. He also made several journeys to the beautiful old chapel of Rosslyn, in the highly ornamented Gothic style, situated some miles to the south of Edinburgh, making careful drawings of the more important parts of that building.

When he had thus improved himself " and studied all that was to be seen in Edinburgh, in returning to the western border," he says, " I visited the justly celebrated Abbey of Melrose." There he was charmed by the delicate and perfect workmanship still visible even in the ruins of that fine old Abbey; and with his folio filled with sketches and drawings, he made his way back to Eskdale and the humble cottage at The Crooks. But not to remain there long. He merely wished to pay a parting visit to his mother and relations before starting upon a longer journey. "Having acquired," he says in his Autobiography, " the rudiments of my profession, I considered that my native country afforded few opportunities of exercising it to any extent, and therefore judged it advisable (like many of my countrymen) to proceed southward, where industry might find more employment and be better remunerated."

Before setting out he called upon all his old friends and acquaintances in the dale—the neighbouring farmers, who had befriended him and his mother when struggling with poverty—his schoolfellows, many of whom were preparing to migrate, like himself, from their native valley—and the many friends and acquaintances he had made whilst working as a mason in Langholm. Everybody knew that Tam was going south, and all wished him God speed. At length the leave-taking was over, and he set out for London in the year 1782, when twenty-five years of age. He had, like the little river Meggat, on the banks of which he was born, floated gradually on towards the outer world: first from the nook in the valley, to Westerkirk school; then to Langholm and its little circle; and now, like the Meggat, which flows with the Esk into the ocean, he was about to be borne away into the wide world. Tam, however, had confidence in himself, and no one had fears for him. As the neighbours said, wisely wagging their heads, "Ah, he's an auld-farran chap is Tam; he'll either mak a spoon or spoil a horn; any how, he's gatten a good trade at his fingers' ends."

Telford had made all his previous journeys on foot; but this one he rode on horseback. It happened that Sir James Johnstone, the laird of Wester Hall, had occasion to send a horse from Eskdale to a member of his family in London; but he had some difficulty in finding a person to take charge of it. It occurred to Mr. Jackson, the laird's factor, that this was a capital opportunity for his cousin Tom, the mason; and it was accordingly arranged that he should ride the horse to town. When a boy, he had learnt rough-riding sufficiently well for the purpose; and the better to fit him for the hardships of the road, Mr. Jackson lent him his buckskin breeches. Thus Tom set out from his native valley well mounted, with his little bundle of "traps" buckled behind him, and, after a prosperous journey, duly reached London, and delivered up the horse as he had been directed. Long after, Mr. Jackson used to tell the story of his cousin's first ride to London with great glee, and he always took care to wind up with—"but Tam forgot to send me back my breeks!"

LOWER VALLEY OF THE MEGGAT, THE CROOKS IN THE DISTANCE.
[By Percival Skelton.]

CHAPTER III.

TELFORD A WORKING MASON IN LONDON, AND FOREMAN OF MASONS AT PORTSMOUTH.

A COMMON working man, whose sole property consisted in his mallet and chisels, his leathern apron and his industry, might not seem to amount to much in "the great world of London." But, as Telford afterwards used to say, very much depends on whether the man has got a head, with brains in it of the right sort, upon his shoulders. In London the weak man is simply a unit added to a vast floating crowd, and may be driven hither and thither, if he do not sink altogether; whilst the strong man will strike out, keep his head above water, and make a course for himself, as Telford did. There is indeed a wonderful impartiality about London. There the capable person usually finds his place. When work of importance is required, nobody cares to ask where the man who can do it best comes from, or what he has been; but what he is, and what he can do. Nor did it ever stand in Telford's way that his father had been a poor shepherd in Eskdale, and that he himself had begun his London career by working for weekly wages with his mallet and chisel.

After duly delivering up the horse, Telford proceeded to present a letter with which he had been charged by his friend Miss Pasley on leaving Langholm. It was addressed to her brother, Mr. John Pasley, an eminent London merchant — brother also of Sir Thomas Pasley, and uncle of the Malcolms. Miss Pasley requested his influence on behalf of the young mason from Eskdale, the bearer of the letter. Mr. Pasley received his countryman kindly, and furnished him with letters of introduction to Sir William Chambers, the architect of Somerset House, then in course of erection. It was the finest architectural work in progress in the metropolis, and Telford, desirous of improving himself by experience of the best kind, wished to be employed upon it. It did not, indeed, need any influence to obtain work there, for good hewers were in great demand; but our mason thought it well to make sure, and accordingly provided himself beforehand with the letter of introduction to the architect. He was employed immediately, and set to work amongst the hewers, receiving the usual wages for his work.

Mr. Pasley also furnished him with a letter to Mr. Robert Adam,[1] another distinguished architect of the time; and Telford seems to have been much gratified by the civility which he received from him. Sir William Chambers he found haughty and reserved, probably being too much occupied to bestow attention on the Somerset House hewer, whilst Adam he described as affable and communicative. "Although I derived no direct advantage from either," Telford says, "yet so powerful is manner, that the latter left the most favourable impression; while the interviews with both convinced me that my safest plan was to endeavour to advance, if by slower degrees, yet by independent conduct."

There was a good deal of fine hewer's work about Somerset House, and from the first Telford aimed at taking the highest place as an artist and tradesman in

---

[1] Robert and John Adam were architects of considerable repute in their day. Among their London erections were the Adelphi Buildings, in the Strand; Lansdowne House, in Berkeley Square; Caen Wood House, near Hampstead (Lord Mansfield's); Portland Place, Regent's Park; and numerous West End streets and mansions. The screen of the Admiralty and the ornaments of Drapers' Hall were also designed by them.

that line.[1]   Diligence, carefulness, and observation will always carry a man onward and upward; and before long we find that Telford had succeeded in advancing himself to the rank of a first-class mason.   Judging by his letters written about this time to his friends in Eskdale, he seems to have been very cheerful and happy; and his greatest pleasure was in calling up recollections of his native valley.   He was full of kind remembrances for everybody.   "How is Andrew, and Sandy, and Aleck, and Davie?" he would say; and "remember me to all the folk of the nook."   He seems to have made a round of the persons from Eskdale in or about London before he wrote, as his letters were full of messages from them to their friends at home; for in those days postage was dear, and as much as possible was necessarily packed within the compass of a working man's letter.   In one, written after more than a year's absence, he says he envies the visit which a young surgeon of his acquaintance was about to pay to the valley; "for the meeting of long absent friends," he adds, "is a pleasure to be equalled by few other enjoyments here below."

He had now been more than a year in London, during which he had acquired much practical information both in the useful and ornamental branches of architecture.   Was he to go on as a working·mason? or what was to be his next move?   He had been quietly making his observations upon his companions, and had come to the conclusion that they very much wanted spirit, and, more than all, fore-thought.   He found very clever workmen about him with no idea whatever beyond their week's wages.   For these they would

---

[1] Long after Telford had become famous, he was passing over Waterloo Bridge one day with a friend, when, pointing to some finely-cut stones in the corner nearest the bridge, he said: "You see those stones there; forty years since I hewed and laid them, when working on that building as a common mason."

make every effort: they would work hard, exert themselves to keep them up to the highest point, and very
readily "strike" to secure an advance; but as for the
next week, or the next year, he thought them exceedingly careless. On the Monday mornings they began
"clean;" and on Saturdays there was the week's earnings to spend. Thus they lived from one week to another
—their limited notion of "the week" seeming to bound
their existence.

Telford, on the other hand, looked upon the week as
only one of the storeys of a building; and upon the
succession of weeks, running on through years, he thought
that the complete life structure should be built up. He
thus describes one of the best of his fellow-workmen
at that time—the only individual he had formed an
intimacy with: "He has been six years at Somerset
House, and is esteemed the finest workman in London,
and consequently in England. He works equally in stone
and marble. He has excelled the professed carvers in
cutting Corinthian capitals and other ornaments about
this edifice, many of which will stand as a monument to
his honour. He understands drawing thoroughly, and
the master he works under looks on him as the principal
support of his business. This man, whose name is Mr.
Hatton, may be half a dozen years older than myself at
most. He is honesty and good nature itself, and is
adored by both his master and fellow-workmen. Notwithstanding his extraordinary skill and abilities, he has
been working all this time as a common journeyman,
contented with a few shillings a-week more than the
rest; but I believe your uneasy friend has kindled a
spark in his breast that he never felt before." [1]

In fact, Telford had formed the intention of inducing
this admirable fellow to join him in commencing business as builders on their own account. "There is nothing

---

[1] Letter to Mr. Andrew Little, Langholm, dated London, July, 1783.

done in stone or marble," he says, "that we cannot do in the completest manner." Mr. Robert Adam, to whom the scheme was mentioned, promised his support, and said he would do all in his power to recommend them. But the great difficulty was money, which neither of them possessed; and Telford, with grief, admitting that this was an "insuperable bar," the scheme went no further.

About this time Telford was consulted by Mr. Pulteney[1] respecting the alterations making in the mansion at Wester Hall, and was often with him on this business. We find him also writing down to Langholm for the prices of roofing, masonry, and timber-work, with a view to preparing estimates for a friend who was building a house in that neighbourhood. Although determined to reach the highest excellence as a manual worker, it is clear that he was already aspiring to be something more. Indeed, his steadiness, perseverance, and general ability pointed him out as one well worthy of promotion.

How he achieved his next step we are not informed; but we find him, in July, 1784, engaged in superintending the erection of a house, after a design by Mr. Samuel Wyatt, intended for the residence of the Commissioner (now occupied by the Port Admiral) at Portsmouth Dockyard, together with a new chapel,

---

[1] Mr., afterwards Sir William Pulteney, was the second son of Sir James Johnstone, of Wester Hall, and assumed the name of Pulteney on his marriage to Miss Pulteney, niece of the Earl of Bath and of General Pulteney, by whom he succeeded to a large fortune; he afterwards succeeded to the baronetcy of his elder brother James, who died without issue in 1797. Sir William Pulteney represented Cromarty, and afterwards Shrewsbury, where he usually resided, in seven successive Parliaments. He was a great patron of Telford's, as we shall afterwards find. The story is told in Eskdale that Miss Pulteney had danced at a public ball with the elder brother, Sir James, and afterwards expressed such admiration of him that it was repeated to him, with the hint that he might do worse than offer his hand to the heiress. " There is only one slight difficulty," said he, " I have got a wife already; but," he added, "if she would like a Johnstone, there's my brother Will—a much better looking fellow than I am." How it may have been brought about is not stated, but Will did make up to the heiress, and married her, assuming the name of her family.

and several buildings connected with the Yard. Telford took care to keep his eyes open to all the other works going forward in the neighbourhood, and he states that he had frequent opportunities of observing the various operations necessary in the foundation and construction of graving-docks, wharf-walls, and such like, which were amongst the principal occupations of his after-life.

The letters written by him from Portsmouth to his Eskdale correspondents about this time were cheerful and hopeful, like those he had sent from London. His principal grievance was that he received so few from home, but he supposed that opportunities for forwarding them by hand had not occurred, postage being so dear as scarcely then to be thought of. To tempt them to correspondence he sent copies of the poems which he still continued to compose in the leisure of his evenings : one of these was a 'Poem on Portsdown Hill.' As for himself, he was doing very well. The buildings were advancing satisfactorily; but, "above all," said he, "my proceedings are entirely approved by the Commissioners and officers here—so much so that they would sooner go by my advice than my master's, which is a dangerous point, being difficult to keep their good graces as well as his. However, I will contrive to manage it." [1]

The following is his own account of the manner in which he was usually occupied during the winter months while at Portsmouth Dock :—"I rise in the morning at 7 (February 1st), and will get up earlier as the days lengthen until it come to 5 o'clock. I immediately set to work to make out accounts, write on matters of business, or draw, until breakfast, which is at 9. Then I go into the Yard about 10, see that all are at their posts, and am ready to advise about any matters that may require attention. This, and going round the several

---

[1] Letter to Andrew Little, Langholm, dated Portsmouth, July 23rd, 1784.

works, occupies until about dinner-time, which is at 2; and after that I again go round and attend to what may be wanted. I draw till 5; then tea; and after that I write, draw, or read until half after 9; then comes supper and bed. This is my ordinary round, unless when I dine or spend an evening with a friend; but I do not make many friends, being very particular, nay, nice to a degree. My business requires a great deal of writing and drawing, and this work I always take care to keep under by reserving my time for it, and being in advance of my work rather than behind it. Then, as knowledge is my most ardent pursuit, a thousand things occur which call for investigation which would pass unnoticed by those who are content to trudge only in the beaten path. I am not contented unless I can give a reason for every particular method or practice which is pursued. Hence I am now very deep in chemistry. The mode of making mortar in the best way led me to inquire into the nature of lime. Having, in pursuit of this inquiry, looked into some books on chemistry, I perceived the field was boundless; but that to assign satisfactory reasons for many mechanical processes required a general knowledge of that science. I have therefore borrowed a MS. copy of Dr. Black's Lectures. I have bought his 'Experiments on Magnesia and Quicklime,' and also Fourcroy's Lectures, translated from the French by one Mr. Elliot, of Edinburgh. And I am determined to study the subject with unwearied attention until I attain some accurate knowledge of chemistry, which is of no less use in the practice of the arts than it is in that of medicine." He adds, that he continues to receive the cordial approval of the Commissioners for the manner in which he performs his duties, and says, " I take care to be so far master of the business committed to me as that none shall be able to eclipse me in that respect." [1] At

---

[1] Letter to Mr. Andrew Little, Langholm, dated Portsmouth Dockyard, Feb. 1, 1786.

the same time he states he is taking great delight in
Freemasonry, and is about to have a lodge-room at
the George Inn fitted up after his plans and under
his direction.   Nor does he forget to add that he has
his hair powdered every day, and puts on a clean shirt
three times a week.   The Eskdale mason is evidently
getting on, as he deserves.   Yet he says that " he
would rather have it said of him that he possessed one
grain of good nature or good sense than shine the finest
puppet in Christendom."   " Let my mother know that
I am well," he writes to Andrew Little, "and that I
will print her a letter soon." [1]   For it was a practice of
this good son, down to the period of his mother's
death, no matter how much burdened he was with
business, to set apart occasional times for the careful
penning of a letter in *printed* characters, that she might
be the more easily able to decipher it with her old and
dimmed eyes by her cottage fireside at The Crooks.
As a man's real disposition usually displays itself most
strikingly in small matters—like light, which gleams
the most brightly when seen through narrow chinks—it
will probably be admitted that this trait, trifling though
it may appear, was truly characteristic of the simple and
affectionate nature of the hero of our story.

The buildings at Portsmouth were finished by the end
of 1786, when Telford's duties at that place being at an
end, and having no engagement beyond the termination
of the contract, he prepared to leave, and began to look
about him for other employment.

---

[1] Letter to Mr. Andrew Little, Langholm, dated Portsmouth Dockyard,
Feb. 1, 1786.

CHAPTER IV.

BECOMES SURVEYOR FOR THE COUNTY OF SALOP.

MR. PULTENEY, member for Shrewsbury, was the owner
of extensive estates in that neighbourhood by virtue
of his marriage with the niece of the last Earl of Bath.
Having resolved to fit up the Castle there as a residence,
he bethought him of the young Eskdale mason, who had,
some years before, advised him as to the repairs of the
Johnstone mansion at Wester Hall. Telford was soon
found, and engaged to go down to Shrewsbury to
superintend the necessary alterations. Their execution
occupied his attention for some time, and during their
progress he was so fortunate as to obtain the appointment of Surveyor of Public Works for the county of
Salop, most probably through the influence of his patron.
Indeed, Telford was known to be so great a favourite
with Mr. Pulteney that at Shrewsbury he usually went
by the name of "Young Pulteney."

Much of his attention was from this time occupied
with the surveys and repairs of roads, bridges, and gaols,
and the supervision of all public buildings under the
control of the magistrates of the county. He was also
frequently called upon by the corporation of the borough
to furnish plans for the improvement of the streets and
buildings of that fine old town ; and many alterations
were carried out under his directions during the period
of his residence there.

While the Castle repairs were in course of execution,
he was called upon by the justices to superintend the
erection of a new gaol, the plans for which had
already been prepared and settled. The benevolent

Howard, who devoted himself with such zeal to gaol improvement, on hearing of the intentions of the magistrates, made a visit to Shrewsbury for the purpose of
examining the plans; and the circumstance is thus ad-
verted to by Telford in one of his letters to his Eskdale
correspondent:—"About ten days ago I had a visit
from the celebrated John Howard, Esq.   I say *I*, for he
was on his tour of gaols and infirmaries; and those of
Shrewsbury being both under my direction, this was, of
course, the cause of my being thus distinguished.   I
accompanied him through the infirmary and the gaol.
I showed him the plans of the proposed new buildings,
and had much conversation with him on both subjects.
In consequence of his suggestions as to the former, I
have revised and amended the plans, so as to carry out
a thorough reformation; and my alterations having been
approved by a general board, they have been referred
to a committee to carry out.   Mr. Howard also took objection to the plan of the proposed gaol, and requested
me to inform the magistrates that, in his opinion, the
interior courts were too small, and not sufficiently venti-
lated; and the magistrates, having approved his suggestions, ordered the plans to be amended accordingly.   You
may easily conceive how I enjoyed the conversation of
this truly good man, and how much I would strive to
possess his good opinion.   I regard him as the guardian
angel of the miserable.   He travels into all parts of
Europe with the sole object of doing good, merely for
its own sake, and not for the sake of men's praise.   To
give an instance of his delicacy, and his desire to avoid
public notice, I may mention that, being a Presbyterian,
he attended the meeting-house of that denomination in
Shrewsbury on Sunday morning, on which occasion I
accompanied him; but in the afternoon he expressed a
wish to attend another place of worship, his presence in
the town having excited considerable curiosity, though
his wish was to avoid public recognition.   Nay, more,

he assures me that he hates travelling, and was born to be a domestic man. He never sees his country-house but he says within himself, 'Oh! might I but rest here, and never more travel three miles from home; then should I be happy indeed!' But he has become so committed, and so pledged himself to his own conscience to carry out his great work, that he says he is doubtful whether he will ever be able to attain the desire of his heart—life at home. He never dines out, and scarcely takes time to dine at all: he says he is growing old, and has no time to lose. His manner is simplicity itself. Indeed, I have never yet met so noble a being. He is going abroad again shortly on one of his long tours of mercy "[1] The journey to which Telford here refers was Howard's last. In the following year he left England to return no more; and the great and good man died at Cherson, on the shores of the Black Sea, less than two years after his interview with the young engineer at Shrewsbury.

Telford writes to his Langholm friend at the same time, that he is working very hard, and studying to improve himself in branches of knowledge in which he feels himself deficient. He is practising very temperate habits: for half a year past he has taken to drinking water only, avoiding all sweets, and eating no "nick-nacks." He has "sowens and milk" (oatmeal flummery) every night for his supper. His friend having asked his opinion of politics, he says he really knows nothing about them; he had been so completely engrossed by his own business that he has not had time to read even a newspaper. But, though an ignoramus in politics, he has been studying *linne*, which is more to his purpose. If his friend can give him any information about that, he will promise to read a newspaper now and then in the ensuing session of Parliament, for the purpose of forming

---

[1] Letter to Mr. Andrew Little, Langholm, dated Shrewsbury Castle, 21st Feb., 1788.

some opinion of politics: he adds, however, "not if it interfere with my business—mind that!" His friend had told him that he proposed translating a system of chemistry. "Now you know," said he, "that I am chemistry mad; and if I were near you, I would make you promise to communicate any information on the subject that you thought would be of service to your friend, especially about calcareous matters and the mode of forming the best composition for building with, as well above as below water. But not to be confined to that alone, for you must know I have a book for the pocket,[1] which I always carry with me, into which I have extracted the essence of Fourcroy's Lectures, Black on Quicklime, Scheele's Essays, Watson's Essays, and various points from the letters of my respected friend Dr. Irving.[2] So much for chemistry. But I have also crammed into it facts relating to mechanics, hydrostatics, pneumatics, and all manner of stuff, to which I keep continually adding, and it will be a charity to me if you will kindly contribute your mite."[3] He says it has been, and will continue to be, his aim to endeavour to unite those "two frequently jarring pursuits, literature and business;" and he does not see why a man should be less efficient in the latter capacity because he has well informed, stored, and humanized his mind by the cultivation of letters. There was both good sense and sound practical wisdom in this view of Telford.

While the gaol was in course of erection, after the improved plans suggested by Howard, a variety of important matters occupied the county surveyor's attention. During

---

[1] This practice of noting down information, the result of reading and observation, was continued by Mr. Telford until the close of his life; his last pocket memorandum book, containing a large amount of valuable information on mechanical subjects—a sort of engineer's vade mecum—

being printed in the appendix to the 4to. 'Life of Telford' published by his executors in 1838. Pp. 663-90.

[2] A medical man, a native of Eskdale, of great promise, who died comparatively young.

[3] Letter to Mr. Andrew Little, Langholm.

the summer of 1788 he says he is very much occupied, having about ten different jobs on hand : roads, bridges, streets, drainage-works, gaol, and infirmary. Yet he had time to write verses, copies of which he forwarded to his Eskdale correspondent, inviting his criticism. Several of these were elegiac lines, somewhat exaggerated in their praise of the deceased, though doubtless sincere. One poem was in memory of George Johnstone, Esq., a member of the Wester Hall family, and another on the death of William Telford, an Eskdale farmer's son, an intimate friend and schoolfellow of our engineer.[1] These, however, were but the votive offerings of private friendship, persons more immediately about him knowing nothing of his stolen pleasures in versemaking. He continued to be shy of strangers, and was very " nice," as he calls it, as to those whom he admitted to his bosom.

Two circumstances of considerable interest occurred in the course of the same year (1788), which are worthy of passing notice. The one was the fall of the· church of St. Chad's, at Shrewsbury ; the other was the discovery of the ruins of the Roman city of Uriconium, in the immediate neighbourhood. The church of St. Chad's was about four centuries old, and stood greatly

---

[1] It would occupy unnecessary space to cite these poems. The following, from the verses in memory of William Telford, relates to schoolboy days. After alluding to the lofty Fell Hills, which formed part of the sheep farm of his deceased friend's father, the poet goes on to say :—

" There, 'mongst those rocks I'll form a rural seat,
And plant some ivy with its moss compleat ;
I'll benches form of fragments from the stone,
Which, nicely pois'd, was by our hands o'erthrown,—

A simple frolic, but now dear to me,
Because, my Telford, 'twas performed with thee.
There, in the centre, sacred to his name,
I'll place an altar, where the lambent flame
Shall yearly rise, and every youth shall join
The willing voice, and sing the enraptured line.
But we, my friend, will often steal away
To this lone seat, and quiet pass the day ;
Here oft recall the pleasing scenes we knew
In early youth, when every scene was new,
When rural happiness our moments blest,
And joys untainted rose in every breast."

in need of repairs. The roof let in the rain upon the congregation, and the parish vestry met to settle the plans for mending it; but they could not agree about the mode of procedure. In this emergency Telford was sent for by the churchwardens, and requested to advise them what was best to be done. He accordingly examined the building, and found that not only the roof but the walls of the church were in a most decayed state. It appeared that, in consequence of graves having been dug in the loose soil close to the shallow foundation of the north-west pillar of the tower, it had sunk so as to endanger the whole structure. "I discovered," says he, "that there were large fractures in the walls, on tracing which I found that the old building was in a most shattered and decrepit condition, though until then it had been scarcely noticed. Upon this I declined giving any recommendation as to the repairs of the roof unless they would come to the resolution to secure the more essential parts, as the fabric appeared to me to be in a very alarming condition. I sent in a written report to the same effect." [1] The parish vestry again met, and the report was read; but the meeting exclaimed against so extensive a proposal, imputing mere motives of self-interest to the surveyor. "Popular clamour," says Telford, "overcame my report. 'These fractures,' exclaimed the vestrymen, 'have been there from time immemorial;' and there were some otherwise sensible persons, who remarked that professional men always wanted to carve out employment for themselves, and that the whole of the necessary repairs could be done at a comparatively small expense." [2] Telford at length left the meeting, advising that, if they wished to discuss anything besides the alarming state of the church, they had better adjourn to some other place, where there was no danger of its falling on their heads. The suggestion was received

---

[1] Letter to Mr. Andrew Little, Langholm, dated 16th July, 1788.   [2] Ibid.

with ridicule, and the vestry called in another person, a mason of the town, directing him to cut away the injured part of the pillar, in order to underbuild it. On the second evening after the commencement of these operations, the sexton was alarmed at the fall of lime-dust and mortar when he attempted to toll the great bell, on which he immediately desisted and left the church. Early next morning (on the 9th of July), while the workmen were waiting at the church door for the key, the bell struck four, and the vibration at once brought down the tower, which overwhelmed the nave, demolishing all the pillars along the north side, and shattering the rest. "The very parts I had pointed out," says Telford, "were those which gave way, and down tumbled the tower, forming a very remarkable ruin, which astonished and surprised the vestry, and roused them from their infatuation, though they have not yet recovered from the shock." [1]

The other circumstance to which we have above referred was the discovery of the Roman city of Uriconium, near Wroxeter, about five miles from Shrewsbury, in the year 1788. The situation of the place is extremely beautiful, the river Severn flowing along its western margin, and forming a barrier against what were once the hostile districts of West Britain. For many centuries the dead city had slept under the irregular mounds of earth which covered it, like those of Mossul and Nineveh. Farmers raised heavy crops of turnips and grain from the surface; and they scarcely ever ploughed or harrowed the ground without turning up Roman coins or pieces of pottery. They also observed that in certain places the corn was more apt to be scorched in dry weather than in others— a sure sign· to them that there were ruins underneath; and their practice, when they wished to find stones for building, was to set a mark upon the scorched places

---

[1] Letter to Mr. Andrew Little, Langholm, dated 16th July, 1788.

when the corn was on the ground, and after harvest to
dig down, sure of finding the store of stones which they
wanted for walls, cottages, or farm-houses.   In fact, the
place came to be regarded in the light of a quarry, rich
in ready-worked materials for building purposes.   About
this time a quantity of stone was wanted for the purpose
of erecting a blacksmith's shop, and on digging down
upon one of the marked places, the labourers came
upon some ancient works of a more perfect appearance
than usual.   Curiosity was excited—antiquarians made
their way to the spot—and lo ! they pronounced the ruins
to be neither more nor less than a Roman bath, in a
remarkably perfect state of preservation.   Mr. Telford
was requested to apply to Mr. Pulteney, the lord of the
manor, to prevent the destruction of these interesting
remains, and also to permit the excavations to proceed,
with a view to the buildings being completely explored.
This was readily granted, and Mr. Pulteney authorised
Telford himself to conduct the necessary excavations
at his expense.   This he promptly proceeded to do,
and the result was, that an extensive hypocaust apartment was brought to light, with baths, sudatorium,
dressing-room, and a number of tile pillars—all forming
parts of a Roman floor—sufficiently perfect to show
the manner in which they had been constructed and
used.[1]

Among Telford's less agreeable duties about the same
time was that of keeping the felons at work.   He had
to devise the ways and means of employing them without
risk of their escaping, which gave him much trouble
and anxiety.   "Really," he says, "my felons are a very
troublesome family.   I have had a great deal of plague
from them, and I have not yet got things quite in the
train that I could wish.   I have had a dress made for

---

[1] The discovery formed the subject
of a paper read before the Society of
Antiquaries in London on the 7th of
May, 1789, published in the 'Archæ-
ologia,' together with a drawing of the
remains supplied by Mr. Telford.

them of white and brown cloth, in such a way that they
are pyebald. They have each a light chain about one
leg. Their allowance in food is a penny loaf and a halfpenny worth of cheese for breakfast; a penny loaf, a
quart of soup, and half a pound of meat for dinner; and
a penny loaf and a halfpenny worth of cheese for supper;
so that they have meat and clothes at all events. I
employ them in removing earth, serving masons or brick-
layers, or in any common labouring work on which they
can be employed; during which time, of course, I have
them strictly watched."

Much more pleasant was his first sight of Mrs. Jordan
at the Shrewsbury theatre, where he seems to have
been worked up to a pitch of rapturous enjoyment. She
played for six nights there at the race time, during which
there were various other entertainments. On the second
day there was what was called an Infirmary Meeting,
or an assemblage of the principal county gentlemen in
the infirmary, at which, as county surveyor, Telford
was present. They proceeded thence to church to hear a
sermon preached for the occasion; after which there was
a dinner, followed by a concert. He attended all. The
sermon was preached in the new pulpit, which had just
been finished after his designs, in the Gothic style; and
he confidentially informed his Langholm correspondent
that he believed the pulpit secured greater admiration
than the sermon. With the concert he was completely
disappointed, and he then became convinced that he could
have no ear for music. Other people seemed very much
pleased; but for the life of him he could make nothing
of it. The only difference that he recognised between
one tune and another was that there was a difference of
noise. "It was all very fine," he says, "I have no
doubt; but I would not give a song of Jock Stewart[1] for

---

[1] An Eskdale crony. His son,
Colonel Jonas Stewart, rose to eminence in the East India Company's
service, having been for many years
Resident at Gwalior and Indore.

the whole of them.   The melody of sound is thrown away upon me.   One look, one word of Mrs. Jordan, has more effect upon me than all the fiddlers in England.   Yet I sat down and tried to be as attentive as any mortal could be.   I endeavoured, if possible, to get up an interest in what was going on; but it was all of no use.   I felt no emotion whatever, excepting only a strong inclination to go to sleep.   It must be a defect; but it is a fact, and I cannot help it.   I suppose my ignorance of the subject, and the want of musical experience in my youth, may be the cause of it." [1]

Telford's mother was still living in her old cottage at The Crooks.   Since he had parted from her he had written many printed letters to keep her informed of his progress; and he never wrote to any of his friends in the dale without including some message or other to his mother.   Like a good and dutiful son, he had taken care out of his means to provide for her comfort in her declining years.   "She has been a good mother to me," he said, "and I will try and be a good son to her."   In a letter written from Shrewsbury about this time, enclosing a ten pound note, seven pounds of which was to be given to his mother, he said, " I have from time to time written William Jackson [his cousin] and told him to furnish her with whatever she wants to make her comfortable ; but there may be many little things she may wish to have, and yet not like to ask him for.   You will therefore agree with me that it is right she should have a little cash to dispose of in her own way. . . I am not rich yet ; but it will ease my mind to set my mother above the fear of want.   That has always been my first object ; and next to that, to be the *somebody* which you have always encouraged me to believe I might aspire to become. Perhaps after all there may be something in it ! " [2]

---

[1] Letter to Mr. Andrew Little, Langholm, dated 3rd Sept., 1788.
[2] Letter to Mr. Andrew Little, Langholm, dated Shrewsbury, 8th October, 1789.

He now seems to have occupied much of his leisure
hours in miscellaneous reading. Amongst the numerous
books which he read, he expressed the highest admiration for Sheridan's 'Life of Swift.' But his Langholm friend, who was a great politician, having invited
his attention to politics, Telford's reading gradually
extended in that direction. Indeed the exciting events
of the French Revolution then tended to make all
men more or less politicians. The capture of the
Bastille by the people of Paris in 1789 passed like
an electric thrill through Europe. Then followed the
Declaration of Rights; after which, in the course of six
months, all the institutions which had before existed in
France were swept away, and the reign of justice was
fairly inaugurated upon earth! In the spring of 1791
the first part of Paine's 'Rights of Man' appeared, and
Telford, like many others, read it, and was at once carried
away by it. Only a short time before, he had admitted
with truth that he knew nothing of politics; but no
sooner had he read Paine than he felt completely en-
lightened. He now suddenly discovered how much
reason he and everybody else in England had for being
miserable. Whilst residing at Portsmouth he had quoted
to his Langholm friend the lines from Cowper's 'Task,'
then just published, beginning "Slaves. cannot breathe
in England;" but lo! Mr. Paine had filled his imagination
with the idea that England was nothing but a nation of
bondmen and aristocrats. To his natural mind the
kingdom had appeared to be one in which a man had
pretty fair play, could think and speak, and do the
thing he would,—tolerably happy, tolerably prosperous,
and enjoying many blessings. He himself had felt free
to labour, to prosper, and to rise from manual to head
work. No one had hindered him; his personal liberty
had never been interfered with; and he had freely employed his earnings as he thought proper. But now the
whole thing appeared a delusion. Those rosy-cheeked

old country gentlemen who came riding in to Shrewsbury to quarter sessions, and were so fond of their young
Scotch surveyor — occupying themselves in building
bridges, maintaining infirmaries, making roads, and regulating gaols—those county magistrates and members of
parliament, aristocrats all, were the very men who,
according to Paine, were carrying the country headlong
to ruin.

If Telford could not offer an opinion on politics before,
because he " knew nothing about them," he had now no
such difficulty. Had his advice been asked about the
foundations of a bridge, or the security of an arch, he
would have read and studied much before giving it ; he
would have carefully inquired into the chemical qualities
of different kinds of lime—into the mechanical principles of weight and resistance, and such like ; but he had
no such hesitation in giving an opinion about the foundations of a constitution of more than a thousand years'
growth. Here, like other young politicians, with Paine's
book before him, he felt competent to pronounce a
decisive judgment at once. " I am convinced," said he,
writing to his Langholm friend, " that the situation of
Great Britain is such, that nothing short of some signal
revolution can prevent her from sinking into bankruptcy,
slavery, and insignificancy." He held that the national
expenditure was so enormous,[1] arising from the corrupt
administration of the country, that it was impossible the
" bloated mass " could hold together any longer ; and as
he could not expect that " a hundred Pulteneys," such as
his employer, could be found to restore it to health, the
conclusion he arrived at was that ruin was " inevitable." [2]

---

[1] It was then under seventeen millions sterling, or about a fourth of
what it is now.

[2] Letter to Mr. Andrew Little,
Langholm, dated 28th July, 1791.
Notwithstanding the theoretical ruin
of England which pressed so heavy
on his mind at this time, we find

Telford strongly recommending his
correspondent to send any good wrights
he can find in his neighbourhood to
Bath, where they would be enabled
to earn twenty shillings or a guinea
a week at piece-work—the wages paid
at Langholm for similar work being
only about half those amounts.

Fortunately for Telford, his intercourse with the towns-people of Shrewsbury was so small that his views on these subjects were never known; and we very shortly find him employed by the clergy themselves in building a new church in the town of Bridgenorth. His patron and employer, Mr. Pulteney, however, knew of his extreme views, and the knowledge came to him quite accidentally. He found that Telford had made use of his frank to send through the post a copy of Paine's 'Rights of Man' to his Langholm correspondent,[1] where the pamphlet excited as much fury in the minds of some of the people of that town as it had done in that of Telford himself. The " Langholm patriots" broke out into drinking revolutionary toasts at the Cross, and so disturbed the peace of the little town that some of them were confined for six weeks in the county gaol. Mr. Pulteney was very indignant at the liberty Telford had taken with his frank, and a rupture between them seemed likely to ensue; but the former was forgiving, and the matter went no further. It is only right to add, that as Telford grew older and wiser, he became more careful in jumping at conclusions on political topics. The events which shortly occurred in France tended in a great measure to heal his mental distresses as to the future of England. When the " liberty" won by the Parisians ran into riot, and the " Friends of Man" occupied themselves in taking off the heads of those who differed from them, he became wonderfully reconciled to the

---

[1] The writer of a memoir of Telford, in the 'Encyclopedia Britannica,' says :—" Andrew Little kept a private and very small school at Langholm. Telford did not neglect to send him a copy of Paine's 'Rights of Man;' and, as he was totally blind, he employed one of his scholars to read it in the evenings. Mr. Little had received an academical education before he lost his sight; and, aided by a memory of uncommon powers, he taught the classics, and particularly Greek, with much higher reputation than any other schoolmaster within a pretty extensive circuit. Two of his pupils read all the Iliad, and all or the greater part of Sophocles. After hearing a long sentence of Greek or Latin distinctly recited, he could generally construe and translate it with little or no hesitation. He was always much gratified by Telford's visits, which were not infrequent, to his native district."

enjoyment of the substantial freedom which, after all, was secured to him by the English Constitution. At the same time he was so much occupied in carrying out his important works, that he found but little time to devote either to political speculation or to verse-making.

Whilst living at Shrewsbury, he had his poem of 'Eskdale' reprinted for private circulation. We have also seen several MS. verses by him, written about the same period, which do not appear ever to have been printed. One of these—the best—is entitled 'Verses to the Memory of James Thomson, author of " Liberty, a poem ;" ' another is a translation from Buchanan, 'On the Spheres;' and a third, written in April, 1792, is entitled 'To Robin Burns, being a postscript to some verses addressed to him on the establishment of an Agricultural Chair in Edinburgh.' It would unnecessarily occupy our space to print these effusions ; and, to tell the truth, they exhibit few if any indications of poetic power. No amount of perseverance will make a poet of a man in whom the divine gift is not born. The true line of Telford's genius lay in building and engineering, in which direction we now propose to follow him.

SHREWSBURY CASTLE. [By Percival Skelton.]

CHAPTER V.

Telford's First Employment as an Engineer.

As surveyor for the county, Telford was frequently called upon by the magistrates to advise them as to the improvement of roads and the building or repairs of bridges. His early experience of bridge-building in his native district now proved of much service to him, and he used often to congratulate himself, even when he had reached the highest rank in his profession, upon the circumstances which had compelled him to begin his career by working with his own hands. To be a thorough judge of work, he held that a man must himself have been practically engaged in it. "Not only," he said, "are the natural senses of seeing and feeling requisite in the examination of materials, but also the practised eye, and the hand which has had experience of the kind and qualities of stone, of lime, of iron, of timber, and even of earth, and of the effects of human ingenuity in applying and combining all these substances, are necessary for arriving at mastery in the profession; for, how can a man give judicious directions unless he possesses personal knowledge of the details requisite to effect his ultimate purpose in the best and cheapest manner? It has happened to me more than once, when taking opportunities of being useful to a young man of merit, that I have experienced opposition in taking him from his books and drawings, and placing a mallet, chisel, or trowel in his hand, till, rendered confident by the solid knowledge which experience only can bestow, he was qualified to insist on the due performance of workmanship, and to

judge of merit in the lower as well as the higher depart-
ments of a profession in which no kind or degree of
practical knowledge is superfluous." [1]

The first bridge designed and built under Telford's
superintendence was one of no great magnitude, across the
river Severn at Montford, about four miles west of Shrewsbury. It was a stone bridge of three elliptical arches, one
of 58 feet and two of 55 feet span each. The Severn at
that point is deep and narrow, and its bed and banks are
of alluvial earth. It was necessary to make the foundations very secure, as the river is subject to high floods;
but this was effectually accomplished by means of coffer-
dams. The building was substantially executed in red
sandstone, and proved a very serviceable bridge, forming
part of the great high road from Shrewsbury into Wales.
It was finished in the year 1792.

In the same year we find Telford engaged as an architect in preparing the designs and superintending the
construction of the new parish church of St. Mary Magdalen at Bridgenorth. It stands at the end of Castle
Street, near to the old ruined fortress perched upon the
bold red sandstone bluff on which the upper part of the
town is built. The situation of the church is very fine,
and an extensive view of the beautiful vale of the Severn
is obtained from it. Telford's design is by no means
striking; " being," as he said, " a regular Tuscan elevation; the inside is as regularly Ionic : its only merit is
simplicity and uniformity; it is surmounted by a Doric
tower, which contains the bells and a clock." A graceful
Gothic church would have been more appropriate to the
situation, and a much finer object in the landscape ; but
Gothic was not then in fashion—only a mongrel mixture
of many styles, without regard to either purity or grace-
fulness. The church, however, proved comfortable and

---

[1] Telford's ' Autobiography,' p. 3.

commodious, and these were doubtless the points to which the architect paid most attention.

His completion of the church at Bridgenorth to the satisfaction of the inhabitants brought Telford a commission, in the following year, to erect a similar edifice at Coalbrookdale. But in the mean time, to enlarge his knowledge and increase his acquaintance with the best forms of architecture, he determined to make a journey to London and through some of the principal towns of the south of England. He accordingly visited Gloucester, Worcester, and Bath, remaining several days in the last-mentioned city. He was charmed beyond expression by his journey through the manufacturing districts of Gloucestershire. The whole seemed to him a smiling scene of prosperous industry and middle class comfort. But passing out of this " Paradise," as he styles it, another stage brought him into a region the very opposite. " We stopped," says he, " at a little alehouse on the side of a rough hill to water the horses, and lo! the place was full of drunken blackguards, bellowing out ' Church and King!' A poor ragged German Jew happened to come up, whom those furious loyalists had set upon and accused of being a Frenchman in disguise. He protested that he was only a poor German who ' cut de corns,' and that all he wanted was to buy a little bread and cheese. Nothing would serve them but that he must be carried before the Justice. The great brawny fellow of a landlord swore he should have nothing in his house, and, being a constable, told him that he would carry him to gaol. I interfered, and endeavoured to pacify the assailants of the poor man; when suddenly the landlord, snatching up a long knife, sliced off about a pound of raw bacon from a ham which hung overhead, and, presenting it to the Jew, swore that if he did not swallow it down at once he should not be allowed to go. The man was in a worse plight than

ever. He said he was a 'poor Shoe,' and durst not
eat that. In the midst of the uproar Church and King
were forgotten, and eventually I prevailed upon the
landlord to accept from me as much as enabled poor
little Moses to get his meal of bread and cheese; and
by the time the coach started they seemed all perfectly
reconciled." [1]

Telford was much gratified by his visit to Bath, and
inspected its fine buildings with admiration. But he
thought that Mr. Wood, who, he says, "created modern
Bath," had left no worthy successor. In the buildings
then in progress he saw clumsy designers at work, "blun-
dering round about a meaning"—if there was meaning
at all in their designs, which, indeed, he failed to see.
From Bath he went to London by coach, making the
journey in safety, "although," he says, "the collectors
had been doing duty on Hounslow Heath." During
his stay in London he carefully examined the principal public buildings by the light of the experience
which he had gained since he last saw them. He also
spent a good deal of his time in studying rare and
expensive works on architecture—the use of which he
could not elsewhere procure—at the libraries of the
Antiquarian Society and the British Museum. There
he perused the various editions of Vitruvius and Pal-
ladio, as well as Inigo Jones's 'Parentalia.' He
found a rich store of ancient architectural remains in
the British Museum, which he studied with great care:
antiquities from Athens, Baalbec, Palmyra, and Her-
culaneum; "so that," he says, "what with the information I was before possessed of, and that which I have
now accumulated, I think I have obtained a tolerably
good general notion of architecture."

From London he proceeded to Oxford, where he

---

[1] Letter to Mr. Andrew Little, Langholm, dated Shrewsbury, 10th March,
1793.

carefully inspected its colleges and churches, afterwards
expressing the great delight and profit which he had
derived from his visit. He was entertained while there
by Mr. Robertson, an eminent mathematician, then
superintending the publication of an edition of the
works of Archimedes. The architectural designs of
buildings that most pleased him were those of Dr.
Aldrich, Dean of Christchurch about the time of Sir
Christopher Wren. He tore himself from Oxford with
great regret, proceeding by Birmingham on his way
home to Shrewsbury. "Birmingham," he says, "famous
for its buttons and locks, its ignorance and barbarism—
its prosperity increases with the corruption of taste and
morals. Its nicknacks, hardware, and gilt gimcracks
are proofs of the former; and its locks and bars,
and the recent barbarous conduct of its populace,[1] are
evidences of the latter." His principal object in visiting the place was to call upon a stained glass maker
respecting a window for the new church at Bridgenorth.

On his return to Shrewsbury, Telford proposed to
proceed with his favourite study of architecture; but
this, said he, "will probably be very slowly, as I must
attend to my every day employment," namely, the
superintendence of the county road and bridge repairs,
and the direction of the convicts' labour. "If I keep
my health, however," he added, "and have no unforeseen hindrance, it shall not be forgotten, but will be
creeping on by degrees." An unforeseen circumstance,
though not a hindrance, did very shortly occur, which
launched Telford upon a new career, for which his unremitting study, as well as his carefully improved experience, eminently fitted him : we refer to his appointment
as engineer to the Ellesmere Canal Company.

---

[1] Referring to the burning of Dr. Priestley's library.

The conscientious carefulness with which Telford had performed the duties entrusted to him, and the skill with which he directed the works placed under his charge, secured the cordial approbation of the gentlemen of the county. His straightforward and outspoken manner had further obtained for him the friendship of many of them. At the meetings of quarter-sessions his. plans had often to encounter considerable opposition, and, when called upon to defend them, he did so with such firmness, persuasiveness, and good temper, that he usually carried his point. " Some of the magistrates are ignorant," he wrote in 1789, " and some are obstinate ; though I must say that on the whole there is a very respectable bench, and with the sensible part I believe I am on good terms." This was amply proved some four years later, when it became necessary to appoint an engineer to the Ellesmere Canal, on which occasion the magistrates, who were mainly the promoters of that undertaking, almost unanimously solicited their Surveyor to accept the office.

Indeed, Telford had become a general favourite in the county. He was cheerful and cordial in his manner, though somewhat brusque. Though now thirty-five years old, he had not lost the humorousness which had procured for him the sobriquet of " Laughing Tam." He laughed at his own jokes as well as at others. He was spoken of as jolly—a word then much more rarely as well as more choicely used than it is now. Yet he had a manly spirit, and was very jealous of his independence. All this made him none the less liked by free-minded men. Speaking. of the friendly support which he had throughout received from Mr. Pulteney, he said, " His good opinion has always been a great satisfaction to me ; and the more so, as it has neither been obtained nor preserved by deceit, cringing, nor flattery. On the contrary, I believe I am almost the only man that speaks out fairly to him, and who contradicts him the most. In fact, between us, we sometimes quarrel like tinkers; but I hold my ground, and when he sees I am right he quietly gives in."

Although Mr. Pulteney's influence had no doubt assisted Telford in obtaining the appointment of surveyor, it had nothing to do with the unsolicited invitation which now emanated from the county gentlemen. Telford was not even a candidate for the engineership, and had not dreamt of offering himself, so that the proposal came upon him entirely by surprise. Though he admitted he had self-confidence, he frankly confessed that he had not a sufficient amount of it to justify him in aspiring to the office of engineer to one of the most important undertakings of the day. The following is his own account of the circumstance :—

" My literary project[1] is at present at a stand, and may be retarded for some time to come, as I was last Monday appointed sole agent, architect, and engineer to the canal which is projected to join the Mersey, the Dee, and the Severn. It is the greatest work, I believe, now in hand in this kingdom, and will not be completed for many years to come. You will be surprised that I have not mentioned this to you before; but the fact is that I had no idea of any such appointment until an application was made to me by some of the leading gentlemen, and I was appointed, though many others had made much interest for the place. This will be a great and laborious undertaking, but the line which it opens is vast and noble; and coming as the appointment does in this honourable way, I thought it too great an opportunity to be neglected, especially as I have stipulated for, and been allowed, the privilege of carrying on my architectural profession. The work will re-

---

[1] The preparation of some translations from Buchanan which he had contemplated.

quire great labour and exertions, but it is worthy of them all." [1]

Telford's appointment was duly confirmed by the next general meeting of the shareholders of the Ellesmere Canal. An attempt was made to get up a party against him, but it failed. "I am fortunate," he said, "in being on good terms with most of the leading men, both of property and abilities; and on this occasion I had the decided support of the great John Wilkinson, king of the ironmasters, himself a host. I travelled in his carriage to the meeting, and found him much disposed to be friendly." [2]   The salary at which Telford

---

[1] Letter to Mr. Andrew Little, Langholm, dated Shrewsbury, 29th September, 1793.

[2] John Wilkinson was the first of the great ironmasters, of whom there are now so many. His father was a small farmer in Cumberland, who afterwards went to work at the iron furnace and forge at Blackbarrow, in Cartmel, where he became foreman of the works. His two sons, John and William, were employed there. The first iron furnace and forge erected by John was at a place called Wilson House, in the same neighbourhood, for the purpose of smelting the hæmatite iron ore of Furness. The patent which he took out for the manufacture of box smoothing irons proved very successful, and he gradually extended his operations. The two brothers erected iron forges at Bersham, near Chester, at Bradley, Brimbo, Merthyr Tydvil, and other places; and became by far the largest iron manufacturers of their day. John was particularly celebrated for his castings and borings. Watt never could get cylinders cast true for his condensing steam-engines until Wilkinson supplied them. He himself invented and introduced a new boring machine, since become common, which was a great improvement on that formerly in use. The Wilkinsons cast the whole of the tubes, pipes, cylinders, and ironwork required for the great Paris waterworks—the most formid-

able undertaking of the kind at that day. The *first iron vessel ever built* was erected by them at Willey in South Wales, and traded upon the Severn before the year 1790. John Wilkinson retired late in life to his native county with a handsome fortune, and he built the mansion of Castlehead, in the parish of Cartmel, where he died. When the brothers gave up business, labourers were em-

ployed to break up the machinery at the Welsh works with sledge-hammers, in order that the materials might be equally divided between them. Many thought this an exceedingly insane act; yet it was not entirely so. Both were extremely stubborn men, and knew each other's temper; and perhaps they concluded that, though sledge-hammers might be very destructive when wielded by labourers amongst their fine machinery, the corrosive though more tedious process of a Chancery suit, managed by skilful lawyers, might be still more damaging

was engaged was 500*l*. a year, out of which he had to pay one clerk and one confidential foreman, besides defraying his own travelling expenses. It would not appear that after making these disbursements much would remain for Telford's own labour; but in those days engineers were satisfied with comparatively small pay, and did not dream of making large fortunes. Though he intended to continue his architectural business, he decided to give up his county surveyorship and other minor matters, which, he said, "give a great deal of very unpleasant labour for very little profit; in short they are like the calls of a country surgeon." One part of his former business which he did not give up was what related to Mr. Pulteney and Lady Bath, with whom he continued on intimate and friendly terms. He incidentally mentions in a letter a graceful and charming act of her Ladyship. On going into his room one day he found that, before setting out for Buxton, she had left upon his table a copy of Ferguson's 'Roman Republic,' in three quarto volumes, superbly bound and gilt.

He now looked forward with anxiety to the commencement of the canal, the execution of which would necessarily call for great exertion on his part, as well as unremitting attention and industry; "for," said he,

---

to the interests of both; so the machinery was all broken up. John had great faith in iron, and in its applicability to nearly every purpose for which durable material was required. Having made his fortune by its manufacture, he determined that his body should lie encased by his favourite metal when he died. In his will he directed that he should be buried in his garden in an iron coffin with an iron monument over him of twenty tons weight; and he was so buried within thirty yards of his mansion at Castlehead. He had the coffin made long before his death, and used to take pleasure in showing it to his visitors, very much to the horror of many of them. He would also offer a present of an iron coffin to any one who might desire to possess one. When he came to be placed in his narrow bed, it was found that the coffin he had provided was too small, so he was temporarily interred until another could be made. When placed in the ground a second time, the coffin was found to be too near the surface; accordingly it was taken up, and an excavation cut in the rock, after which it was buried a third time; and on the Castlehead estates being sold in 1828, the family directed the coffin again to be taken up and removed to the neighbouring chapel yard of Lindale, where it now lies. A man is still living at the latter place who assisted at all the four interments.

" besides the actual labour which necessarily attends so extensive a public work, there are contentions, jealousies, and prejudices, stationed like gloomy sentinels from one extremity of the line to the other.   But, as I have heard my mother say that an honest man might look the Devil in the face without being afraid, so we must just trudge along in the old way." [1]

---

[1] Letter to Mr. Andrew Little, Langholm, dated Shrewsbury, 3rd November, 1793.

ST. MARY MAGDALEN BRIDGENORTH.

[By Percival Skelton]

CHAPTER VI.

THE ELLESMERE CANAL.

THE ELLESMERE CANAL consists of a series of navigations proceeding from the river Dee in the vale of Llangollen. One branch passes northward, near the towns of Ellesmere, Whitchurch, Nantwich, and the city of

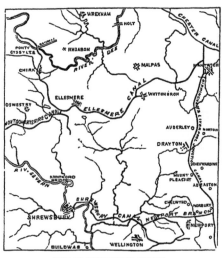

MAP OF ELLESMERE CANAL.

Chester, to Ellesmere Port on the Mersey; another, in a southeasterly direction, through the middle of Shropshire towards Shrewsbury on the Severn; and a third, in a south-westerly direction, by the town of Oswestry, to the Montgomeryshire Canal near Llanymynech; its whole extent, including the Chester Canal, incorporated with it, being about 112 miles.

The success of the Duke of Bridgewater's Canal had awakened the attention of the landowners throughout England, but more especially in the districts immediately adjacent to the scene of the Duke's operations, as they saw with their own eyes the extraordinary benefits which had followed the opening up of the navigations. The resistance of the landed gentry, which many of these schemes had originally to encounter, had now completely given way, and, instead of opposing canals, they were everywhere found anxious for their construction. The navigations brought lime, coal, manure, and merchandise almost to the farmers' doors, and provided them at the same time with ready means of conveyance for their produce to good markets. Farms in remote situations were thus placed more on an equality with those in the neighbourhood of large towns; rents rose in consequence, and the owners of land everywhere became the advocates and projectors of canals. The dividends paid by the first companies were also very high, and it was well known that the Duke's property was bringing him in immense wealth. There was, therefore, no difficulty in getting the shares in new projects readily subscribed for: indeed Mr. Telford relates that at the first meeting of the Ellesmere projectors, so eager were the public, that four times the estimated expense was subscribed without hesitation. Yet this navigation passed through a difficult country, necessarily involving very costly works; and as the district was but thinly inhabited, it did not present a very inviting prospect of dividends.[1] But the mania had fairly set in, and it was determined that the canal should be made. And whether the investment repaid the immediate proprietors or not, it unquestionably proved of immense advantage to the population of the districts through which it passed, and contributed to enhance the value of most of the adjoining property.

The Act authorising the construction of the canal was obtained in 1793, and Telford commenced operations very shortly after his appointment in October of the same year. His first business was to go carefully over the whole of the proposed line, and make a careful working survey, settling the levels of the different lengths, and the position of the locks, embankments, cuttings, and aqueducts. In all matters of masonry work he felt

---

[1] The Ellesmere Canal now pays about 4 per cent. dividend.

himself master of the necessary details; but having had comparatively small experience of earthwork, and none of canal-making, he determined to take the advice of Mr. William Jessop on that part of the subject; and he cordially acknowledges the obligations he was under to that eminent engineer for the kind assistance which he received from him on many occasions.

The heaviest and most important part of the works occurred in carrying the canal through the rugged hill country between the rivers Dee and Ceriog, in the vale of Llangollen. From Nantwich to Whitchurch the distance is 16 miles, and the rise 132 feet, involving nineteen locks; and from thence to Ellesmere, Chirk, Pont-Cysylltau, and the river Dee, 1¾ mile above Llangollen, the distance is 38¼ miles, and the rise 13 feet, involving only two locks. The latter part of the undertaking presented the greatest difficulties; as, in order to avoid the expense of constructing numerous locks, which would also involve serious delay and heavy expense in working the navigation, it became necessary to contrive means for carrying the canal on the same level from one side of the respective valleys of the Dee and the Ceriog to the other, and hence the magnificent aqueducts of Chirk and Pont-Cysylltau, characterised by Phillips as "among the boldest efforts of human invention in modern times." [1]

The Chirk Aqueduct carries the canal across the valley of the Ceriog, between Chirk Castle and the village of that name. At this point the valley is above 700 feet wide; the banks are steep, with a flat alluvial meadow between them, through which the river flows. The country is finely wooded. Chirk Castle stands on an eminence on its western side, with the Welsh mountains and Glen Ceriog as a background; the whole composing

---

[1] 'A General History of Inland Navigation, Foreign and Domestic,' &c. By J. Phillips. Fourth edition. London, 1803.

a landscape of great beauty, in the centre of which Telford's aqueduct forms a highly picturesque object.

CHIRK AQUEDUCT.

[By Percival Skelton, after his original Drawing]

The aqueduct consists of ten arches of 40 feet span each. The level of the water in the canal is 65 feet above the meadow, and 70 feet above the level of the river Ceriog. The proportions of this work far exceeded everything of the kind that had up to that time been attempted in England. It was a very costly structure ; but Telford, like Brindley, thought it better to incur a considerable capital outlay in maintaining the uniform level of the canal, than to raise and lower it up and down the sides of the valley by locks at a heavy expense in works, and a still greater cost in time and water. The aqueduct is a splendid specimen of the finest class of masonry, and Telford

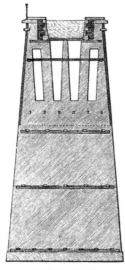

showed himself a master of his profession by the manner in which he carried out the whole details of the undertaking. The piers were carried up solid to a certain height, above which they were built hollow, with cross walls. The spandrels also, above the springing of the arches, were constructed with longitudinal walls, and left hollow.[1] The first stone was laid on the 17th of June, 1796, and the work was completed in the year 1801; the whole remaining in a perfect state to this day.

SECTION OF PIER.

The other great aqueduct on the Ellesmere Canal, named Pont-Cysylltau, is of even

---

[1] Telford himself thus modestly describes the merit of this original contrivance : " Previously to this time such canal aqueducts had been uniformly made to retain the water necessary for navigation by means of puddled earth retained by masonry ; and in order to obtain sufficient breadth for this superstructure, the masonry of the piers, abutments, and arches was of massive strength ; and after all this expense, and every imaginable precaution, the frosts, by swelling the moist puddle, frequently created fissures, which burst the masonry, and suffered the water to escape —nay, sometimes actually threw down the aqueducts; instances of this kind having occurred even in the works of the justly celebrated Brindley. It was evident that the increased pressure of the puddled earth was the chief cause of such failures : I therefore had recourse to the following scheme in order to avoid using it. The spandrels of the stone arches were constructed with longitudinal walls, instead of being filled in with earth (as at Kirkcudbright Bridge), and across these the canal bottom was formed by cast iron plates at each side, infixed in square stone masonry. These bottom plates had flanches on their edges, and were secured by nuts and screws at every juncture. The sides of the canal were made waterproof by ashlar masonry, backed with hard burnt bricks laid in Parker's cement, on the outside of which was rubble stone work, like the rest of the aqueduct. The towing path had a thin bed of clay under the gravel, and its outer edge was protected by an iron railing. The width of the water-way is 11 feet ; of the masonry on each side, 5 feet 6 inches; and the depth of the water in the canal, 5 feet. By this mode of construction the quantity of masonry is much diminished, and the iron bottom plate forms a continuous tie, preventing the sidewalls from separation by lateral pressure of the contained water."—' Life of Telford,' p. 40.

greater dimensions, and a far more striking object in the landscape. Sir Walter Scott spoke of it to Southey as " the most impressive work of art he had ever seen." [1] It is situated about four miles to the north of Chirk, at the crossing of the Dee, in the romantic vale of Llangollen. The north bank of the river is very abrupt; but on the south side the acclivity is more gradual. The lowest part of the valley in which the river runs is 127 feet beneath the water-level of the canal; and it became a question with the engineer whether the valley was to be crossed, as originally intended, by locking down one side and up the other —which would have involved seven or eight locks on each side—or by carrying it directly across by means of an aqueduct. The execution of the proposed locks would have been very costly, and the working of them in carrying on the navigation would necessarily involve a great waste of water, which was a serious objection, inasmuch as the supply was estimated to be no more than sufficient to provide for the unavoidable lockage and leakage of the summit level. Hence Telford's opinion was strongly in favour of an aqueduct; but, as we have already seen in the case of that at Chirk, the height of the work was such as to render it impracticable to construct it in the usual manner, upon masonry piers and arches of sufficient breadth and strength to afford room for a puddled water-way, which would have been extremely hazardous as well as expensive. He was therefore under the necessity of contriving some more safe and economical method of procedure ; and he again resorted to the practice which he had adopted in the construction of the Chirk Aqueduct, but on a much more formidable scale.

It will be understood that many years elapsed between the period at which Telford was appointed engineer to

---

[1] Selections from the Letters of Southey, vol. iii., p. 502.

the Ellesmere Canal and the designing of these gigantic works. He had in the mean while been carefully gathering experience from a variety of similar undertakings on which he was employed, and bringing his observations of the strength of materials and the different forms of construction to bear upon the plans under his consideration for the great aqueducts of Chirk and Pont-Cysylltau. In 1795 he was appointed engineer to the Shrewsbury Canal, which extended from that town to the collieries and ironworks in the neighbourhood of the Wrekin, crossing the rivers Roden and Tern, and Ketley Brook, after which it joins the Dorrington and Shropshire Canals. Writing to his Eskdale friend, Telford said: "Although this canal is only eighteen miles long, yet there are many important works in its course—several locks, a tunnel about half a mile long, and two aqueducts. For the most considerable of these last, I have just recommended an aqueduct *of iron*. It has been approved, and will be executed under my direction, upon a principle entirely new, and which I am endeavouring to establish with regard to the application of iron." [1]

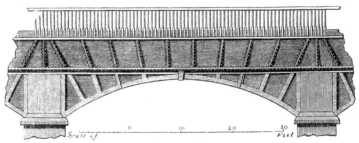

SIDE VIEW OF CAST IRON TROUGH.

It was the same principle which he applied to the great aqueducts of the Ellesmere Canal now under consideration. He had a model made of part of the proposed aqueduct for Pont-Cysylltau, showing the piers, ribs, towing-path, and side railing, with a cast iron

[1] Letter to Mr. Andrew Little, Langholm, dated Shrewsbury, 13th March, 1795.

trough for the canal. The model being approved, the design was completed; the ironwork was ordered for the summit, and the masonry of the piers then proceeded. The foundation-stone was laid on the 25th July, 1795, by Richard Myddelton, Esq., of Chirk Castle, M.P., and the work was not finished until the year 1803,—thus occupying a period of nearly eight years in construction.

The aqueduct is approached on the south side by an embankment 1500 feet in length, extending from the level of the water-way in the canal until its perpendicular height at the " tip " is 97 feet; thence it is carried to the opposite side of the valley, over the river Dee, upon piers supporting nineteen arches, extending for a length of 1007 feet. The height of the piers above the low water in the river is 121 feet. The lower part of each was built solid for 70 feet, all above being hollow, for the purpose of saving masonry as well as ensuring good workmanship. The outer walls of the hollow portion are only two feet thick, with cross inner walls. As each stone was exposed to inspection, and as both Telford and his confidential foreman, Matthew Davidson,[1] kept a vigilant eye upon the work, scamping was rendered impossible, and a first-rate piece of masonry was the result.

Upon the top of the masonry was set the cast iron trough for the canal, with its towing-path and side-rails, all accurately fitted and bolted together, forming a completely water-tight canal, with a water-way of 11 feet 10 inches, of which the towing-path, standing upon iron pillars rising from the bed of the çanal, occupied 4 feet 8 inches, leaving a space of 7 feet 2 inches for the boat.[2] The whole cost of this part of the canal

Section of Top of Pont-Cysylltau Aqueduct.

---

[1] Matthew Davidson had been Telford's fellow workman at Langholm, and was reckoned an excellent mason.

He died at Inverness, where he had a situation on the Caledonian Canal.

[2] Mr. Hughes, C.E., in his ' Me-

was 47,018*l.*, which was considered by Telford a moderate sum compared with what it must have cost if

VIEW OF PONT-CYSYLLTAU AQUEDUCT.

[By Percival Skelton, after his original Drawing]

moir of William Jessop,' published in 'Weale's Quarterly Papers on Engineering,' points out the bold and original idea here adopted, of constructing a water-tight trough of cast iron, in which the water of the canal was to be carried over the valleys, instead of an immense puddled trough, in accordance with the practice until that time in use; and he adds, "the immense importance of this improvement on the old practice is apt to be lost sight of at the present day by those who overlook the enormous size and strength of masonry which would have been required to support a puddled channel at the height of 120 feet." Mr. Hughes, however, claims for Mr. Jessop the merit of having suggested the employment of iron, though, in our opinion, without sufficient reason. Mr. Jessop was, no doubt, consulted by Mr. Telford on the subject; but the whole details of the

executed after the ordinary manner. The aqueduct was formally opened for traffic in 1805. "And thus," says Telford, "has been added a striking feature to the beautiful vale of Llangollen, where formerly was the fastness of Owen Glendower, but which, now cleared of its entangled woods, contains a useful line of intercourse between England and Ireland; and the water drawn from the once sacred Devon furnishes the means of distributing prosperity over the adjacent land of the Saxons." [1]

It is scarcely necessary to refer to the other works upon this canal, some of which were of considerable magnitude, though they may now appear dwarfed by comparison with the works of subsequent engineers. Thus, there were two difficult tunnels cut through hard rock, under the rugged ground which separates the valleys of the Dee and the Ceriog. One of these is 500 and the other 200 yards in length. To ensure a supply of water for the summit of the canal, the lake called Bala Pool was dammed up by a regulating weir, and by its means the water was drawn off at Llandisilio when required for the purposes of the navigation; the navigable feeder being six miles long, carried along the bank of the Llangollen valley. All these works were skilfully executed; and when the undertaking was finished, Mr. Telford may be said to have fairly established his reputation as an engineer of first-rate ability.

We now return to Telford's personal history during this important period of his career. He had long pro-

---

the design, as well as the suggestion of the use of iron (as admitted by Mr. Hughes himself), and the execution of the entire works, rested with the acting engineer. This is borne out by the report published by the Company immediately after the formal opening of the Canal in 1805, in which they state: "Having now detailed the particulars relative to the

Canal, and the circumstances of the concern, the committee, in concluding their report, think it but justice due to Mr. Telford to state that the works have been planned with great skill and science, and executed with much economy and stability, doing him, as well as those employed by him, infinite credit. (Signed) BRIDGEWATER."

[1] Life, p. 45.

mised himself a visit to his dear Eskdale, and the many
friends he had left there; but more especially to see his
infirm mother, who had descended far into the vale of
years, and longed to see her son once more before she
died. He had not forgotten her. He had printed many
letters to her, and had taken constant care that she should
want for nothing. *She* formed the burden of many of
his letters to Andrew Little. " Your kindness in visiting
and paying so much attention to her," said he, "is
doing me the greatest favour which you could possibly
confer upon me." He sends his friend money, which he
urges him to lay out in providing sundry little comforts
for his mother, who seems to have carried her spirit of
independence so far as to have expressed reluctance to
accept money even from her own son. " I must request,"
said he, " that you will purchase and send up what things
may be likely to be wanted, either for her or the person
who may be with her, as her habits of economy will
prevent her from getting plenty of everything, especially
as she thinks that I have to pay for it, which really hurts
me more than anything else."[1] He expresses an anxious
wish to see her; but, what with one urgent matter of
business and another, he fears it will be November before
he can set out. He has to prepare a general statement as
to the navigation affairs for a meeting of the committee;
he must attend the approaching Salop quarter sessions,
and after that a general meeting of the Canal Company;
so that his visit must be postponed for yet another month.
"Indeed," said he, " I am rather distressed at the thoughts
of running down to see a kind parent in the last stage of
decay, on whom I can only bestow an affectionate look,
and then leave her : her mind will not be much consoled
by this parting, and the impression left upon mine will
be more lasting than pleasant."[2]

---

[1] Letter to Mr. Andrew Little, Langholm, dated Shrewsbury, 16th Sept.,
1794. [2] Ibid.

He did, however, contrive to run down to Eskdale in the following November. His mother was alive, but that was all. After doing what he could for her comfort, and providing that all her little wants were properly attended to, he hastened back to his responsible duties in connection with the Ellesmere Canal. When at Langholm he called upon his former friends, and recounted with them the incidents of their youth. He was declared to be the same "canty" fellow as ever, and, though he had risen greatly in the world, he was "not a bit set up." He found one of his old fellow-workmen, Frank Beattie, become the principal innkeeper of the place. "What have you made of your mell and chisels?" asked Telford. "Oh!" replied Beattie, "they are all dispersed—perhaps lost." "I have taken better care of mine," said Telford; "I have them all locked up in a room at Shrewsbury, as well as my old working clothes and leather apron : you know one can never tell what may happen." He was surprised, as most people are who visit the scenes of their youth after a long absence, to see into what small dimensions Langholm had shrunk. That High Street, which before had seemed so big, and that frowning gaol and court-house in the Market Place, were now comparatively paltry to eyes that had been familiar with Shrewsbury, Portsmouth, and London. But he was charmed, as ever, with the sight of the heather hills and the narrow winding valley—

> Where deep and low the hamlets lie
> Beneath their little patch of sky,
> And little lot of stars.

On his return southward he was again delighted by the sight of old Gilnockie Castle and the surrounding scenery. As he afterwards wrote to his friend Little, "Broomholm was in all his glory." Probably one of the results of this visit was the revision of the poem of 'Eskdale,' which he undertook in the course of the following spring, putting in some fresh touches and adding many new lines, whereby the effect of the whole was considerably improved. He had the poem printed privately, merely for distribution amongst friends; "being careful," as he said, that "no copies should be smuggled and sold." Later in the year we find him, on his way to London on business, sparing a day or two for the purpose of visiting the Duke of Buckingham's palace and treasures of art at Stowe; afterwards writing out an eight-page description of it for the perusal of his friend at Langholm. At another time, when engaged upon the viaduct at Pont-Cysylltau, he snatched a few days' leisure to run through North Wales, of which he afterwards gave a glowing account to his correspondent. He passed by Cader Idris, Snowdon, and Penmaen Mawr. "Parts of the country we passed through," he says, "very much resemble the lofty green hills and woody vales of Eskdale. In other parts the magnificent boldness of the mountains, the torrents, lakes, and waterfalls, give a special character to the scenery, unlike everything of the kind I had before seen. The vale of Llanrwst is peculiarly beautiful and fertile. In this vale is the celebrated bridge of Inigo Jones;[1] but what is a much more delightful circumstance, the inhabitants of the vale are the most beautiful race of people I have ever beheld; and I am much astonished that this never seems to have struck the Welsh tourists. The vale of Llangollen is very fine, and not the least interesting object in it, I can assure you, is Davidson's[2] famous aqueduct, which is already reckoned amongst the wonders of Wales. Your old acquaintance thinks nothing of having three or four carriages at his door at a time."[3]

It seems that, besides attending to the construction of the works at this time, Telford had to organise the conduct of the navigation at those points at which the canal was open for traffic. By the middle of 1797 he

---

[1] See Vol. I., p. 252.
[2] See note [1], p. 347.
[3] Letter to Mr. Andrew Meikle, Langholm, dated Salop, 20th August, 1797.

states that twenty miles were in working condition, along which coal and lime were conveyed in considerable quantities, to the profit of the Company and the benefit of the public; the price of these articles having already in some places been reduced twenty-five, and in others as much as fifty, per cent. "The canal affairs," he says in one of his letters, "have required a good deal of exertion, though we are on the whole doing well. But, besides carrying on the works, it is now necessary to bestow considerable attention on the creating and guiding of a trade upon those portions which are executed. This involves various considerations, and many contending and sometimes clashing interests. In short, it is the working of a great machine : in the first place, to draw money out of the pockets of a numerous proprietary to make an expensive canal, and then to make the money return into their pockets by the creation of a business upon that canal."

But, as if all this business were not enough, he was occupied at the same time in writing a book upon the subject of Mills. In the year 1796 he had undertaken to draw up a paper on this topic for the Board of Agriculture, and by degrees it had grown into a large quarto volume, illustrated by upwards of thirty plates. He was also reading extensively in his few leisure moments; and among the solid works which he perused we find him mentioning Robertson's 'Disquisitions on Ancient India,' Stewart's 'Philosophy of the Human Mind,' and Alison's 'Principles of Taste.' But, as a relief from these graver studies, he seems, above all things, to have taken peculiar pleasure in occasionally throwing off a bit of poetry. Thus, when laid up at an hotel in Chester by a blow on his leg, which disabled him for some weeks, he employed part of his time in writing his 'Verses on hearing of the Death of Robert Burns.' On another occasion, when on his way to London, and detained for a night at Stratford-on-Avon, he occupied the evening at his inn in composing some stanzas, entitled 'An Address to the River Avon.' And when on his way back to Shrewsbury, while resting for the night at Bridgenorth, he amused himself with revising and copying out the verses for the perusal of Andrew Little. "There are worse employments," he said, "when one has an hour to spare from business;" and he asked his friend's opinion of the composition. It seems to have been no more favourable than the verses deserved; for, in his next letter, Telford says, "I think your observations respecting the verses to the Avon are correct. It is but seldom I have time to versify; but it is to me something like what a fiddle is to others. I apply to it in order to relieve my mind, after being much fatigued with close attention to business."

It is very pleasant to see the engineer relaxing himself in this way, and submitting cheerfully to unfavourable criticism, which is so trying to even the best of tempers. The time, however, thus taken from his regular work was not loss, but gain. Taking the character of his occupation into account, it was probably the best kind of relaxation he could have indulged in. With his head full of bridges and viaducts, he thus kept his heart open to the influences of beauty in life and nature; and, at all events, the writing of verses, indifferent though they might have been, proved of this value to him—that it cultivated in him the art of writing better prose.

CHAPTER VII.

Iron and other Bridges.

Shrewsbury being situated in the immediate neighbourhood of the Black Country, of which coal and
iron are the principal products, Telford's attention was
naturally directed, at a very early period, to the employment of cast iron in bridge-building. The strength as
well as lightness of a bridge of this material, compared
with one of stone and lime, is of great moment where
headway is of importance, or the difficulties of defective
foundations have to be encountered. The metal can be
moulded in such precise forms and so accurately fitted
together as to give to the arching the greatest possible
rigidity; whilst it defies the destructive influences of
time and atmospheric corrosion with nearly as much
certainty as stone itself.

The Italians and French, who took the lead- in engineering down almost to the end of last century, early
detected the value of this material, and made several
attempts to introduce it in bridge-building; but their
efforts proved unsuccessful, chiefly because of the in-
ability of the early founders to cast large masses of iron,
and also because the metal was then more expensive
than either stone or timber. The first actual attempt to
build a cast iron bridge was made at Lyons in 1755,
and it proceeded so far that one of the arches was put
together in the builder's yard; but the project was
abandoned as too costly, and timber was eventually
used.

It was reserved for English manufacturers to triumph

2 A 2

over the difficulties which had baffled the foreigners. Shortly after the above ineffectual attempt had been made, the construction of an iron bridge over the Severn near Broseley formed the subject of discussion among the iron-masters. It was proposed to substitute a bridge in place of the ferry which then connected the two banks of the river; and Mr. John Wilkinson, who had, as some thought, an extravagant, but, as results have proved, a truly prophetic, appreciation of the extensive uses to which iron might be applied, strongly urged that the structure should be of that material. Everybody knew of Mr. Wilkinson's hobby, and of his prognostication that the time would come when we should live in houses of iron, and even navigate the seas in ships of iron. When he insisted upon an iron bridge being built at Coalbrookdale, people said he was "iron-mad." But as he was a powerful man in his day—"the great iron-master" he was called—his suggestion could not be dismissed without consideration; and the Bridge Company, which had been formed, determined to take the opinion of Mr. Pritchard, of Shrewsbury, on the subject. That architect's opinion was favourable to the suggestion of the iron-master; and he was requested to supply a design of an iron bridge, which was eventually adopted. The work was erected, under contract, by Messrs. Reynolds and Darby, iron-masters at Coalbrookdale, in the year 1777. The bridge has only one semicircular arch of 100 feet span, each of the great ribs consisting of two pieces only. Though it was on the whole a bold design, and well executed, the error was committed of treating the arch as one of equilibrium. There also seems to have been some defect in the abutments, which were forced inwards by the pressure of the earth behind them, and the iron arch was thus partially fractured and raised in the middle. Nevertheless, the first cast iron bridge

THE FIRST IRON BRIDGE, COALBROOKDALE. [By E. M Wimperis.]

ever erected proved a very serviceable structure, and it
remains so to this day.

It is a curious circumstance that the next contriver of
an iron bridge was no other than the same Tom Paine
whose political writings Telford had so much admired.
While residing in America, Paine had proposed to build a
bridge of this material 400 feet in span over the Schuyl-
kill, and he came to England to take out a patent for
his invention,[1] and to order a bridge after his plan, the
materials of which were manufactured at the Rother-
ham Iron-works. They were delivered in London, and
fitted together on a bowling-green at Paddington. But
the French Revolution breaking out, Paine hastened to

---

[1] Specification of Patents, No. 1667, A.D. 1788.

Paris to join the "Friends of Man," leaving his bridge
in the hands of his creditors.   His democratic associates
having incarcerated him in the Luxembourg prison, he
lay there for eleven months, but finally escaped to
America.   In the mean time the materials had been purchased for erection over the river Wear, at Sunderland,
where the second iron bridge in England was erected,
after a design by Mr. T. Wilson, in the year 1796.   Mr.
Robert Stephenson has characterised this bridge as " a
structure which, as regards its proportions and the small
quantity of material employed in its construction, will
probably remain unrivalled." [1]   Indeed, it was long
regarded as the greatest triumph of the art.   Its span

WEAR BRIDGE, AT SUNDERLAND. [By Percival Skelton.]

---

[1] 'Encyclopedia Britannica.' Art.: Iron Bridges.

is 236 feet, with a rise of 34 feet, the springing commencing at 95 feet above the bed of the river ; and its height is such as to allow vessels of 300 tons burden to sail underneath without striking their masts.

The same year in which Paine's bridge was erected at Sunderland, Telford was building his first iron bridge over the Severn at Buildwas, at a point about midway between Shrewsbury and Bridgenorth. An unusually high flood having swept away the old bridge in the year 1795, Telford was called upon, as surveyor for the county, to supply the plan of a new one. Being on intimate terms with Mr. Wilkinson, who had strongly supported his appointment as engineer to the Ellesmere Canal, he was well aware of his sanguine views as to the capabilities of iron. He had also carefully examined the bridge at Coalbrookdale ; and while he had noted the defects in its construction, he fully appreciated its remarkable merits. The conclusion at which he arrived was, to build the proposed bridge at Buildwas of this material; and as the waters came down with great suddenness and fury from the Welsh mountains, he determined to construct it of only one arch, so as to afford the largest possible water-way. He had some difficulty in inducing the Coalbrookdale iron-masters, who undertook the casting of the material, to depart from the plan of the earlier structure ; but he persisted in his own design, which was eventually carried out. It consisted of a single arch of 130 feet span, the segment of a very large circle, calculated to resist that tendency of the abutments to slide inwards which had been the defect of the Coalbrookdale bridge ; the flat arch being itself sustained and strengthened by an outer ribbed one on each side, springing lower than the former and also rising higher, somewhat after the manner of timber-trussing. Although the span of the new bridge was 30 feet wider than the Coalbrookdale bridge, it contained less than half the quantity of iron; Buildwas bridge containing 173, whereas the other contained 378 tons. The new structure was, besides, extremely elegant in form; and when the centres were struck, the arch and abutments stood perfectly firm, and have remained so until this day. But the ingenious yet simple design of this bridge will be better explained by the following representation than by any description in words.[1]

BUILDWAS BRIDGE   [By Percival Skelton.]

This, however, had not been Telford's first employment of iron in bridge-building; for we found him writing to his friend at Langholm—the year before Paine's arching was erected at Sunderland and his own at Buildwas—that he had recommended an iron aqueduct for the Shrewsbury Canal, "on a principle entirely new," and which he was "endeavouring to establish with regard to the

---

[1] Mr. Telford gives the following further details: "Each of the main ribs of the flat arch consists of three pieces, and at each junction they are secured by a grated plate, which connects all the parallel ribs together into one frame. The back of each abutment is in a wedge-shape, so as to throw off laterally much of the pressure of the earth. Under the bridge is a towing path on each side of the river. The bridge was cast in an admirable manner by the Coalbrookdale iron-masters in the year 1796, under contract with the county magistrates. The total cost was 6034l. 13s. 3d."—'Life,' p. 30.

application of iron." [1] This iron aqueduct had now been cast and fixed; and it was found to effect so great a saving in masonry and earthwork, that he was afterwards induced to apply the same principle, as we have already seen, in different forms, in the magnificent aqueducts of Chirk and Pont-Cysylltau.

The uses of cast iron in canal construction became more obvious with every year's successive experience; and Telford was accustomed to introduce it in many cases where formerly only timber or stone had been employed. On the Ellesmere, and afterwards on the Caledonian Canal, he introduced cast iron lock-gates, which were found to answer admirably, being more durable than timber, and not liable like it to shrink and expand with alternate dryness and wet. The turnbridges which he introduced upon his canals, instead of the old drawbridges, were also of cast iron; and in some cases even the locks themselves were of the same material. Thus, on a part of the Ellesmere Canal opposite Beeston Castle, in Cheshire, where a couple of locks, together rising 17 feet, having been built on a stratum of quicksand, were repeatedly undermined, the idea of constructing the entire locks of cast iron was suggested; and this extraordinary application of the new material was successfully accomplished, with entirely satisfactory results.

But Telford's principal application of cast iron was in the construction of road bridges, in which he proved himself a master. His experience in these structures had now become very extensive. During the time that he held the office of surveyor to the county of Salop, he erected no fewer than forty-two, five of which were of iron. Indeed, his success in iron bridge-building so much emboldened him, that in 1801, when Old London Bridge had become so rickety and inconvenient that it was found necessary to take steps to

---

[1] Letter to Mr. Andrew Little, Langholm, dated Shrewsbury, 18th March, 1795.

rebuild or remove it, he proposed the daring plan of a
cast iron bridge of a single arch of not less than 600 feet
span, the segment of a circle 1450 feet in diameter.   In
preparing this design we find that he was associated with
a Mr. Douglas, to whom many allusions are made in his
private letters.[1]   The design of this bridge seems to have
arisen out of a larger subject—the improvement of the

---

[1] Douglas was first mentioned to
Telford, in a letter from Mr. Pasley,
as a young man, a native of Big-
holmes, Eskdale, who had, after serving his time there as a mechanic,
emigrated to America, where he
showed such proofs of mechanical
genius that he attracted the notice
of Mr. Liston, the British Minister,
who paid his expenses home to
England, that his services might not
be lost to his country, and at the
same time gave him a letter of introduction to the Society of Arts in
London. Telford, in a letter to Andrew Little, dated 4th December,
1797, expresses a desire "to know
more of this Eskdale Archimedes."
Shortly after, we find Douglas mentioned as having invented a brick-
machine, a shearing-machine, and a
ball for destroying the rigging of
ships; for the two former of which he
secured patents. Telford having by
this time got rid of his revolutionary
sympathies, expressed the hope that
Douglas's destructive ball might prove
of service against the French war-
ships. In a subsequent letter, dated
Salop, 6th March, 1798, he says:
"We are subscribing most liberally
here to oppose these terrible French-
men, who seem to want more worlds
to subdue, not considering how difficult it is to govern what they already
possess. I have already disapproved
of our meddling with them at all
with regard to their own affairs; but
I am equally averse to their settling
ours. They prove rather troublesome
neighbours, and very much assume
the attitude of our old sturdy beggars.
Upon the whole, I should feel well
satisfied if Douglas would destroy
their floating camps and flat-bottomed

boats." In 1799 Telford says he is
employing Douglas to do some work
for him in connection with the waterworks. In 1802 Douglas seems to
have suddenly absconded, and Telford
writes his friend at Langholm :—
"James Douglas has played us a
pliskey. However, by good manage-
ment, I think there will not be any
great loss—but no thanks to him.
Where he is gone, I know not."
Telford heard nothing more of his
coadjutor who had thus left him
in the lurch until the year 1815,
when he thus writes to Mr. William
Little, postmaster at Langholm:—
"At Inverness," says he, "I was surprised at receiving a letter from James
Douglas, announcing his successful
career in France. I am not surprised
at it: he was peculiarly fitted for that
country. I had never heard of him
since he left Mr. Pasley and myself
in the lurch, which action, with good
reason, soured the mind of that most
excellent of men even to me; although
it was through him I was forced into
the connection, and my injury was
far greater than his. But of this no
more. I wish Douglas may *deserve*
the success which his natural talents
merit, although they might have been
exercised in a more respectable manner. But he was always too impatient for distinction and wealth; in
the race for which, in this country,
he found too many able competitors."
It appears that Douglas introduced
in France machinery for the improved
manufacture of woollen cloth, and,
being patronised by the Government,
he succeeded in realising considerable
wealth, which, however, he did not
live to enjoy.

port of London.   In a private letter of Telford's, dated
the 13th May, 1800, he says :—

" I have twice attended the Select Committee on the
Port of London, Lord Hawkesbury chairman. The subject
has now been agitated for four years, and might have
been so for as many more, if Mr. Pitt had not taken the
business out of the hands of the General Committee and
got it referred to a Select Committee.   Last year they
recommended that a system of docks should be formed
in a large bend of the river opposite Greenwich, called
the Isle of Dogs, with a canal across the neck of the bend.
This part of the contemplated improvements is already
commenced, and is proceeding as rapidly as the nature
of the work will admit.   It will contain ship docks for
large vessels, such as East and West Indiamen, whose
draught of water is considerable.

" There are now two other propositions under consideration.   One is to form another system of docks at
Wapping, and the other to take down London Bridge,
rebuild it of such dimensions as to admit of ships of 200
tons passing under it, and form a new pool for ships of
such burden between London and Blackfriars Bridges,
with a set of regular wharves on each side of the river.
This is with the view of saving lighterage and plunderage,
and bringing the great mass of commerce so much nearer
to the heart of the City.   This last part of the plan has
been taken up in a great measure from some statements
I made while in London last year, and I have been called
before the Committee to explain.   I had previously prepared a set of plans and estimates for the purpose of
showing how the idea might be carried out ; and thus a
considerable degree of interest has been excited on the
subject.   It is as yet, however, very uncertain how far
the plans will be carried out.   It is certainly a matter of
great national importance to render the port of London
as perfect as possible." [1]

---

[1] Letter to Mr. Andrew Little, Langholm, dated London, 13th May, 1800.

Later in the same year he writes that his plans and propositions have been approved and recommended to be carried out, and he expects to have the execution of them. "If they will provide the ways and means," says he, "and give me elbow-room, I see my way as plainly as mending the brig at the auld burn." In November, 1801, he states that his view of London Bridge, as proposed by him, has been published, and admitted to be one of the finest plates lately produced. On the 14th of April, 1802, he writes, "I have got into mighty favour with the Royal folks. I have received notes written by order of the King, the Prince of Wales, Duke of York, and Duke of Kent, about the bridge print, and in future it is to be dedicated to the King."

TELFORD'S PROPOSED ONE-ARCHED BRIDGE OVER THE THAMES.

The bridge in question was one of the boldest of Telford's designs. He proposed by his one arch to provide a clear headway of 65 feet above high water. The arch was to consist of seven cast iron ribs, in segments as large as possible, and they were to be connected by diagonal cross-bracing, disposed in such a manner that any part of the ribs and braces could be taken out and replaced without injury to the stability of the bridge or interruption to the traffic over it. The roadway was to be

45 feet wide in the centre, increasing to double that width at the abutments. It was to contain 6500 tons of iron, and the cost of the whole was to be 262,289*l*.

The originality of the design was greatly admired, though there were many who received with incredulity the proposal of bridging the Thames by a single arch, and it was sarcastically said of Telford that he might as well think of " setting the Thames on fire." Before any outlay was incurred in its construction, it was determined to submit the design to the most eminent scientific and practical men of the day ; after which evidence was taken at great length before a Select Committee which sat on the subject. Amongst those examined on the occasion were the venerable James Watt of Birmingham, Mr. John Rennie, Professor Hutton of Woolwich, Professors Playfair and Robison of Edinburgh, Mr. Jessop, Mr. Southern, and Dr. Maskelyne. Their evidence will still be found interesting as indicating the state at which constructive science had at that time arrived in England.[1] There was a considerable diversity of opinion amongst the witnesses, as might have been expected ; for experience was as yet very limited as to the resistance of cast iron to extension and compression. Some of them anticipated immense difficulty in casting pieces of metal of the necessary size and exactness, so as to secure that the radiated joints should be all straight and bearing. Others laid down certain ingenious theories of the arch, which did not quite square with the plan proposed by the engineer. But, as was candidly observed by Professor Playfair in concluding his report—" It is not from theoretical men that the most valuable information in such a case as the present is to be expected. When a mechanical arrangement becomes in a certain degree complicated, it baffles the efforts of the geometer, and refuses to submit to even the most approved methods

---

[1] The evidence is fairly set forth in ' Cresy's Encyclopedia of Civil Engineering,' p. 475.

of investigation. This holds good particularly of bridges, where the principles of mechanics, aided by all the resources of the higher geometry, have not yet gone further than to determine the equilibrium of a set of smooth wedges acting on one another by pressure only, and in such circumstances as, except in a philosophical experiment, can hardly ever be realised. It is, therefore, from men educated in the school of daily practice and experience, and who to a knowledge of general principles have added, from the habits of their profession, a certain feeling of the justness or insufficiency of any mechanical contrivance, that the soundest opinions on a matter of this kind can be obtained."

It would appear that the Committee came to the general conclusion that the construction of the proposed bridge was practicable and safe; for the river was contracted to the requisite width, and the preliminary works were actually begun. Mr. Stephenson says the design was abandoned, owing more immediately to the difficulty of constructing the approaches with such a headway, which would have involved the formation of extensive inclined planes from the adjoining streets, and thereby led to serious inconvenience, and depreciation of much valuable property on both sides of the river.[1] From that time we hear no more of the subject in Telford's private letters.

Besides his iron bridges, Telford was busily engaged, at this early period of his career, in designing and erecting bridges of stone of considerable magnitude and importance. In the spring of 1795 a thaw, after a long continued fall of snow, caused a sudden flood to sweep down the Severn, which carried away many bridges—amongst others one at Bewdley, in Worcestershire. Telford was called upon to supply a design for a new structure. He was very busily occupied on similar works

---

[1] Article on Iron Bridges, in the 'Encyclopedia Britannica.' Edinburgh, 1857.

about the same time—amongst others, in planning a new bridge for the town of Bridgenorth; "in short," he says, "I have been at it night and day." So uniform a success had heretofore attended the execution of his designs, that his reputation as a bridge-builder was universally acknowledged. "Last week," he says, "Davidson and I struck the centre of an arch of 76 feet span, and this is the third which has been thrown this summer, none of which have shrunk a quarter of an inch."

Bewdley Bridge is a handsome and substantial piece of masonry. The streets upon each side being on low ground, it was necessary to provide land arches at both ends for the passage of the flood waters; and as the Severn was navigable at the point crossed, it was considered necessary to provide much greater width in the river arches than had been

BEWDLEY BRIDGE. [By Percival Skelton.]

the case in the former structure. The arches were three in number—one of 60 feet span and two of 52 feet, the land arches being 9 feet span. The works were proceeded with and the bridge was completed during the summer of 1798, Telford writing to his friend, in December of that year—"We have had a remarkably dry summer and autumn; after that an early fall of snow and some frost, followed by rain. The drought of the summer was unfavourable to our canal working; but it has enabled us to raise Bewdley

Bridge as if by enchantment. We have thus built a magnificent bridge over the Severn in one season, which is no contemptible work for John Simpson [1] and your humble servant, amidst so many other great undertakings. John Simpson is a treasure—a man of great talents and integrity. I met with him here by chance, employed and recommended him, and he has now under his charge all the works of any magnitude in this great and rich district."

Another of our engineer's early stone bridges, which may be mentioned in this place, was erected by him in 1805, over the river Dee at Tongueland in the county of Kirkcudbright. It is a bold and picturesque bridge, situated in a lovely locality. The river is very deep at high water there, the tide rising no less than 20 feet;

TONGUELAND BRIDGE.

[By R. P. Leitch, after a Drawing by J. S Smiles.]

---

[1] His foreman of masons at Bewdley Bridge, and afterwards his assistant in numerous important works.

and as the banks were steep and rocky, he determined to bridge the stream by a single arch of 112 feet span. The rise being considerable, high wingwalls and deep spandrels were requisite; but the weight of the structure was much lightened by the expedient which he adopted of perforating the wings, and building a number of longitudinal walls in the spandrels, instead of filling them with earth or inferior masonry, as had until then been the usual practice. The ends of these walls, connected and steadied by the insertion of tee-stones, were built so as to abut against the back of the arch-stones and the cross walls of each abutment. Thus great strength as well as lightness was secured, and a very graceful and at the same time substantial bridge was provided for the accommodation of the district.[1]

In his letters written about this time, Telford seems to have been very full of employment, which required him to travel about a great deal. "I have become," said he, "a very wandering being, and am scarcely ever two days in one place, unless detained by business, which, however, occupies my time very completely." At another time he says, "I am tossed about like a tennis ball: the

---

[1] The work is thus described in Robert Chambers's 'Picture of Scotland':—" Opposite Compston there is a magnificent new bridge over the Dee. It consists of a single arch, the span of which is 112 feet; and it is built of vast blocks of freestone brought from the Isle of Arran. The cost of this work was somewhere about 7000*l.* sterling; and it may be mentioned, to the honour of the Stewartry, that this sum was raised by the private contributions of the gentlemen of the district. From Tongueland Hill, in the immediate vicinity of the bridge, there is a view well worthy of a painter's eye, and which is not inferior in beauty and magnificence to any in Scotland. The distant horizon is bounded by the everlasting ocean and the mountains of Man, which are seen in their whole extent. The centre of the picture is occupied by the high and rocky island of Little Bess, the woods of St. Mary's Isle, the town of Kirkcudbright, with its raised and castellated outline, and the windings of the Dee, which flows to the foot of the eminence upon which the spectator is supposed to stand. On the left, the Bay of Kirkcudbright and the course of the river are bounded by a long range of wooded hills, for the beauty of which we are mainly indebted to the patriotic improvements of Basil Lord Daer, the eldest brother of the late Earl of Selkirk. The right of the picture, again, is filled by a similar range of hills, extending along the west side of the bay, and terminating in the ornamented grounds of Compston, which are seen to the greatest advantage from Tongueland Hill."

other day I was in London, since that I have been in
Liverpool, and in a few days I expect to be in Bristol.
Such is my life ; and to tell you the truth, I think it suits
my disposition."

One of the projects on which he was engaged at this
time was a plan for supplying the town of Liverpool
with water, conveyed through pipes, in the same manner
as had long before been adopted in London. He was
much struck by the activity and enterprise apparent in
that town compared with Bristol. " Liverpool," he said,
" has taken firm root in the country by means of the
canals : it is young, vigorous, and well situated. Bristol
is sinking in commercial importance : its merchants are
rich and indolent, and in their projects they are always
too late. Besides, the place is badly situated. There
will probably arise another port somewhere near the
Severn ; but Liverpool will nevertheless continue of
the first commercial importance, and their water will be
turned into wine. We are making rapid progress in this
country—I mean from Liverpool to Bristol, and from
Wales to Birmingham. This is an extensive and rich
district, abounding in coal, lime, iron, and lead. Agriculture too is improving, and manufactures are advancing
at rapid strides towards perfection. Think of such a
mass of population, industrious, intelligent, and ener-
getic, in continual exertion ! In short, I do not believe
that any part of the world, of like dimensions, ever
exceeded Great Britain, as it now is, in regard to the
production of wealth and the practice of the useful
arts." [1]

At the same time, he thought there was a prospect
for Ireland too, amidst all this improvement. " There
is a board of five members, appointed by Parliament, to
act as a board of control over all the inland navigations,
&c., of Ireland. One of the members is a particular

---

[1] Letter to Mr. Andrew Little, Langholm, dated Salop, 13th July, 1799.

friend of mine, and at this moment a pupil, as it were, anxious for information. This is a noble object: the field is wide, the ground new and capable of vast improvement. To take up and manage the water of a fine island is like a fairy tale, and, if properly conducted, it would render Ireland truly a jewel among the nations." [1] It does not, however, appear that Telford was ever employed by the board to carry out the grand scheme which thus fired his engineering imagination.

Mixing freely with men of all classes, our engineer seems to have made many new friends and acquaintances about this time. Whilst on his journeys north and south, he frequently took the opportunity of looking in upon the venerable James Watt, at his house at Heathfield, near Birmingham,—"the steam-engine man from Glasgow, a great and good man," he terms him. At London, he says, he is "often with old Brodie and Black, each the first in his profession, though they walked up together to the great city on foot,[2] more than half a century ago—Gloria!" About the same time we find Telford taking interest in the projects of a deserving person, named Holwell, a coal-master in Staffordshire, and assisting him to take out a patent for boring wooden pipes; "he being a person," says Telford, "little known, and not having capital, interest, or connections, to bring the matter forward. I have had the machine at work for about a year, and it answers the purpose well."

He also kept up his literary friendships and preserved his love for poetical reading. At Shrewsbury, one of

---

[1] Letter to Mr. Andrew Little, Langholm, dated Liverpool, 9th September, 1800.

[2] Letter to Mr. Andrew Little, Langholm, dated Salopian Coffee House, Charing Cross, London, 26th February, 1799. Mr. Brodie was originally a blacksmith. He was a man of great ingenuity and industry, and introduced many improvements in iron work. He invented stoves for chimneys, ships' hearths, &c. He had above a hundred men working in his London shop, besides carrying on an iron work at Coalbrookdale. He afterwards established a woollen manufactory near Peebles.

his most intimate friends was Dr. Darwin, son of the
author of the 'Botanic Garden.' At Liverpool, he
made the acquaintance of Dr. Currie, and was favoured
with a sight of his manuscript of the 'Life of Burns,'
then in course of publication. Curiously enough, Dr.
Currie had found amongst Burns's papers a copy of
some verses, addressed to the poet, which Telford recognised as his own, written many years before whilst
working as a mason at Langholm. Their purport was
to urge Burns to devote himself to the composition of
poems of a serious character, such as the 'Cotter's
Saturday Night.' With Telford's permission, several
extracts from his Address to Burns were published in
1800 in Currie's Life of the poet. Another of his literary friendships, formed about the same time, was that
with Thomas Campbell, then a very young man, whose
'Pleasures of Hope' had just made its appearance.
Telford, in one of his letters, says, "I will not leave a
stone unturned to try to serve the author of that charming poem." In a subsequent communication [1] he says,
"The author of the 'Pleasures of Hope' has been here
for some time. I am quite delighted with him. He is
the very spirit of poetry. On Monday I introduced him
to the King's librarian, and I imagine some good may
result to him from the introduction."

In the midst of his plans of docks, canals, and bridges,
he writes letters about the peculiarities of Goethe's
poems and Kotzebue's plays, Buonaparte's campaign in
Egypt, and the merits of sundry new books recently
published. He confesses, however, that he has now
very little leisure for reading, but has purchased the
'Encyclopedia Britannica,' which he finds "a perfect
treasure, containing everything, and being always at
hand." He gives his correspondent an idea of the
manner in which his time is engrossed. "A few days

---

[1] Dated London, 14th April, 1802.

since, I attended a general assembly of the canal proprietors in Shropshire. I have to be at Chester again in a week, upon an arbitration business respecting the rebuilding of the county hall and gaol ; but previous to that I must visit Liverpool, and after that proceed into Worcestershire. So you see what sort of a life I have of it. It is something like Buonaparte, when in Italy, fighting battles at fifty or a hundred miles distance every other day. However, plenty of employment is what every professional man is seeking after, and my various occupations now require of me great exertions, which they certainly shall have so long as life and health are spared to me." [1]

Amidst all his engagements, Telford found time to make particular inquiry about many poor families formerly known to him in Eskdale, for some of whom he paid house-rent, whilst he transmitted the means of supplying others with coals, meal, and other necessaries, during the severe winter months,—a practice which he continued to the close of his life.

---

[1] Letter to Mr. Andrew Little, Langholm, dated Salop, 30th November, 1799.

CHAPTER VIII.

HIGHLAND ROADS AND BRIDGES.

IN our introduction to the Life of Rennie, we gave a
rapid survey of the state of Scotland about the middle
of last century. We found a country without roads, fields
lying uncultivated, mines unexplored, and all branches of
industry languishing, in the midst of an idle, miserable,
and haggard population. Fifty years passed, and the
state of the Lowlands had become completely changed.
Roads had been made, canals dug, coal-mines opened up,
iron works established; manufactures were extending in
all directions; and Scotch agriculture, instead of being the
worst, was admitted to be the best in the island.

"I have been perfectly astonished," wrote Romilly
from Stirling, in 1793, "at the richness and high cultivation of all the tract of this calumniated country
through which I have passed, and which extends quite
from Edinburgh to the mountains where I now am. It
is true, however, that almost everything which one sees
to admire in the way of cultivation is due to modern
improvements; and now and then one observes a few
acres of brown moss, contrasting admirably with the
corn-fields to which they are contiguous, and affording
a specimen of the dreariness and desolation which, only
half a century ago, overspread a country now highly
cultivated and become a most copious source of human
happiness."

It must, however, be admitted, that the industrial
progress to which we have above referred was confined

---

[1] Romilly's ' Autobiography.'

mainly to the Lowlands, and had scarcely penetrated the mountainous regions lying towards the north-west. The rugged nature of that part of the country interposed a formidable barrier to improvement, and the district had thus remained very imperfectly opened up. The only practicable roads were those which had been made by the soldiery after the rebellions of 1715 and '45, through counties which before had been inaccessible except by dangerous footpaths across high and rugged mountains. One was formed along the Great Glen of Scotland, in the line of the present Caledonian Canal, connected with the Lowlands by the road through Glencoe by Tyndrum down the western banks of Loch Lomond; another, more northerly, connected Fort Augustus with Dunkeld by Blair Athol; whilst a third, still further to the north and east, connected Fort George with Cupar-in-Angus by Badenoch and Braemar. These roads were about eight hundred miles in extent, and maintained at the public expense. But they were laid out for purposes of military occupation rather than for the convenience of the districts which they traversed. Hence they were comparatively little used, and the Highlanders, in passing from one place to another, for the most part continued to travel by the old cattle tracks along the mountain sides. But the population were so poor and so spiritless, and industry was as yet in so backward a state all over the Highlands, that the want of communication was little felt. The state of agriculture may be inferred from the fact that an instrument called the cas-chrom [1]—literally, the " crooked-foot "—

---

[1] The cas-chrom was a rude combination of a lever for the removal of rocks, a spade to cut the earth, and a foot-plough to turn it. We annex an illustration of this curious and now obsolete instrument. It weighed about eighteen pounds. In working it, the upper part of the handle, to which the left hand was applied, reached the workman's shoulder, and being slightly elevated, the point, shod with iron, was pushed into the ground horizontally; the soil being turned over by inclining the handle to the furrow side, at the same time making the heel act as a fulcrum to raise the point of the instrument. In turning up unbroken ground, it was first em-

the use of which had been forgotten for hundreds of years in every other country in Europe, was almost the only tool employed in tillage in those parts of the Highlands which were separated by impassable roads from the rest of the United Kingdom.   The country lying on the west of the Great Glen was absolutely without a road.   The native population were by necessity peaceful. Old feuds were restrained by the strong arm of the law, if indeed the spirit of the clans had not been completely broken by the upshot of the rebellion of Forty-five. But the people had not yet learnt to bend their backs, like the Sassenach, to the stubborn soil, and they sat gloomily by their turf-fires at home, or wandered away to settle in other lands beyond the seas.   It even began to be feared that the country would become entirely depopulated ; and it became a matter of national concern to devise methods of opening up the district so as to develope its industry and afford improved means of sustenance for its population.   The poverty of the inhabitants rendered the attempt to construct roads—even had they desired them—beyond their scanty means ; but the ministry of the day entertained the opinion that by contributing a certain proportion of the necessary expense, the proprietors of Highland estates might be induced to advance the remainder ; and on this principle the construction of the new roads in those districts was undertaken.

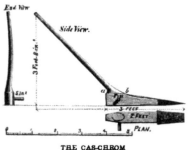

THE CAS-CHROM

ployed with the heel uppermost, with pushing strokes to cut the breadth of the sward to be turned over ; after which, it was used horizontally as above described.   We are indebted to a Parliamentary Blue Book for our representation of this interesting relic of ancient agriculture.   It is given in the appendix to the 'Ninth Report of the Commissioners for Highland Roads and Bridges,' ordered by the House of Commons to be printed, 19th April, 1821.

In 1802 Mr. Telford was called upon by the Government to make a survey of Scotland, and report as to the measures which were necessary for the improvement of the roads and bridges of that part of the kingdom, and also on the means of promoting the fisheries on the east and west coasts, with the object of better opening up the country and preventing further extensive emigration. Previous to this time he had been employed by the British Fisheries Society—of which his friend Sir William Pulteney was Governor—to inspect the harbours at their several stations, and to devise a plan for the establishment of a fishery on the coast of Caithness. He accordingly made an extensive tour of Scotland, examining, amongst other harbours, that of Annan; from which he proceeded northward by Aberdeen to Wick and Thurso, returning to Shrewsbury by Edinburgh and Dumfries.[1] He accumulated a large mass of data for his report, which was sent in to the Fishery Society, with charts and plans, in the course of the following year.

In July, 1802, he was requested by the Lords of the Treasury, most probably in consequence of the preceding report, to make a further survey of the interior of the Highlands, the result of which he communicated in his report presented to Parliament in the following year. Although full of important local business, " kept running," as he says, " from town to country, and from country to town, never when awake, and perhaps not always when asleep, have my Scotch surveys been absent from my mind." He had worked very hard at

---

[1] He was accompanied on this tour by Colonel Dixon, with whom he returned to his house at Mount Annan, in Dumfries. Telford says of him: " The Colonel seems to have roused the county of Dumfries from the lethargy in which it has slumbered for centuries. The map of the county, the mineralogical survey, the new roads, the opening of lime works, the competition of ploughing, the improving harbours, the building of bridges, are works which bespeak the exertions of no common man."—Letter to Mr. Andrew Little, dated Shrewsbury, 30th November, 1801.

his report, and hoped that it might be productive of some good.

The report was duly presented, printed,[1] and approved, and it formed the starting-point of a system of legislation with reference to the Highlands which extended over many years, and had the effect of completely opening up that romantic but rugged district of country, and extending to its inhabitants the advantages of improved intercourse with the other parts of the kingdom. Mr. Telford pointed out that the military roads which had been made there were altogether insufficient, and that the use of them was in many places very much circumscribed by the want of bridges over some of the principal rivers. For instance, the route from Edinburgh to Inverness, by the Central Highlands, was seriously interrupted at Dunkeld, where the Tay is broad and deep, and not always easy to be crossed by means of a boat. The route to the same place by the east coast was in like manner broken at Fochabers, where the rapid Spey could only be crossed by a dangerous ferry.

The difficulties encountered by the Bar, in travelling the north circuit about this time, are well described by Lord Cockburn in his 'Memorials.' "Those who are born to modern travelling," he says, "can scarcely be made to understand how the previous age got on. The state of the roads may be judged of from two or three facts. There was no bridge over the Tay at Dunkeld, or over the Spey at Fochabers, or over the Findhorn at Forres. Nothing but wretched pierless ferries, let to poor cottars, who rowed, or hauled, or pushed a crazy boat across, or more commonly got their wives to do it. There was no mail-coach north of Aberdeen till, I think, after the battle of Waterloo. What it must have been

---

[1] Ordered to be printed 5th of April, 1803.

a few years before my time may be judged of from
Bozzy's ' Letter to Lord Braxfield,' published in 1780.
He thinks that, besides a carriage and his own carriage-
horses, every judge ought to have his sumpter-horse,
and ought not to travel faster than the waggon which
carried the baggage of the circuit. I understood from
Hope that, after 1784, when he came to the Bar, he and
Braxfield *rode* a whole north circuit; and that, from the
Findhorn being in a flood, they were obliged to go up
its banks for about twenty-eight miles to the bridge of
Dulsie before they could cross. I myself rode circuits
when I was Advocate-Depute between 1807 and 1810.
The fashion of every Depute carrying his own shell on
his back, in the form of his own carriage, is a piece of
very modern antiquity." [1]

North of Inverness, matters were still worse, if possible. There was no bridge over the Beauley or the
Conan. The drovers coming south swam the rivers
with their cattle. There being no roads, there was little
use for carts. In the whole county of Caithness there
was scarcely a farmer who owned a wheel-cart. Burdens
were conveyed usually on the backs of ponies, but quite
as often on the backs of women.[2] The interior of the
county of Sutherland being almost inaccessible, the only
track lay along the shore, amongst rocks and sand,
covered by the sea at every tide. "The people lay
scattered in inaccessible straths and spots among the
mountains, where they lived in family with their pigs
and kyloes (cattle), in turf cabins of the most miserable
description; they spoke only Gaelic, and spent the whole
of their time in indolence and sloth. Thus they had
gone on from father to son, with little change, except
what the introduction of illicit distillation had wrought,
and making little or no export from the country beyond

---

[1] ' Memorials of his Time,' by
Henry Cockburn, pp. 341-3.
[2] ' Memoirs of the Life and Writings

of Sir John Sinclair, Bart.,' vol. i., p.
339.

the few lean kyloes, which paid the rent and produced
wherewithal to pay for the oatmeal imported." [1]

Telford's first recommendation was, that a bridge
should be thrown across the Tay at Dunkeld, to connect
the improved lines of road proposed to be made on each
side of the river. He regarded this measure as of the
first importance to the Central Highlands; and as the
Duke of Athol was willing to pay one-half of the cost of
the erection, if the Government would defray the other—
the bridge to be free of toll after a certain period—it appeared to the engineer that this was a reasonable and just
mode of providing for the contingency. In the next
place, he recommended a bridge over the Spey, which
drained a great extent of mountainous country, and,
being liable to sudden inundations, was very dangerous
to cross. Yet this ferry formed the only link of communication between the whole of the northern counties.
The site pointed out for the proposed bridge was adjacent to the town of Fochabers, and here also the Duke
of Gordon and other county gentlemen were willing to
provide one-half the means for its erection.

Mr. Telford further described in detail the roads
necessary to be constructed in the north and west Highlands, with the object of opening up the western parts
of the counties of Inverness and Ross, and affording a
ready communication from the Clyde to the fishing
lochs in the neighbourhood of the Isle of Skye. As to
the means of executing these improvements, he suggested
that Government would be justified in dealing with the
Highland roads and bridges as exceptional and extraordinary works, and extending the public aid towards
carrying them into effect, as but for such assistance the
country must remain, perhaps for ages to come, imperfectly connected. His report further embraced certain
improvements in the harbours of Aberdeen and Wick,

---

[1] Extract of a letter from a gentleman residing in Sutherland, quoted in
' Life of Telford,' p. 465.

and a description of the country through which the proposed line of the Caledonian Canal would necessarily pass—a canal which had long been the subject of inquiry, but had not as yet emerged from a state of mere speculation.

The new roads and bridges, and other improvements suggested by the engineer, excited much interest in the north.   The Highland Society voted him their thanks by acclamation ; the counties of Inverness and Ross followed ; and he had letters of thanks and congratulation from many of the Highland chiefs.   " If they will persevere," says he, " with anything like their present zeal, they will have the satisfaction of greatly improving a country that has been too long neglected.   Things are greatly changed now in the Highlands.   Even were the chiefs to quarrel, de'il a Highlandman would stir for them.   The lairds have transferred their affections from their people to flocks of sheep, and the people have lost their veneration for the lairds.   It seems to be the natural progress of society ; but it is not an altogether satisfactory change.   There were some fine features in the former patriarchal state of society ; but now clanship is gone, and chiefs and people are hastening into the opposite extreme.   This seems to me to be quite wrong." [1]

In the same year Telford was elected a member of the Royal Society of Edinburgh, on which occasion he was proposed and supported by three professors ; so that the former Edinburgh mason was rising in the world and receiving due honour in his own country.   The effect of his report was such, that in the session of 1803 a Parliamentary Commission was appointed, under whose direction a series of practical improvements was commenced, which issued in the construction of not less than 920 additional miles of roads and bridges throughout the Highlands, one-half of the cost of which was defrayed by

---

[1] Letter to Mr. Andrew Little, Langholm, dated Salop, 18th February, 1803.

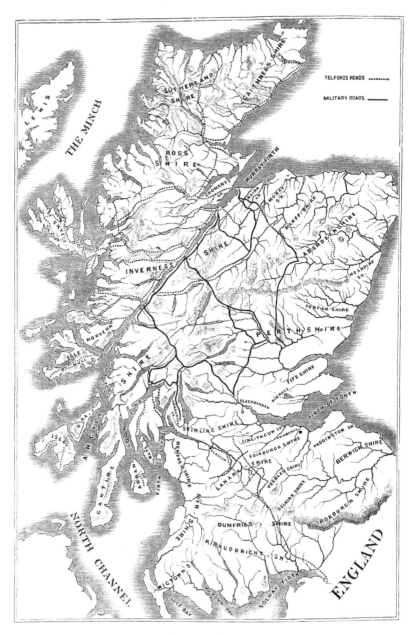

MAP OF TELFORD'S ROADS.

the Government and the other half by local assessment. But in addition to these main lines of communication, numberless county roads were formed by statute labour, under local road Acts and by other means; the land-owners of Sutherland alone having formed nearly 300 miles of district roads at their own cost.

By the end of the session of 1803 Telford received his instructions from Mr. Vansittart as to the survey he was forthwith to enter upon, with a view to commencing practical operations; and he again proceeded to the High-lands to lay out the roads and plan the bridges which were most urgently needed. The district of the Solway was, at his representation, included, with the object of improving the road from Carlisle to Portpatrick—the nearest point at which Great Britain meets the Irish coast, and where the sea passage forms only a sort of wide ferry.

It would occupy too much space, and indeed it is alto-gether unnecessary, to describe in detail the operations of the Commission and of their engineer in opening up the communications of the Highlands. Suffice it to say, that one of the first things taken in hand was the con-nection of the new lines of road by means of bridges at the more important points; such as at Dunkeld over the Tay, and near Dingwall over the Conan and Orrin. That at Dunkeld was the most important, as being the portal to the Central Highlands; and at the second meeting of the Commissioners Mr. Telford submitted his plan and estimates of the proposed bridge. In conse-quence of some difference with the Duke of Athol as to his share of the expense—which proved to be greater than he had taken into account—some delay occurred in the commencement of the work; but at length it was fairly begun, and after being three years in hand the structure was finished and opened for traffic in 1809.

The bridge is a handsome one of five river and two land arches. The span of the centre arch is 90 feet,

of the two adjoining it 84 feet, and of the two side
arches 74 feet; affording a clear waterway of 446 feet.
The total breadth of the roadway and footpaths is 28 feet
6 inches.   The cost of the structure was about 14,000*l.*,
one-half of which was defrayed by the Duke of Athol.
It forms a fine feature in a landscape not often surpassed, presenting within a comparatively small compass
a great variety of character and beauty.

DUNKELD BRIDGE.

[By Percival Skelton, after a sketch by J. S. Smiles.]

The communication by road north of Inverness was
also perfected by the construction of a bridge of five arches
over the Beauley, and·another of the same number over
the Conan, the central arch being 65 feet span; and the
formerly wretched bit of road between these points having
been put in good repair, the town of Dingwall was there-
upon rendered easily approachable from the south.   At
the same time a beginning was made with the construction of new roads through the districts most in need of
them.   The first contracted for was the Loch-na-Gaul

road, from Fort William to Arasaig, on the western coast,
nearly opposite the island of Egg. Another was begun
from Loch Oich, on the line of the Caledonian Canal,
across the middle of the Highlands, through Glengarry,
to Loch Hourn on the western sea. Other roads were
opened north and south; through Morvern to Loch Moi-
dart; through Glen Morrison and Glen Sheil, and through
the entire Isle of Skye; from Dingwall, eastward, to Loch-
carron and Loch Torridon, quite through the county of
Ross; and from Dingwall, northward, through the county
of Sutherland as far as Tongue on the Pentland Frith;
whilst another line, striking off at the head of the Dornoch
Frith, proceeded along the coast in a north-easterly direction to Wick and Thurso, in the immediate neighbourhood
of John o' Groats. There were numerous other subordi-
nate lines, which it is unnecessary to specify in detail;
but some idea may be formed of their extent, as well as
of the rugged character of the country through which
they were carried, when we state that they involved the
construction of no fewer than twelve hundred bridges.
Several important bridges were also erected at other
points to connect existing roads, such as those at Ballater
and Potarch over the Dee; at Alford over the Don; and
at Craig-Ellachie over the Spey.

The last-named bridge is a remarkably elegant structure, thrown over the Spey at a point where the river,
rushing obliquely against the lofty rock of Craig-Ellachie,[1]
has formed for itself a deep channel not exceeding fifty
yards in breadth. Only a few years before, there had not
been any provision for crossing this river at its lower
parts except the very dangerous ferry at Fochabers. The
Duke of Gordon had, however, erected a suspension bridge
at that town, and the inconvenience was in a great

---

[1] The names of Celtic places are
highly descriptive. Thus *Craig-Ellachie* literally means, the rock of separation; *Badenoch*, bushy or woody;
Cairngorm, the blue cairn; *Lochinet*,
the lake of nests; *Balknockan*, the
town of knolls; *Dalnasealg*, the hunt-
ing dale; *All'n dater*, the burn of
the horn-blower; and so on.

measure removed. Its utility was so generally felt, that the demand arose for a second bridge across the river; for there was not another by which it could be crossed for a distance of nearly fifty miles up Strath Spey. It was a difficult stream to span by a bridge at any place, in consequence of the violence with which the floods descended at particular seasons. Sometimes, even in summer, when not a drop of rain had fallen, the flood would come down the Strath in great fury, sweeping everything before it; this remarkable phenomenon being accounted for by the prevalence of a strong south-westerly wind, which blew the loch waters from their beds into the Strath, and thus suddenly filled the valley of the Spey.[1] The same phenomenon, similarly caused, is also frequently observed in the neighbouring river, the Findhorn, cooped up in its deep rocky bed, where the water sometimes comes down in a wave six feet high, like a liquid wall, sweeping everything before it. To meet such a contingency, it was deemed necessary to provide abundant waterway, and to build a bridge offering as little resistance as possible to the passage of the Highland floods. Telford accordingly designed for the passage of the river at Craig-Ellachie a light cast iron arch of 150 feet span, with a rise of 20 feet, the arch being composed of four ribs, each consisting of two concentric arcs forming panels, which are filled in with diagonal bars. The roadway is 15 feet wide, and is formed of another arc of greater radius, attached to which is the iron railing; the spandrels being filled by diagonal ties, forming trellis-work. Mr. Robert Stephenson took objection to the two dissimilar arches, as liable to subject the structure, from variations of temperature, to very unequal strains. Nevertheless this bridge, as well as many others constructed by Mr. Telford after a similar plan, has stood perfectly

[1] Sir Thomas Dick Lauder has vividly described the destructive character of the Spey-side inundations in his capital book on the 'Morayshire Floods.'

well, and to this day remains a very serviceable structure. Its appearance is highly picturesque. The scattered pines and beech trees on the side of the impending mountain, the meadows along the valley of the Spey, and the western approach road to the bridge cut deeply into the face of the rock, combine, with the slender appearance of the iron arch, in rendering this spot one of the most remarkable in Scotland.[1]

CRAIG-ELLACHIE BRIDGE. [By Percival Skelton]

An iron bridge of a similar span to that at Craig-Ellachie had previously been constructed across the head of the Dornoch Frith at Bonar, near the point where the waters of the Shin join the sea. The very severe trial which this structure sustained from the tremendous blow of an irregular mass of fir-tree logs,

[1] 'Report of the Commissioners on Highland Roads and Bridges.' Appendix to 'Life of Telford,' p. 400.

consolidated by ice, as well as, shortly after, from the blow
of a schooner which drifted against it on the opposite
side, and had her two masts knocked off by the collision,
gave him every confidence in the strength of this form
of construction, and he accordingly repeated it in several
of his subsequent bridges, though none of them are com-
parable in beauty with that of Craig-Ellachie.

Thus, in the course of eighteen years, 920 miles of
capital roads, connected together by no fewer than 1200
bridges, were added to the road communications of the
Highlands, at an expense defrayed partly by the localities
immediately benefited, and partly by the nation. The
effects of these twenty years' operations were such as
follow the making of roads everywhere—development
of industry and increase of civilization. In no districts
were the benefits derived from them more marked than
in the remote northern counties of Sutherland and
Caithness. The first stage-coaches that ran northward
from Perth to Inverness were tried in 1806, and became
regularly established in 1811; and by the year 1820
no fewer than forty arrived at the latter town in the
course of every week, and the same number departed
from it. Others were established in various directions
through the Highlands, which became as accessible as
any English county.

Agriculture made rapid progress. The use of carts
became practicable, and manure was no longer carried
to the field on women's backs. Sloth and idleness
disappeared before the energy, activity, and industry
which were called into life by the improved communications; better built cottages took the place of
the old mud biggins with a hole in the roof to let
out the smoke; the pigs and cattle were treated to a
separate table; the dunghill was turned to the outside
of the house; tartan tatters gave place to the produce
of Manchester and Glasgow looms; and very soon few
young persons were to be found who could not both

read and write English.   But not less remarkable were
the effects of the road-making upon the industrial habits
of the people.   Before Telford went into the Highlands,
they did not know how to work, having never been
accustomed to labour continuously and systematically.
Telford himself thus describes the moral influences of his
Highland contracts :—" In these works," says he, " and
in the Caledonian Canal, about three thousand two
hundred men have been annually employed.   At first,
they could scarcely work at all; they were totally unac-
quainted with labour; they could not use the tools.
They have since become excellent labourers, and of
the above number we consider about one-fourth left us
annually, taught to work.   These undertakings may,
indeed, be regarded in the light of a working academy,
from which eight hundred men have annually gone
forth improved workmen.   They have either returned
to their native districts with the advantage of having
used the most perfect sort of tools and utensils (which
alone cannot be estimated at less than ten per cent. on
any sort of labour), or they have been usefully distributed through the other parts of the country.   Since
these roads were made accessible, wheelwrights and
cartwrights have been established, the plough has been
introduced, and improved tools and utensils are generally used.   The plough was not previously employed;
in the interior and mountainous parts they used crooked
sticks, with iron on them, drawn or pushed along.   The
moral habits of the great masses of the working classes
are changed; they see that they may depend on their
own exertions for support : this goes on silently, and is
scarcely perceived until apparent by the results.   I consider these improvements among the greatest blessings
ever conferred on any country.   About two hundred
thousand pounds has been granted in fifteen years.   It
has been the means of advancing the country at least a
century."

CHAPTER IX.

Telford's Scotch Harbours.

No sooner were the Highland roads and bridges in progress than attention was directed to the improvement of the harbours around the coast. Very little had as yet been done for them beyond what nature had effected. Happily, there was a public fund at disposal—the accumulation of rents and profits derived from the estates forfeited in consequence of the rebellion of 1745—which was available for the purpose. The suppression of the rebellion produced good in many ways. It broke the feudal spirit, which had lingered in the Highlands long after it had ceased in every other part of Britain; it led to the effectual opening up of the country by a system of good roads; and now the accumulated rents of the defeated Jacobite chiefs were about to be applied to the improvement of the Highland harbours for the benefit of the industrial population.

The harbour of Wick was one of the first to which Mr. Telford's attention was directed. It will be remembered that Mr. Rennie had reported on the subject of its improvement as early as the year 1793, and his plans were not adopted only because their execution was beyond the means of the locality at that time. The place had now, however, become of increased importance. It was largely frequented by Dutch fishermen during the herring season; and it was hoped that, if they could be induced to form a settlement there by improving the accommodation of the harbour, their example might exercise a beneficial influence upon the population generally. Mr. Telford reported that, by the expenditure of about 5890*l.*, a capacious and well-protected tide-basin might be formed, capable of containing about two hundred herring-busses. The Commission adopted his plan, and voted the requisite funds for carrying out the works, which were begun in 1808. The new station was named Pulteney Town, in compliment to Sir William Pulteney, the Governor of the Fishery Society; and the harbour was built at a cost of about 12,000*l.*, of which 8500*l.* was granted from the balance of forfeited estates. A handsome stone bridge, erected over the river Wick in 1805, after the design of our engineer, connects these improvements with the older town : it is formed of three arches, having a clear waterway of 156 feet. The money was well expended, as the result proved; and Wick is now, we believe, the greatest fishing station in the world. The place has increased from the dimensions of a little poverty-stricken village to those of a large and thriving town, which swarms during the fishing season with lowland Scotchmen, fair Northmen, broad-built Dutchmen, and kilted Highlanders. The bay is then frequented by upwards of a thousand fishing boats, and the take of herrings in some years amounts to more than a hundred thousand barrels. The harbour has of late years been considerably improved to meet the growing requirements of the herring trade, the principal additions having been carried out, in 1823, by Mr. Bremner,[1] a native engineer of great ability.

---

[1] Hugh Millar, in his 'Cruise of the Betsy,' attributes the invention of columnar pier-work to Mr. Bremner, whom he terms "the Brindley of Scotland." He has acquired great fame for his skill in raising sunken ships, having warped the *Great Britain* steamer off the shores of Dundrum Bay. But we believe Mr. Telford had adopted the practice of columnar pier-work before Mr. Bremner, in forming the little harbour of Folkestone, built in 1808, where the work is still to be seen quite perfect. The most solid mode of laying stone on land is in flat courses; but in open pier work the reverse process is adopted. The blocks are laid on end in columns, like upright beams jammed together. Thus laid, the wave which dashes against them is broken, and spends itself on the interstices; whereas, if it struck the broad solid blocks, the tendency would be to lift them from their beds and set the work afloat; and in a furious

Improvements of a similar kind were carried out by the Fishery Board at various other parts of the coast, and many snug and convenient harbours were provided at the principal fishing stations in the Highlands and Western Islands. Where the local proprietors were themselves found actively at work expending money in carrying out piers and harbours, the Board assisted them with grants to enable them to complete the works in the most substantial manner and after the most approved plans. Thus, along that part of the bold northern coast of the mainland of Scotland which projects into the German Ocean, many old harbours were improved or new ones constructed—as at Peterhead, Frazerburgh, Banff, Cullen, Burgh Head, and Nairn. At Fortrose, in the Murray Frith; at Dingwall, in the Cromarty Frith; at Portmaholmac, within Tarbet Ness, the remarkable headland of the Frith of Dornoch; at Kirkwall, the principal town and place of resort in the Orkney Islands, so well known from Sir Walter Scott's description of it in the 'Pirate;' at Tobermory, in the island of Mull; and at other points of the coast, piers were erected and other

---

storm such blocks would be driven about almost like pebbles. The rebound from flat surfaces is also very heavy, and produces violent commo-tion; whereas these broken, upright, columnar-looking piers seem to absorb the fury of the sea, and render its wildest waves comparatively innocuous.

FOLKESTONE HARBOUR.   [By Percival Skelton]

improvements carried out to suit the convenience of the growing traffic and trade of the country.

The principal works were those connected with the harbours situated upon the line of coast extending from the harbour of Peterhead, in the county of Aberdeen, round to the head of the Murray Frith. The shores there are exposed to the full force of the seas rolling in from the Northern Ocean; and safe harbours were especially necessary for the protection of the shipping passing from north to south. Wrecks had become increasingly frequent, and harbours of refuge were loudly called for. At one part of the coast as many as thirty wrecks had occurred within a very short time, chiefly for want of shelter. The situation of Peterhead was peculiarly well adapted for a haven of refuge, and the improvement of its port was early regarded as a matter of national importance. Not far from it, on the south, are the famous Bullars or Boilers of Buchan—bold rugged rocks, some 200 feet high, against which the sea beats with great fury, boiling and churning in the deep caves and recesses with which they are perforated. Peterhead stands on the most easterly part of the entire mainland of Scotland, the town standing on the north-east side of the bay, and being connected with the country on the north-west by an isthmus only 800 yards in breadth. In Cromwell's time the port possessed no more than twenty tons of boat tonnage, and its only harbour was a small basin dug out of the rock. Even down to the close of the sixteenth century the place was but an insignificant fishing village. It is now a town bustling with trade, having long been the principal seat of the whale fishery, 1500 men of the port being engaged in that pursuit alone; and it sends out ships of its own building to all parts of the world, its handsome and commodious harbours being accessible at all winds to vessels of almost the largest burden.

It may be mentioned that about sixty years since

PETERHEAD. [By R. P. Leitch.]

the port was formed by the island called Keith Island,
situated a small distance eastward from the shore, between which and the mainland an arm of the sea formerly passed. A causeway had, however, been formed
across this channel, thus dividing it into two small bays;
after which the southern one had been converted into
a harbour by means of two rude piers erected along
either side of it. The north inlet remained without any
pier, and being very inconvenient and exposed to the
north-easterly winds, it was little used.

The first works carried out at Peterhead were of a

comparatively limited character, the old piers of the south harbour having been built by Smeaton; but improvements proceeded apace with the enterprise and wealth of the inhabitants. Mr. Rennie, and after him Mr. Telford, fully reported as to the capabilities of the port and the best means of improving it. Mr. Rennie recommended the deepening of the south harbour and the extension of the jetty of the west pier, at the same time cutting off all projections of rock from Keith Island on the eastward, so as to render the access

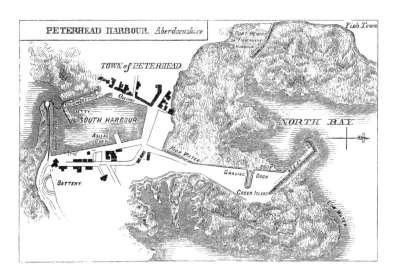

more easy. The harbour, when thus finished, would, he estimated, give about 17 feet depth at high water of spring tides. He also proposed to open a communication across the causeway between the north and south harbours, and form a wet dock between them, 580 feet long and 225 feet wide, the water being kept in by gates at each end. He further proposed to provide an entirely new harbour, by constructing two extensive piers for the effectual protection of the northern part of the channel, running out one from a rock north of the Green Island, about 680 feet long, and another

from the Roan Head, 450 feet long, leaving an opening
between them of 70 yards. This sagacious design unhappily could not be carried out at the time for want
of funds; but it may be said to have formed the ground-
work of all that has been subsequently done for the
improvement of the port of Peterhead.

It was resolved, in the first place, to commence operations by improving the south harbour, and protecting it
more securely from south-easterly winds. The bottom
of the harbour was accordingly deepened by cutting
out 30,000 cubic yards of rocky ground; and part of
Mr. Rennie's design was carried out by extending the
jetty of the west pier, though only for a distance of
twenty yards. These works were executed under Mr.
Telford's directions; they were completed by the end of
the year 1811, and proved to be of very great public
convenience. The trade of the town, however, so much
increased, and the port was found of so great importance
as a place of refuge for vessels frequenting the north
seas, that in 1816 it was determined to proceed with the
formation of a harbour on the northern part of the old
channel; and the inhabitants having agreed amongst
themselves to contribute to the extent of 10,000*l.* towards
carrying out the necessary works, they applied for the
grant of a like sum from the Forfeited Estates Fund,
which was eventually voted for the purpose. The plan
adopted was on a more limited scale than that proposed
by Mr. Rennie; but in the same direction and contrived
with the same object—so that, when completed, vessels
of the largest burden employed in the Greenland fishery
might be able to enter one or other of the two harbours
and find safe shelter, from whatever quarter the wind
might blow.

The works were vigorously proceeded with, and had
made considerable progress, when, in October, 1819, a
violent hurricane from the north-east, which raged along
the coast for days and inflicted heavy damage on many

of the northern harbours, destroyed a large part of the
unfinished masonry and hurled the heaviest blocks into
the sea, tossing them about as if they had been pebbles.
The finished work had, however, stood well, and the
foundations of the piers under low water were ascertained
to have remained comparatively uninjured. There was
no help for it but to repair the damaged work, though
it involved a heavy additional cost, one-half of which
was borne by the Forfeited Estates Fund and the remainder by the inhabitants. Increased strength was
also given to the more exposed parts of the pierwork,
and the slope at the sea side of the breakwater was
considerably extended.[1] These alterations in the design
were carried out, together with a spacious graving-dock,
as shown in the preceding plan, and they proved completely successful, enabling Peterhead to offer an amount
of accommodation for shipping of a more effectual kind
than was at that time to be met with along the whole
eastern coast of Scotland.

The old harbour of Frazerburgh, situated on a projecting point of the coast at the foot of Mount Kennaird,
about twenty miles north of Peterhead, had become so
ruinous that vessels lying within it received almost as
little shelter as if they had been exposed in the open sea.
Mr. Rennie had prepared a plan for its improvement by
running out a substantial north-eastern pier; and this was
eventually carried out by Mr. Telford in a modified form,
proving of substantial service to the trade of the port.
Since then a large and commodious new harbour has
been formed at the place, partly at the public expense
and partly at that of the inhabitants, rendering Frazerburgh a safe retreat for vessels of war as well as merchantmen; and the place has thus become thriving, wealthy,
and populous.

---

[1] 'Memorials from Peterhead and
Banff, concerning Damage occasioned
by a Storm.' Ordered by the House
of Commons to be printed, 5th July,
1820. [242.]

Among the other important harbour works on the north-east coast carried out by Mr. Telford under the Commissioners appointed to administer the funds of the Forfeited Estates, were those at Banff, the execution of which extended over many years; but though costly, they did not prove of anything like the same convenience as those executed at Peterhead. The old harbour at the end of the ridge running north and south, on which what is called the "sea town" of Banff[1] is situated, was completed in 1775, when the place was already considered of some importance as a fishing station.

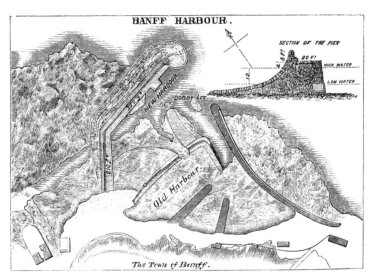

BANFF HARBOUR.

SECTION OF THE PIER

HIGH WATER

LOW WATER

NEW HARBOUR

BOBBY LEE

Old Harbour

The Town of Banff.

This harbour occupies the triangular space at the north-eastern extremity of the projecting point of land, at the opposite side of which, fronting the north-west, is the little town and harbour of Macduff. In 1816 Mr. Telford furnished a plan of a new pier and breakwater, covering the old entrance, which presented an opening to the N.N.E., with a basin occupying the

---

[1] The old harbour is shown at the right extremity of the distant view of Banff, given at p. 2 of this volume.

intermediate space. The inhabitants agreed to defray one-half of the necessary cost, and the Commissioners the other; and the plans having been approved, the works were commenced in 1818. They were in full progress when, unhappily, the same hurricane which in 1819 did so much injury to the works at Peterhead, also fell upon those at Banff, and carried away a large part of the unfinished pier. This accident had the effect of interrupting the work, as well as increasing its cost; but the whole was successfully completed by the year 1822. Although the new harbour did not prove very safe, and exhibited a tendency to become silted up with sand, it proved of use in some respects, more particularly by preventing all swell and agitation in the old harbour, which was thus rendered probably the safest artificial haven in the Murray Firth.

It is unnecessary to specify the alterations and improvements of a similar character, adapted to the respective localities, which were carried out by Mr. Telford at Burgh Head, Nairn, Kirkwall, Tarbet, Tobermory, Portmaholmac, Dingwall (with its canal two thousand yards long, connecting the town in a complete manner with the Frith of Cromarty), Cullen, Fortrose, Ballintraed, Portree, Jura, Gourdon, Invergordon, and other places. Down to the year 1823 the Commissioners had expended 108,530*l.* on the improvements of these several ports, in aid of the local contributions of the inhabitants and adjoining proprietors to a considerably greater extent; the result of which was a great increase in the shipping accommodation of the coast towns, to the benefit of the local population, and of shipowners and navigators generally.

Mr. Telford's principal harbour works in Scotland, however, were those of Aberdeen and Dundee, which, next to Leith, the port of Edinburgh, formed the principal havens along the east coast. The neighbourhood of Aberdeen was originally so wild and barren that Telford expressed his surprise that any class of men should ever have settled there. An immense shoulder of the Grampian mountains there extends itself down to the sea-coast, where it terminates in a bold, rude promontory. The country on either side of the Dee, which flows past the town, was originally covered with innumerable granite blocks, one, called Craig Metellan, lying right in the river's mouth, which, with the sand, formed an almost effectual bar to its navigation. Although, in ancient times, a little cultivable land lay immediately outside the town, the region beyond was as sterile as it is possible for land to be in such a latitude. " Any wher," says an ancient writer, " after yow pas a myll without the toune, the countrey is barren lyke, the hills craigy, the plaines full of marishes and mosses, the feilds are covered with heather or peeble stons, the corne feilds mixt with thes bot few. The air is temperat and healthful about it, and it may be that the citizens owe the acuteness of their wits thereunto and their civill inclinations; the lyke not easie to be found under northerlie climats, damped for the most pairt with air of a grosse consistence." [1]

But the old inhabitants of Aberdeen and its neighbourhood were really as rough as their soil. Judging from their records, they must have been dreadfully haunted by witches and sorcerers down to a comparatively recent period, witch-burning having been common in the town until the end of the sixteenth century. We find that, in one year, no fewer than twenty-three women and one man were burnt; the Dean of Guild Records containing the detailed accounts of the " loads of peattis, tar barrellis," and other combustibles used in burning them. The lairds of the Garioch, a district in the immediate neighbourhood, seem to have been still more terrible

---

[1] ' A Description of Bothe Touns of Aberdeene.' By James Gordon, Parson of Rothiemay. Reprinted in Gavin Turreff's 'Antiquarian Gleanings from Aberdeenshire Records.' Aberdeen, 1859.

than the witches, being accustomed to enter the place and
make an onslaught upon the citizens, according as local
rage and thirst for spoil might incline them.   On one
of such occasions, eighty of the inhabitants were killed
and wounded.[1]   Down even to the middle of last century the Aberdonian notions of personal liberty seem to
have been very restricted; for between 1740 and 1746
we find that persons of both sexes were kidnapped, put
on board ships, and despatched to the American planta-
tions, where they were sold for slaves.   Strangest of
all, the men who carried on this slave trade were local
dignitaries, one of them being a town's baillie, another
the town-clerk depute.   Those kidnapped were openly
" driven in flocks through the town, like herds of sheep,
under the care of a keeper armed with a whip." [2]
So open was the traffic that the public workhouse was
used for their reception until the ships sailed, and when
that was filled, the tolbooth or common prison was made
use of.   The vessels which sailed from the harbour for
America in 1743 contained no fewer than sixty-nine
persons; and it is supposed that, in the six years during
which the Aberdeen slave trade was at its height, about
six hundred were transported for sale, very few of whom
ever returned.[3]

   This slave traffic was doubtless stimulated by the
foreign ships now beginning to frequent the port; for

---

[1] Robertson's ' Book of Bon-Accord.'
[2] Ibid., quoted in Turreff's ' Antiquarian Gleanings,' p. 222.
[3] One of them, however, did return—Peter Williamson, a native of
the town, sold for a slave in Penn-
sylvania, "a rough, ragged, humle-
headed, long, stowie, clever boy,"
who, reaching York, published an
account of the infamous traffic, in a
pamphlet which excited extraordinary
interest at the time, and met with
a rapid and extensive circulation.
But his exposure of kidnapping gave
very great offence to the magistrates,
who dragged him before their tribunal
as having "published a scurrilous and
infamous libel on the corporation,"
and he was sentenced to be imprisoned until he should sign a denial of
the truth of his statements.   He
brought an action against the corporation for their proceedings, and obtained a verdict and damages; and
he further proceeded against Baillie
Fordyce (one of his kidnappers), and
others, from whom he obtained 200*l.*
damages, with costs.   The system was
thus effectually put a stop to.

the inhabitants were industrious, and their plaiding, linen, and worsted stockings were in much request as articles of merchandise. Cured salmon was also exported in large quantities. As early as 1659 a quay was formed along the Dee towards the village of Foot Dee. "Beyond Futty," says an old writer, "lyes the fisherboat heavne; and after that, towards the promontorie called Sandenesse, ther is to be seen a grosse bulk of a building, vaulted and flatted above (the Blockhous they call it), begun to be builded anno 1513, for guarding the entree of the harboree from pirats and algarads; and cannon wer planted ther for that purpose, or, at least, that from thence the motions of pirats might be tymouslie foreseen. This rough piece of work was finished anno 1542, in which yer lykewayes the mouth of the river Dee was locked with cheans of iron and masts of ships crossing the river, not to be opened bot at the citizens' pleasure."[1]

After the Union, but more especially after the rebellion of 1745, the trade of Aberdeen made considerable progress. Although Burns, in 1787, briefly described the place as a "lazy toun," the inhabitants were displaying much energy in carrying out improvements in their port.[2] In 1775 the foundation-stone of the new pier designed by Mr. Smeaton was laid with great ceremony, and, the works proceeding to completion, a new pier, twelve hundred feet long, terminating in a round head, was finished in less than six years. The trade of the place was, however, as yet too small to justify anything beyond a tidal harbour, and the engineer's views were

---

[1] 'A Description of Bothe Touns of Aberdeene.' By James Gordon, Parson of Rothiemay. Quoted by Turreff, p. 109.

[2] Communication with London was as yet by no means frequent, and far from expeditious, as the following advertisement of 1778 will show:—
"For London: To sail positively on Saturday next, the 7th November, wind and weather permitting, the *Aberdeen* smack. Will lie a short time at London, and, if no convoy is appointed, will sail under care of a fleet of colliers—the best convoy of any. For particulars apply," &c., &c.

limited to that object. He found the river meandering over an irregular space about five hundred yards in breadth; and he applied the only practicable remedy, by confining the channel as much as the limited means placed at his disposal enabled him to do, and directing the land floods so as to act upon and diminish the bar. Opposite the north pier, on the south side of the river, Smeaton constructed a breast-wall about half the length of the pier. Owing, however, to a departure from that engineer's plans, by which the pier was placed too far to the north, it was found that a heavy swell entered the harbour, and, to obviate this formidable inconvenience, a bulwark was projected from it, so as to occupy about one-third of the channel entrance.

The trade of the place continuing to increase, Mr. Rennie was called upon, in 1797, to examine and report upon the best means of improving the harbour, when he recommended the construction of floating docks upon the sandy flats called Foot Dee. Nothing was done at the time, as the scheme was very costly and considered beyond the available means of the locality. But the magistrates kept the subject in mind; and when Mr. Telford reported as to the improvement of the harbour in 1801, he intimated that the inhabitants were ready to co-operate with the Government in rendering it capable of accommodating ships of war, so far as their circumstances would permit. In 1807 the south pier-head, built by Mr. Smeaton, was destroyed by a storm, and the time had arrived when something must be done, not only to improve but even to preserve the port. The magistrates accordingly proceeded, in 1809, to rebuild the pier-head of cut granite, and at the same time they applied to Parliament for authority to carry out further improvements after the plan recommended by Mr. Telford; and the necessary powers were conferred in the following year. The new works comprehended a large extension of the wharfage accommodation, the construction

2 D 2

of floating and graving docks, increased means of scouring the harbour and ensuring greater depth of water
on the bar across the river's mouth, and the provision of
a navigable communication between the Aberdeenshire
Canal and the new harbour.

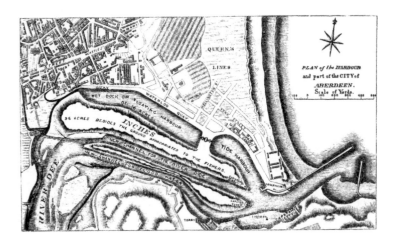

The extension of the north pier was first proceeded
with, under the superintendence of Mr. John Gibb, the
resident engineer; and by the year 1811 the whole
length of 300 additional feet had been completed.   The
beneficial effects of this extension were so apparent, that
a general wish was expressed that it should be carried
further; and it was eventually determined to extend the
pier 780 feet beyond Mr. Smeaton's head, by which not
only was much deeper water secured, but vessels were
better enabled to clear the Girdleness Point.   This extension was successfully carried out by the end of the
year 1812, whilst a strong breakwater, about 800 feet
long, was run out from the south shore, leaving a space
of about 250 feet as an entrance, thereby giving greater
protection to the shipping in the harbour, whilst the
contraction of the channel, by increasing the " scour,"
tended to give a much greater depth of water on
the bar.

ABERDEEN HARBOUR. [By R. P. Leitch.]

The outer head of the pier was seriously injured
by the heavy storms of the two succeeding winters, which
rendered it necessary to alter its formation to a very
flat slope of about five to one all round the head.[1]   New

---

[1] "The bottom under the foundations," says Mr. Gibb, in his description of the work, "is nothing better
than loose sand and gravel, constantly
thrown up by the sea on that stormy
coast, so that it was necessary to con-
solidate the work under low water by
dropping large stones from lighters,
and filling the interstices with smaller
ones, until it was brought within about

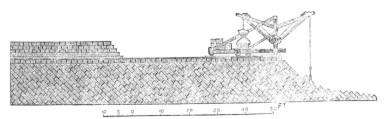

SECTION OF PIER-HEAD WORK.

wharves were at the same time constructed inside the
harbour; a new channel for the river was excavated,
which further enlarged the floating space and wharf
accommodation; wet and dry docks were added; until
at length the quay berthage amounted to not less than
6290 feet, or nearly a mile and a quarter in length.

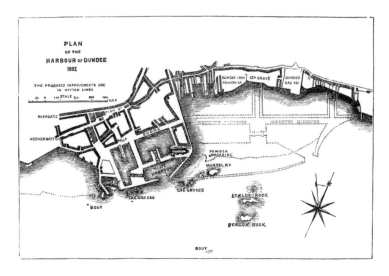

By these combined improvements an additional extent
of quay room was obtained of about 4000 feet; an
excellent tidal harbour was formed, in which, at spring
tides, the depth of water is about 15 feet; whilst on the
bar it was increased to about 19 feet. The prosperity of
Aberdeen had meanwhile been advancing apace. The

---

a foot of the level of low water, when
the ashlar work was commenced; but
in place of laying the stones horizontally in their beds, each course was
laid at an angle of 45 degrees, to
within about 18 inches of the top,
when a level coping was added. This
mode of building enabled the work to
be carried on expeditiously, and rendered it while in progress less liable
to temporary damage, likewise affording three points of bearing; for while
the ashlar walling was carrying up
on both sides, the middle or body of
the pier was carried up at the same
time by a careful backing throughout
of large rubble-stone, to within 18
inches of the top, when the whole
was covered with granite coping and
paving 18 inches deep, with a cut
granite parapet wall on the north side
of the whole length of the pier, thus
protected for the convenience of those
who might have occasion to frequent
it."—Mr. Gibb's 'Narrative of Aberdeen Harbour Works.'

city had been greatly beautified and enlarged; ship-
building had made rapid progress; Aberdeen clippers
became famous, and Aberdeen merchants carried on a
trade with all parts of the world; manufactures of wool,
cotton, flax, and iron were carried on with great success;
its population rapidly increased; and, as a maritime city,
Aberdeen took rank as the third in Scotland, the tonnage
entering the port having increased from 50,000 tons in
1800 to about 300,000 in 1860.

DUNDEE HARBOUR.　[By R. P. Leitch.]

Improvements of an equally important character were
carried out by Mr. Telford in the port of Dundee, also
situated on the east coast of Scotland, at the entrance to
the Frith of Tay. There are those still living at the
place who remember its former haven, consisting of a
crooked wall, affording shelter to only a few fishing boats
or smuggling vessels—its trade being then altogether
paltry, scarcely deserving the name, and its population
not one-fifth of what it now is. Helped by its commodious and capacious harbour, it has become one of the
most populous and thriving towns on the east coast.

The trade of the place took a great start forward at

the close of the war, and Mr. Telford was called upon to supply the plans of a new harbour. His first design, which he submitted in 1814, was of a comparatively limited character; but it was greatly enlarged during the progress of the works. Floating docks were added, as well as graving docks for large vessels. The necessary powers were obtained in 1815; the works proceeded vigorously under the Harbour Commissioners, who superseded the old obstructive corporation; and in 1825 the splendid new floating dock—750 feet long by 450 feet broad, having an entrance-lock 170 feet long and 40 feet wide—was opened to the shipping of all countries.

CHAPTER X.

Caledonian and other Canals.

The formation of a navigable highway through the chain of lochs occupying the Great Glen of the Highlands, extending diagonally across Scotland from the Atlantic to the North Sea, had long been regarded as a work of national importance. As early as 1773, James Watt, when following the business of a land-surveyor at Glasgow, made a survey of the country at the instance of the Commissioners of Forfeited Estates. He pronounced the canal practicable, and pointed out how it could best be constructed. There was certainly no want of water, for Watt was drenched with rain during most of his survey, and had difficulty in preserving even his journal book. " On my way home," he says, " I passed through the wildest country I ever saw, and over the worst conducted roads."

Twenty years later, in 1793, Mr. Rennie was consulted as to the canal, and he also prepared a scheme ; but nothing was done. The project was, however, revived in 1801, during the war with Napoleon, when various inland ship canals—such as those from London to Portsmouth, and from Bristol to the English Channel—were under consideration, with the view of protecting British shipping against French privateers. But there was another reason for urging the formation of the canal through the Great Glen of Scotland, which was regarded as of importance before the introduction of steam enabled vessels to set the winds and tides at comparative defiance. It was this : vessels sailing from the eastern ports to America had to beat up the Pentland Frith often against adverse winds and stormy seas, which rendered the navigation both tedious and dangerous. Thus it was cited by Sir Edward Parry, in his evidence before Parliament in favour of completing the Caledonian Canal, that of two vessels despatched from Newcastle on the same day—one bound for Liverpool by the north of Scotland, and the other for Bombay by the English Channel and the Cape of Good Hope—the latter reached its destination first! Another case may be mentioned, of an Inverness vessel, which sailed for Liverpool on a Christmas Day, reached Stromness Harbour, in Orkney, on the 1st of January, and lay there windbound, with a fleet of other traders, until the middle of April following! In fact the Pentland Frith, which is the throat connecting the Atlantic and German Oceans, and through which the former rolls its long majestic waves with tremendous force, was the dread of mariners, and it was considered an object of national importance to mitigate the dangers of the passage towards the western seas.

As the lochs occupying the chief part of the bottom of the Great Glen were of sufficient depth to be navigable by large vessels, it was thought that if they could be connected by a ship canal, so as to render the line of navigation continuous, it would be used by shipping to a large extent, and prove of great public service. Five hundred miles of dangerous navigation by the Orkneys and Cape Wrath would thus be saved, and ships of war, were this track open to them, might reach the north of Ireland in two days from Fort George, near Inverness. When the scheme of the proposed canal was revived in 1801, Mr. Telford was requested to make a survey and send in his report on the subject. He immediately wrote to his friend James Watt, saying, " I have so long accustomed myself to look with a degree of reverence to your work, that I am particularly anxious to learn what occurred to you in this business while the whole was fresh in your mind. The object appears to

me so great and so desirable, that I am convinced you
will feel a pleasure in bringing it again under investigation, and I am very desirous that the thing should be
fully and fairly explained, so that the public may be
made aware of its extensive utility. If I can accomplish
this, I shall have done my duty ; and if the project is not
executed now, some future period will see it done, and I
shall have the satisfaction of having followed you and
promoted its success." [1]   We may here state that Telford's
survey agreed with Watt's in the most important particulars, and that he largely cited his descriptions of the
intended scheme in his own report.

Mr. Telford's first inspection of the district was made
in 1801, and his report was sent in to the Treasury in
the course of the following year. Lord Bexley, then
Secretary to the Treasury, took a warm personal interest
in the project, and lost no opportunity of actively promoting it. A board of commissioners was eventually
appointed to carry out the formation of the canal.
Mr. Telford, on being appointed principal engineer of
the undertaking, was requested at once to proceed to
Scotland and prepare the necessary working survey.
He was accompanied on the occasion by Mr. Jessop
as consulting engineer. Twenty thousand pounds were
granted under the provisions of the 43 Geo. III.
(chap. cii.), and the works were commenced, in the
beginning of 1804, by the formation of a dock or
basin adjoining the intended tide-lock at Corpach, near
Bannavie.

The basin at Corpach formed the southernmost point
of the intended canal. It is situated at the head of
Loch Eil, amidst some of the grandest scenery of the
Highlands. Across the Loch is the little town of Fort
William, one of the forts established at the end of the

---

[1] 'The Origin and Progress of the | By J. P. Muirhead, Esq., M.A.
Mechanical Inventions of James Watt.' | Vol. i., p. cxxvi.

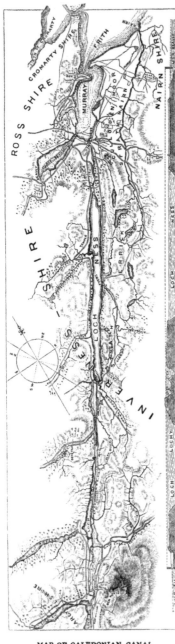

MAP OF CALEDONIAN CANAL.

seventeenth century to keep the wild Highlanders in subjection. Above it, hills over hills arise, of all forms and sizes, and of all hues, from grass-green below to heather-brown and purple above, capped with heights of weather-beaten grey; whilst towering over all stands the rugged mass of Ben Nevis —a mountain almost unsurpassed for picturesque grandeur. Along the western foot of the range, which extends for some six or eight miles, lies a long extent of brown bog, on the verge of which, by the river Lochy, stand the ruins of Inverlochy Castle.

The works at Corpach involved great labour, and extended over a long series of years. The difference between the level of Loch Eil and Loch Lochy is ninety feet, whilst the distance between them was less than eight miles, and it was therefore necessary to climb up the side of the hill by a flight of eight gigantic locks, clustered together, and which Telford named Neptune's Staircase. The ground passed over was in some places very difficult, requiring large masses of embankment, the slips of which in the course of the work frequently occasioned serious embarrassment. The basin on Loch Eil, on the other hand, was constructed amidst rocks, and considerable difficulty was experienced in getting in the necessary coffer-dam for the construction of the entrance to the sea-lock, the entrance-sill of which was laid upon the rock itself, so that there was a depth of 21 feet of water upon it at high water of neap tides.[1]

At the same time that the works at Corpach were commenced, the dock or basin at the north-eastern extremity of the canal, situated at Clachnagarry, on the shore of Loch Beauly, was also laid out, and the excavations and embankments were carried on with considerable activity. This dock was constructed about 967 yards long, and upwards of 162 yards in breadth, giving an area of about 32 acres,—forming, in fact, a harbour for the vessels using the canal. The dimensions of the artificial waterway were of unusual size, as the intention was to adapt it throughout for the passage of a 32-gun frigate of that day, fully equipped and laden with stores. The canal, as originally resolved upon, was designed to be 110 feet wide at the surface, and 50 feet at the bottom, with a depth in the middle of 20 feet; though these dimensions were somewhat modified in the execution of the work. The locks were of corresponding large dimensions, each being from 170 to 180 feet long, 40 broad, and 20 deep.

Between these two extremities of the canal—Corpach on the south-west and Clachnagarry on the north-east— extends the chain of fresh-water lochs: Loch Lochy on the south; next Loch Oich; then Loch Ness; and lastly, furthest north, the small Loch of Doughfour. The whole length of the navigation is 60 miles 40 chains, of which the navigable lochs constitute about 40 miles,

---

[1] For professional details of this work, see ' Life of Telford,' p. 305.

leaving only about 20 miles of canal to be constructed, but of unusually large dimensions and through a very difficult country.

The summit loch of the whole is Loch Oich, the surface of which is exactly a hundred feet above high water-mark, both at Inverness and Fort William; and to this sheet of water the navigation climbs up by a series of locks from both the eastern and western seas. The whole number of these is twenty-eight: the entrance-lock at Clachnagarry, constructed on piles, at the

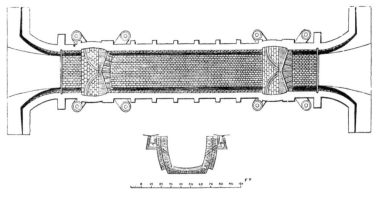

LOCK, CALEDONIAN CANAL.

end of huge embankments, forced out into deep water, at Loch Beauly;[1] another at the entrance to the capacious artificial harbour above mentioned at Muirtown; four connected locks at the southern end of this basin; a

[1] In the 'Sixteenth Report of the Commissioners of the Caledonian Canal,' the following reference is made to this important work, which was finished in 1812:—"The depth of the mud on which it may be said to be artificially seated is not less than 60 feet; so that it cannot be deemed superfluous, at the end of seven years, to state that no subsidence is discoverable; and we presume that the entire lock, as well as every part of it, may now be deemed as immovable, and as little liable to destruction, as any other large mass of masonry. This was the most remarkable work performed under the immediate care of Mr. Matthew Davidson, our superintendent at Clachnagarry, from 1804 till the time of his decease. He was a man perfectly qualified for the employment by inflexible integrity, unwearied industry, and zeal to a degree of anxiety, in all the operations committed to his care."

regulating lock a little to the north of Loch Doughfour;
five contiguous locks at Fort Augustus, at the south
end of Loch Ness; another, called the Kytra Lock,
about midway between Fort Augustus and Loch Oich;
a regulating lock at the north-east end of Loch Oich;
two contiguous locks between Lochs Oich and Lochy; a
regulating lock at the south-west end of Loch Lochy;
next, the grand series of locks, eight in number, called
" Neptune's Staircase," at Bannavie, within a mile and
a quarter of the sea; two locks, descending to Corpach
basin; and lastly, the great entrance or sea-lock at
Corpach.

As may naturally be supposed, the execution of these
great works involved vast labour and anxiety. They
were designed with much skill, and executed with equal
ability. There were lock-gates to be constructed, principally of cast iron, sheathed with pine planking; public
road bridges crossing the line of the canal, of which
there were eight, constructed of cast iron and swung
horizontally. There were many mountain streams,
swollen to torrents in winter, crossing under the canal,
for which abundant accommodation had to be provided,
involving the construction of numerous culverts, tunnels,
and under-bridges of large dimensions. There were
also powerful sluices provided to let off the excess
of water sent down from the adjacent mountains into
the canal during winter. Three of these, of large dimensions, high above the river Lochy, are constructed at a
point where the canal is cut through the solid rock; and
the sight of the mass of waters rushing down into the
valley beneath, gives an impression of power which,
once seen, is never forgotten.

These great works were only brought to a completion
after the labours of many years, during which the difficulties encountered in their construction had swelled the
cost of the canal far beyond the original estimate. The
rapid advances which had taken place in the interval

in the prices of labour and materials also tended greatly
to increase the expenses, and, after all, the canal, when
completed and opened, was comparatively little used.
This was doubtless owing, in a great measure, to the
rapid changes which occurred in the system of navigation shortly after the projection of the undertaking.
For these Telford was not responsible. He was called
upon to make the canal, and he did so in the best
manner. Engineers are not required to speculate as
to the commercial value of the works they are required
to construct; and there were circumstances connected
with the scheme of the Caledonian Canal which removed
it from the category of mere commercial adventures.
It was a Government project, and it proved a failure.
Hence it formed a prominent topic for discussion in
the journals of the day; but the attacks made upon the
Government because of their expenditure on the hapless
undertaking were perhaps more felt by Telford, who
was its engineer, than by all the ministers of state con-
joined.

"The unfortunate issue of this great work," writes
the present engineer of the canal, to whom we are indebted for many of the preceding facts, "was a grievous
disappointment to Mr. Telford, and was in fact the one
great bitter in his otherwise unalloyed cup of happiness
and prosperity. The undertaking was maligned by
thousands who knew nothing of its character. It became 'a dog with a bad name,' and all the proverbial
consequences followed. The most absurd errors and
misconceptions were propagated respecting it from year
to year, and it was impossible during Telford's lifetime
to stem the torrent of popular prejudice and objurga-
tion. It must, however, be admitted, after a long
experience, that Telford was greatly over-sanguine in
his expectations as to the national uses of the canal, and
he was doomed to suffer acutely in his personal feelings,
little though he may have been personally to blame,

the consequences of what in this commercial country is regarded as so much worse than a crime, namely, a financial mistake." [1]

Mr. Telford's great sensitiveness made him feel the ill success of this enterprise far more than most other men would have done. He was accustomed to throw himself into the projects on which he was employed with an enthusiasm almost poetic. He regarded them not merely as so much engineering, but as works which were to be instrumental in opening up the communications of the country and extending its civilization. Viewed in this light, his canals, roads, bridges, and harbours were unquestionably of great national importance, though their commercial results might not in all cases justify the estimates of their projectors. To refer to like instances—no one can doubt the immense value and public uses of Mr. Rennie's Waterloo Bridge or Mr. Robert Stephenson's Britannia and Victoria Bridges, though every one knows that, commercially, they have been failures. But it is probable that neither of these eminent engineers gave himself anything like the anxious concern that Telford did about the financial issue of his undertaking. Were railway engineers to fret and vex themselves about the commercial value of the schemes in which they have been engaged, there

---

[1] The misfortunes of the Caledonian Canal did not end with the life of Telford. The first vessel passed through it from sea to sea in October, 1822, by which time it had cost about a million sterling, or double the original estimate. Notwithstanding this large outlay, it appears that the canal was opened before the works had been properly completed; and the consequence was that they very shortly fell into decay. It even began to be considered whether the canal ought not to be abandoned. In 1838, Mr. James Walker, C.E., an engineer of the highest eminence, examined it, and reported fully on its then state, strongly recommending its completion as well as its improvement. His advice was eventually adopted, and the canal was finished accordingly, at an additional cost of about 200,000l., and the whole line was re-opened in 1847, since which time it has continued in useful operation. The passage from sea to sea at all times can now be depended on, and it can usually be made in forty-eight hours. As the trade of the North increases, the uses of the canal will probably become much more decided than they have heretofore proved.

are few of them but would be so haunted by the ghosts of wrecked speculations that they could scarcely lay their heads upon their pillows for a single night in peace.

While the Caledonian Canal was in progress, Mr. Telford was occupied in various works of a similar kind in England and Scotland, and also one in Sweden. In 1804, while on one of his journeys to the north, he was requested by the Earl of Eglinton and others to examine a project for making a canal from Glasgow to Saltcoats and Ardrossan, on the north-western coast of the county of Ayr, passing near the important manufacturing town of Paisley. A new survey of the line was made, and the works were carried on during several successive years until a very fine capacious canal was completed, on the same level, as far as Paisley and Johnstown. But the funds of the company falling short, the works were stopped, and the canal was carried no further. Besides, the measures so actively employed by the Clyde Trustees to deepen the bed of that river and enable ships of large burden to pass up as high as Glasgow, had proved so successful that the ultimate extension of the canal to Ardrossan, so as to avoid the shoals of the Clyde, was no longer necessary, and the prosecution of the work was accordingly abandoned. But as Mr. Telford has observed, no person suspected, when the canal was laid out in 1805, " that steamboats would not only monopolise the trade of the Clyde, but penetrate into every creek where there is water to float them, in the British Isles and the continent of Europe, and be seen in every quarter of the world."

Another of the navigations on which Mr. Telford was long employed was that of the river Weaver in Cheshire. It was only twenty-four miles in extent, but of considerable importance to the country through which it passed, accommodating the salt-manufacturing districts, of which the towns of Nantwich, Northwich, and

Frodsham are the centres. The channel of the river was extremely crooked and much obstructed by shoals, when Telford took the navigation in hand in the year 1807, and a number of essential improvements were made in it, by means of new locks, weirs, and side cuts, which had the effect of greatly improving the communications of these important districts.

In the following year we find our engineer consulted, at the instance of the King of Sweden, on the best mode of constructing the Gotha Canal, between Lake Wenern and the Baltic, to complete the communication with the North Sea. In 1808, at the invitation of Count Platen, Mr. Telford visited Sweden and made a careful survey of the district. The service occupied him and his assistants two months, after which he prepared and sent in a series of detailed plans and sections, together with an elaborate report on the subject. His plans having been adopted, he again visited Sweden in 1810, to inspect the excavations which had already been begun, when he supplied the drawings for the locks and bridges. With the sanction of the British Government, he at the same time furnished the Swedish contractors with patterns of the most improved tools used in canal making, and took with him a number of experienced lock-makers and navvies for the purpose of instructing the native workmen. The construction of the Gotha Canal was an undertaking of great magnitude and difficulty, similar in many respects to the Caledonian Canal, though much more extensive. The length of artificial canal was 55 miles, and of the whole navigation, including the lakes, 120 miles. The locks are 120 feet long and 24 feet broad; the width of the canal at bottom being 42 feet, and the depth of water 10 feet. The results, so far as the engineer was concerned, were much more satisfactory than in the case of the Caledonian Canal. Whilst in the one case he had much obloquy to suffer for the services he had given,

in the other he was honoured and fêted as a public
benefactor, the King conferring upon him the Swedish
order of knighthood, and presenting him with his portrait set in diamonds.

Among the various canals throughout England which
Mr. Telford was employed to construct or improve,
down to the commencement of the railway era, were the
Gloucester and Berkeley Canal, in 1818; the Grand
Trunk Canal, in 1822; the Harecastle Tunnel, which he
constructed anew, in 1824-7; the Birmingham Canal, in
1824; and the Macclesfield, and Birmingham and Liverpool Junction Canals, in 1825. The Gloucester and
Berkeley Canal Company had been unable to finish their
works, begun some thirty years before; but with the
assistance of a loan of 160,000*l.* from the Exchequer
Bill Loan Commissioners, they were enabled to proceed
with the completion of their undertaking. A capacious
canal was cut from Gloucester to Sharpness Point, about
eight miles down the Severn, which had the effect of
greatly improving the convenience of the port of
Gloucester; and by means of this navigation, ships
of large burden can now avoid the circuitous and difficult passage of the higher part of the river, very much
to the advantage of the trade of the place.

The formation of a new tunnel through Harecastle
Hill, for the better accommodation of the boats passing
along the Grand Trunk Canal, was a formidable work.
The original tunnel, it will be remembered, was laid
out by Brindley, about fifty years before, and occupied
eleven years in construction. But the engineering
appliances of those early days were very limited; the
pumping powers of the steam-engine had not been
fairly developed, and workmen were as yet only half-
educated in the expert use of tools. The tunnel, no
doubt, answered the purpose for which it was originally
intended, but it was very soon found too limited for
the traffic passing along the navigation. It was little

larger than a sewer, and admitted the passage of only
one narrow boat, seven feet wide, at a time, involving
very heavy labour on the part of the men who worked
it through. This was performed by what was called
legging. The Leggers lay upon the deck of the vessel,
or upon a board slightly projecting from either side
of it, and, by thrusting their feet against the slimy
roof or sides of the tunnel—walking horizontally as it
were—they contrived to push it through. But it was
no better than horse-work; and after "legging" Harecastle Tunnel, which is more than a mile and a half
long, the men were usually completely exhausted, and
as wet from perspiration as if they had been dragged
through the canal itself. The process occupied about
two hours, and by the time the passage of the tunnel
was made, there was usually a collection of boats at the
other end waiting their turn to pass. Thus much contention and confusion took place amongst the boatmen
—a very rough class of labourers,—and many furious
battles were fought by the claimants for the first turn
"through." Regulations were found of no avail to
settle these disputes, still less to accommodate the large
traffic which continued to keep flowing along the line
of the Grand Trunk, and steadily increased with the
advancing trade and manufactures of the country. Loud
complaints were made by the public, but they were disregarded for many years; and it was not until the proprietors were threatened with rival canals and railroads
that they determined on—what they could no longer
avoid if they desired to retain the carrying trade of the
district—the enlargement of the Harecastle Tunnel.

Mr. Telford was requested to advise the Company what
course was most proper to be adopted in the matter,
and after examining the place, he recommended that an
entirely new tunnel should be constructed, nearly parallel
with the old one, but of much larger dimensions. The
work was begun in 1824, and completed in 1827, in

less than three years.   There were at that time throughout the country plenty of skilled labourers and contractors, many of them trained by their experience upon
Telford's own works, whereas Brindley had in a great
measure to make his workmen out of the rawest material.
Telford also had the advantage of greatly improved
machinery and an abundant supply of money—the Grand
Trunk Canal Company having become prosperous and
rich, paying large dividends.   It is therefore meet, while
eulogising the despatch with which he was enabled to
carry out the work, to point out that the much greater
period occupied in the earlier undertaking is not to be set
down to the disparagement of Brindley, who had difficulties
to encounter which the later engineer knew nothing of.

The length of the new tunnel is 2926 yards; it is
16 feet high and 14 feet broad, 4 feet 9 inches of
the breadth being occupied by the towing-path—for
" legging " was now dispensed with, and horses hauled
along the boats instead of their being thrust through by
men.   The tunnel is in so perfectly straight a line that
its whole length can be seen through at one view; and
though it was constructed by means of fifteen different
pitshafts sunk to the same
line along the length of
the tunnel, the workman-
ship is so perfect that the
joinings of the various
lengths of brickwork are
scarcely discernible.   The
convenience afforded by
the new tunnel was very
great, and Telford men-
tions that, on surveying it
in 1829, he asked a boat-
man coming out of it how

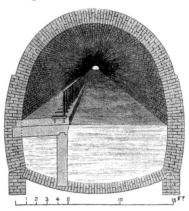

CROSS SECTION OF HARECASTLE TUNNEL.

he liked it ?   " I only wish," he replied, " that it reached
all the way to Manchester ! "

At the time that Mr. Telford was engaged upon the tunnel at Harecastle, he was employed to improve and widen the Birmingham Canal, another of Brindley's works. Though the accommodation provided by it had been sufficient for the traffic when originally constructed, the expansion of the trade of Birmingham and the neighbourhood, accelerated by the formation of the canal itself, had been such as completely to outgrow its limited convenience and capacity, and its enlargement and improvement now became absolutely necessary. Brindley's Canal, for the sake of cheapness of construction—money being much scarcer and more difficult to be raised in the early days of canals—was also winding and crooked; and it was considered desirable to shorten and straighten it by cutting off the bends at different points. At the point at which the canal entered Birmingham, it had become "little better than a crooked ditch, with scarcely the appearance of a towing-path, the horses frequently sliding and staggering in the water, the hauling-lines sweeping the gravel into the canal, and the entanglement at the meeting of boats being incessant; whilst at the locks at each end of the short summit at Smethwick crowds of boatmen were always quarrelling, or offering premiums for a preference of passage; and the mine-owners, injured by the delay, were loud in their just complaints." [1]

Mr. Telford proposed an effective measure of improvement, which was taken in hand without loss of time, and carried out, greatly to the advantage of the trade of the district. The numerous bends in the canal were cut off, the water-way was greatly widened, the summit at Smethwick was cut down to the level on either side, and a straight canal, forty feet wide, without a lock, thus formed as far as Bilston and Wolverhampton; whilst the length of the main line between Birmingham and Auther-

---

[1] 'Life of Telford,' p. 82, 83.

ley, along the whole extent of the "Black country," was reduced from twenty-two to fourteen miles. At the same time the obsolete curvatures in Brindley's old canal were converted into separate branches or basins, for the accommodation of the numerous mines and manufactories on either side of the main line. In consequence of the alterations which had been made in the canal, it was found necessary to construct numerous large bridges. One of these—a cast iron bridge, at Galton, of 150 feet span

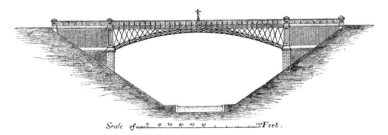

Scale of ....0 10 20 30 40 50 ....100 Feet.

GALTON BRIDGE, BIRMINGHAM CANAL.

—has been much admired for its elegance, lightness, and economy of material. Several others of cast iron were constructed at different points, and at one place the canal itself is carried along on an aqueduct of the same material as at Pont - Cysylltau. The whole of these extensive improvements were carried out in the short space of two years; and the result was highly satisfactory, " proving," as Mr. Telford himself observes, " that where business is extensive, liberal expenditure of this kind is true economy."

In 1825 Mr. Telford was called upon to lay out a canal to connect the Grand Trunk, at the north end of Harecastle Tunnel, with the rapidly improving towns of Congleton and Macclesfield. The line was twenty-nine miles in length, ten miles on one level from Harecastle to beyond Congleton; then, ascending 114 feet by eleven locks, it proceeded for five miles on a level past Macclesfield, and onward to join the Peak Forest Canal at Marple. The navigation was thus conducted upon two levels, each of considerable length; and it so happened that the trade of each was in a measure distinct, and required separate accommodation. The traffic of the whole of the Congleton district had ready access to the Grand Trunk system, without the labour, expense, and delay involved by passing the boats through locks; whilst the coals brought to Macclesfield to supply the mills there were carried throughout upon the upper level, also without lockage. The engineer's arrangement proved highly judicious, and furnishes an illustration of the tact and judgment which he usually displayed in laying out his works for practical uses. Mr. Telford largely employed cast iron in the construction of this canal, using it in the locks and gates, as well as in an extensive aqueduct which it was necessary to construct over a deep ravine, after the plan pursued by him at Pont-Cysylltau and other places.

The last canal constructed by Mr. Telford was the Birmingham and Liverpool Junction, extending from the Birmingham Canal, near Wolverhampton, in nearly a direct line, by Market Drayton, Nantwich, and through the city of Chester by the Ellesmere Canal, to Ellesmere Port on the Mersey. The proprietors of canals were becoming alarmed at the numerous railways projected through the districts heretofore served by their waterways; and amongst other projects one was set on foot, as early as 1825, for constructing a line of railway from London to Liverpool. Mr. Telford was consulted as to the best means of protecting existing investments, and his advice was to render the canal system as complete as it could be made; for he entertained the conviction, which has been justified by experience, that such navigations possessed peculiar advantages for the conveyance of heavy goods, and that, if the interruptions presented by locks could be done away with, or materially reduced, a large portion of the trade of the country must continue to be carried by the water roads. The new line recommended by him was approved and adopted, and the works were commenced in 1826. A second complete route was thus opened up between Birmingham and Liverpool, and Manchester, by which the distance was shortened twelve miles, and the delay occasioned by 320 feet of upward and downward lockage was done away with.

Telford was justly proud of his canals, which were the finest works of their kind that had yet been executed in England. Capacious, convenient, and substantial, they embodied his most ingenious contrivances, and his highest engineering skill. Hence we find him writing to a friend at Langholm, that, so soon as he could find "sufficient leisure from his various avocations in his own unrivalled and beloved island," it was his intention to visit France and Italy, for the purpose of ascertaining what foreigners had been able to accomplish, compared with ourselves, in the construction of canals, bridges, and harbours. "I have no doubt," said he, "as to their inferiority. During the war just brought to a close, England has not only been able to guard her own head and to carry on a gigantic struggle, but at the same time to construct canals, roads, harbours, bridges—magnificent works of peace—the like of which are probably not to be found in the world. Are not these things worthy of a nation's pride?"

CHAPTER XI.

TELFORD AS A ROAD-MAKER.

MR. TELFORD'S extensive practice as a bridge-builder led his friend Southey to designate him " Pontifex Maximus." Besides the numerous bridges erected by him in the West of England, we have found him furnishing designs for about twelve hundred in the Highlands, of various dimensions, some of stone and others of iron. His practice in bridge-building had, therefore, been of an unusually extensive character, and Southey's sobriquet was not ill applied. But besides being a great bridge-builder, Telford was also a great road-maker. With the progress of industry and trade, the easy and rapid transit of persons and goods had come to be regarded as an increasing object of public interest. Fast coaches now ran regularly between all the principal towns of England; every effort being made, by straightening and shortening the roads, cutting down hills, and carrying embankments across valleys and viaducts over rivers, to render travelling by the main routes as easy and expeditious as possible.

Attention was especially turned to the improvement of the longer routes, and to perfecting the connection of London with the chief towns of Scotland and Ireland. Telford was early called upon to advise as to the repairs of the road between Carlisle and Glasgow, which had been allowed to fall into a wretched state; as well as the formation of a new line from Carlisle, across the counties of Dumfries, Kirkcudbright, and Wigton, to Port Patrick, for the purpose of ensuring a more rapid communication with Belfast and the northern

parts of Ireland. Although Glasgow had become a place
of considerable wealth and importance, the roads to it,
north of Carlisle, continued in a very unsatisfactory
state. It was only in July, 1788, that the first mail-coach
from London had driven into Glasgow by that route,
when it was welcomed by a procession of the citizens
on horseback, who went out several miles to meet it.
But the road had been shockingly made, and before long
had become almost impassable. Robert Owen states
that, in 1795, it took him two days and three nights'
incessant travelling to get from Manchester to Glasgow,
and he mentions that the coach had to cross a well-
known dangerous mountain at midnight, called Trick-
stone Bar, which was then always passed with fear and
trembling.[1]

As late as the year 1814 we find a Parliamentary
Committee declaring the road between Carlisle and
Glasgow to be in so ruinous a state as often seriously to
delay the mail and endanger the lives of travellers.
The bridge over Evan Water was so much decayed, that
one day the coach and horses fell through it into the
river, when " one passenger was killed, the coachman
survived only a few days, and several other persons were
dreadfully maimed ; two of the horses being also killed." [2]
The remaining part of the bridge continued for some time
unrepaired, just space enough being left for a single
carriage to pass. The road trustees seemed to be help-
less, and did nothing ; a local subscription was tried and
failed, the district passed through being very poor ; but
as the road was absolutely required for more than merely
local purposes, it was eventually determined to undertake its reconstruction as a work of national importance,
and 50,000l. was granted by Parliament with this object,
under the provisions of the Act passed in 1816. The

---

[1] 'Life of Robert Owen,' by himself.
[2] 'Report from the Select Committee on the Carlisle and Glasgow
Road,' 28th June, 1815.

works were placed under Mr. Telford's charge; and
an admirable road was very shortly under construction
between Carlisle and Glasgow. That part of it between Hamilton and Glasgow, eleven miles in length,
was however left in the hands of local trustees, as was
the diversion of thirteen miles at the boundary of the
counties of Lanark and Dumfries, for which a previous
Act had been obtained.

The length of new line constructed by Mr. Telford
was sixty-nine miles, and it was probably the finest piece
of road which up to that time had been made. The
engineer paid especial attention to two points: first, to
lay it out as nearly as possible upon a level, so as to
reduce the draught to horses dragging heavy vehicles,—one in thirty being about the severest gradient
at any part of the road. The next point was to make
the working, or middle portion of the road, as firm and
substantial as possible, so as to bear, without shrinking,
the heaviest weight likely to be brought over it. With
this object he specified that the metal bed was to be formed
in two layers, rising about four inches towards the centre
—the bottom course being of stones (whinstone, limestone, or hard freestone), seven inches in depth. These
were to be carefully set by hand, with the broadest ends
downwards, all crossbonded or jointed, no stone being
more than three inches wide on the top. The spaces between them were then to be filled up with smaller stones,
packed by hand, so as to bring the whole to an even and
firm surface. Over this a top course was to be laid,
seven inches in depth, consisting of properly broken hard
whinstones, none exceeding six ounces in weight, and
each to be able to pass through a circular ring, two inches
and a half in diameter; a binding of gravel, about an
inch in thickness, being placed over all. A drain crossed
under the bed of the bottom layer to the outside ditch in
every hundred yards. The result was an admirably easy,
firm, and dry road, capable of being travelled upon in

all weathers, and standing in comparatively small need of repairs.[1]

Owing to the mountainous nature of the country through which this road passes, the bridges are unusually numerous and of large dimensions. Thus, the

---

[1] A similar practice was introduced in England about the same time by Mr. Macadam; and, though his method was not so thorough as that of Telford, it was usefully employed on nearly all the leading high roads of the kingdom. Mr. Macadam's atten-

J. L. MACADAM.

tion was first directed to the subject while acting as one of the trustees of a road in Ayrshire. Afterwards, while employed as Government agent for victualling the navy in the western parts of England, he continued the study of road-making, keeping in view the essential conditions of a compact and durable substance and a smooth surface. At that time road legislation was principally directed to the breadth of the wheels of vehicles; whilst Macadam was of opinion that the main point was to attend to the nature of the roads on which they were to travel. Most roads were then made with gravel, or flints tumbled upon them in their natural state, and so rounded that they had no points of contact, and rarely consolidated. When a heavy vehicle of

any sort passed over them, their loose structure presented no resistance; the roads were thus constantly standing in need of repair, and they were bad even at the best. He pointed out that the defect did not arise from the want of materials, which were not worn out by the traffic, but merely displaced. The practice he urged was this: to break the stones into angular fragments, so that a bed several inches in depth should be formed, the best adapted for the purpose being fragments of granite, greenstone, or basalt; to watch the repairs of the road carefully during the process of consolidation, filling up the inequalities caused by the traffic passing over it, until a hard and level surface had been obtained, when the road would last for years without further attention. In 1815 Mr. Macadam devoted himself with great enthusiasm to roadmaking as a profession, and being appointed surveyor-general of the Bristol roads, he had full opportunities of exemplifying his system. It proved so successful that the example set by him was quickly followed over the entire kingdom. Even the streets of many large towns were *Macadamised*. In carrying out his improvements, however, Mr. Macadam spent several thousand pounds from his own resources, and in 1825, having proved this expenditure before a committee of the House of Commons, the amount was reimbursed to him, together with an honorary tribute of two thousand pounds. Mr. Macadam died poor, but, as he himself said, "at least an honest man." By his indefatigable exertions and his success as a roadmaker, by greatly saving animal labour, facilitating commercial intercourse, and rendering travelling easy and expeditious, he entitled himself to the reputation of a public benefactor.

CARTLAND CRAGS BRIDGE.

[By R. P. Leitch, after a Drawing by J. S. Smiles]

Fiddler's Burn Bridge is of three arches, one of 150 and two of 105 feet span each. There are fourteen other bridges, presenting from one to three arches, of from 20 to 90 feet span. But the most picturesque and remarkable bridge constructed by Telford in that district was upon another line of road subsequently carried out by him, in the upper part of the county of Lanark, and crossing the main line of the Carlisle and Glasgow road almost at right angles. Its northern and eastern part formed a direct line of communication

between the great cattle-markets of Falkirk, Crief, and
Doune, and Carlisle and the West of England. It was
carried over deep ravines by several lofty bridges, the
most formidable of which was that across the Mouse
Water at Cartland Crags, about a mile to the westward of Lanark. The stream here flows through a deep
rocky chasm, the sides of which are in some places
about four hundred feet high. At a point where
the height of the rocks is considerably less, but still
most formidable, Telford spanned the ravine with the
beautiful bridge represented in the engraving on the
preceding page, its parapet being 129 feet above the
surface of the water beneath.

The reconstruction of the western road from Carlisle
to Glasgow, which Telford had thus satisfactorily carried
out, shortly led to similar demands from the population
on the eastern side of the kingdom. The spirit of road
reform was now fairly on foot. Fast coaches and wheel-
carriages of all kinds had become greatly improved, so
that the usual rate of travelling had advanced from five
or six to nine or ten miles an hour. The desire for the
rapid communication of political and commercial intelligence was found to increase with the facilities for
supplying it; and, urged by the public wants, the Post-
Office authorities were stimulated to unusual efforts in this
direction. Numerous surveys were made and roads laid
out, so as to improve the main line of communication
between London and Edinburgh and the intermediate
towns. The first part of this road taken in hand was
the worst—that lying to the north of Catterick Bridge,
in Yorkshire. A new line was surveyed by West Auck-
land to Hexham, passing over Carter Fell to Jedburgh,
and thence to Edinburgh; but rejected as too crooked
and uneven. Another was tried by Aldstone Moor and
Bewcastle, and rejected for the same reason. The third
line proposed was eventually adopted as the best, passing
from Morpeth, by Wooler and Coldstream, to Edinburgh;

saving rather more than fourteen miles between the two points, and securing a line of road of much more favourable gradients.

The principal bridge on this new highway was at Pathhead, over the Tyne, about eleven miles south of Edinburgh. To maintain the level, so as to avoid the winding of the road down a steep descent on one side of the valley and up an equally steep ascent on the other, Telford ran out a lofty embankment from both sides, connecting their ends by means of a spacious bridge. The structure at Pathhead is of five arches, each 50 feet span, with 25 feet rise from their springing, 49 feet above the bed of the river. Bridges of a similar character were also thrown over the deep ravines of Cranston Dean and Cotty Burn, in the same neighbourhood. At the same time a useful bridge was built on the same line of road at Morpeth, in Northumberland, over the river Wansbeck. It consisted of three arches, of which the centre one was 50 feet span, and two side-arches 40 feet each; the breadth between the parapets being 30 feet.

The advantages derived from the construction of these new roads were found to be so great, that it was proposed to do the like for the remainder of the line between London and Edinburgh; and at the instance of the Post-Office authorities, with the sanction of the Treasury, Mr. Telford proceeded to make detailed surveys of an entire new post-road between London and Morpeth. In laying it out, the main points which he endeavoured to secure were directness and flatness; and 100 miles of the proposed new Great North Road, south of York, were laid out in a perfectly straight line. This survey, which was begun in 1824, extended over several years; and all the requisite arrangements had been made for beginning the works, when the result of the locomotive competition at Rainhill, in 1829, had the effect of directing attention to that new method of travelling, fortunately in time to prevent what would have proved, for the most part, an unnecessary expenditure, on works soon to be superseded by a totally different order of things.

The most important road-improvements actually carried out under Mr. Telford's immediate superintendence were those on the western side of the island, with the object of shortening the distance and facilitating the communication between London and Dublin by way of Holyhead, as well as between London and Liverpool. At the time of the Union the mode of transit between the capital of Ireland and the metropolis of the United Kingdom was tedious, difficult, and full of peril. In crossing the Irish Sea to Liverpool, the packets were frequently tossed about for days together. On the Irish side there was scarcely the pretence of a port, the landing-place being within the bar of the river Liffey, inconvenient at all times, and in rough weather extremely dangerous. To avoid the long voyage to Liverpool, the passage began to be made from Dublin to Holyhead, the nearest point of the Welsh coast. Arrived there, the passengers were landed upon rugged, unprotected rocks, without a pier or landing conveniences of any kind.[1] But the traveller's perils were not at an end,—comparatively speaking they were only begun. From Holyhead, across the island of Anglesea, there was no made road, but only a miserable track, circuitous and craggy, full of terrible jolts, round bogs and over rocks, for a distance of twenty-four miles. Having reached the Menai Strait, the passengers had

---

[1] A diary is preserved of a journey to Dublin from Grosvenor Square, London, 12th June, 1787, in a coach and four, accompanied by a post-chaise and pair, and five outriders. The party reached Holyhead in four days, at a cost of 75l. 11s. 3d. The state of intercourse between this country and the sister island at this part of the account is strikingly set forth in the following entries: "Ferry at Bangor, 1l. 10s.; expenses of the yacht hired to carry the party across the channel, 28l. 7s. 9d.; duty on the coach, 7l. 13s. 4d.; boats on shore, 1l. 1s.: total, 114l. 3s. 4d."—Roberts's 'Social History of the Southern Counties,' p. 504.

again to take to an open ferry-boat before they could gain the main land. The tide ran with great rapidity through the Strait, and, when the wind blew strong, the boat was liable to be driven far up or down the channel, and was sometimes swamped altogether. The perils of the Welsh roads had next to be encountered, and these were in as bad a condition at the beginning of the present century as those of the Highlands above described. Through North Wales they were rough, narrow, steep, and unprotected, mostly unfenced, and in winter almost impassable. The whole traffic on the road between Shrewsbury and Bangor was conveyed by a small cart, which passed between the two places once a week in summer. As an illustration of the state of the roads in South Wales, which were quite as bad as those in the North, we may state that, in 1803, when the late Lord Sudeley took home his bride from the neighbourhood of Welshpool to his residence only thirteen miles distant, the carriage in which the newly married pair rode stuck in a quagmire, and the occupants, having extricated themselves from their perilous situation, performed the rest of their journey on foot.

The first step taken was to improve the landing-places on both the Irish and Welsh sides of St. George's Channel, and for this purpose Mr. Rennie was employed in 1801. The result was, that Howth on the one coast, and Holyhead on the other, were fixed upon as the most eligible sites for packet stations. Improvements, however, proceeded slowly, and it was not until 1810 that a sum of 10,000l. was granted by Parliament to enable the necessary works to be begun. Attention was then turned to the state of the roads, and here Mr. Telford's services were called into requisition. As early as 1808 it had been determined by the Post-Office authorities to put on a mail-coach between Shrewsbury and Holyhead; but it was pointed out that the roads in North Wales were so rude and dangerous that it was

doubtful whether the service could be conducted in
safety. Attempts were made to enforce the law with
reference to their repairs, and no less than twenty-one
townships were indicted by the Postmaster-General.
The route was found too perilous even for a riding
post, the legs of three horses having been broken in one
week.[1] The road across Anglesea was quite as bad.
Sir Henry Parnell mentioned, in 1819, that the coach
had been overturned beyond Gwynder, going down one
of the hills, when a friend of his was thrown a considerable distance from the roof into a pool of water.
Near the post-office of Gwynder, the coachman had
been thrown from his seat by a violent jolt, and broken
his leg. The post-coach, and also the mail, had been
overturned at the bottom of Penmyndd Hill; and the
route was so dangerous that the London coachmen, who
had been brought down to "work" the country, refused
to continue the duty because of its excessive dangers.
Of course, anything like a regular mail-service through
such a district was altogether impracticable.

The indictments of the townships proved of no
use; the localities were too poor to provide the means
required to construct a line of road sufficient for the conveyance of mails and passengers between England and
Ireland. The work was really a national one, to be
carried out at the national cost. How was this best
to be done? Telford recommended that the old road
between Shrewsbury and Holyhead (109 miles long)
should be shortened by about four miles, and made as
nearly as possible on a level; the new line proceeding
from Shrewsbury by Llangollen, Corwen, Bettws-y-
Coed, Capel-Curig, and Bangor, to Holyhead. Mr.
Telford also proposed to cross the Menai Strait by
means of a cast iron bridge, hereafter to be described.

---

[1] 'Second Report from Committee on Holyhead Roads and Harbours,' 1810.
(Parliamentary paper.)

Although a complete survey was made in 1811, nothing was done for several years. The mail-coaches continued to be overturned, and stage-coaches, in the tourist season, to break down as before.[1] The Irish mail-coach took forty-one hours to reach Holyhead from the time of its setting out from St. Martin's-le-Grand; the journey was performed at the rate of only 6¾ miles an hour, the mail arriving in Dublin on the third day. The Irish members made many complaints of the delay and dangers to which they were exposed in travelling up to town. But, although there was much discussion, there was no money voted until the year 1815, when Sir Henry Parnell vigorously took the question in hand and successfully carried it through. A Board of Parliamentary Commissioners was appointed, of which he was chairman, and, under their direction, the new Shrewsbury and Holyhead road was at length commenced and carried to completion, the works extending over a period of about fifteen years. The same Commissioners exercised an authority over the roads between London and Shrewsbury; and numerous improvements were also made in the main line at various points, with the object of facilitating communication between London and Liverpool, as well as between London and Dublin.

---

[1] Many parts of the road are extremely dangerous for a coach to travel upon. At several places between Bangor and Capel-Curig there are a number of dangerous precipices without fences, exclusive of various hills that want taking down. At Ogwen Pool there is a very dangerous place where the water runs over the road, extremely difficult to pass at flooded times. Then there is Dinas Hill, that needs a side fence against a deep precipice. The width of the road is not above twelve feet in the steepest part of the hill, and two carriages cannot pass without the greatest danger. Between this hill and Rhyddlanfair there are a number of dangerous precipices, steep hills, and difficult narrow turnings. From Corwen to Llangollen the road is very narrow, long, and steep; has no side fence, except about a foot and a half of mould or dirt, which is thrown up to prevent carriages falling down three or four hundred feet into the river Dee. Stage-coaches have been frequently overturned and broken down from the badness of the road, and the mails have been overturned; but I wonder that more and worse accidents have not happened, the roads are so bad. — Evidence of Mr. William Akers, of the Post-Office, before Committee of the House of Commons, 1st June, 1815.

The rugged nature of the country through which the new road passed, along the slopes of rocky precipices and across inlets of the sea, rendered it necessary to build many bridges, to form many embankments, and cut away long stretches of rock, in order to secure an easy and commodious route. The line of the valley of the Dee, to the west of Llangollen, was selected, the road proceeding along the scarped sides of the mountains, crossing from point to point by lofty embankments where necessary; and, taking into account the character of the country, it must be acknowledged that a wonderfully level road was secured. Whilst the gradients on the old road had in some cases been as steep as 1 in 6½, passing along the edge of unprotected precipices, the new one was so laid out as to be no more than 1 in 20 at any part, while it was wide and well protected along its whole extent. Mr. Telford pursued the same system that he had adopted in the formation of the Carlisle and Glasgow road, as regards metalling, cross-draining, and fence-walling; for the latter purpose using schistus, or slate rubble-work, instead of sandstone. The largest bridges were of iron; that at Bettws-y-Coed, over the Conway—called the Waterloo Bridge, constructed in 1815—being a very fine specimen of Telford's iron bridge-work.

Those parts of the road which had been the most dangerous were taken in hand first, and, by the year 1819, the route had been rendered comparatively commodious and safe. Angles were cut off, the sides of hills were blasted away, and several heavy embankments run out across formidable arms of the sea. Thus, at Stanley Sands, near Holyhead, an embankment was formed 1300 yards long and 16 feet high, with a width of 34 feet at the top, along which the road was laid. Its breadth at the base was 114 feet, and both sides were coated with rubble stones, as a protection against storms. By the adoption of this expedient, a mile and a half was saved in a distance of six miles. Heavy embankments were also run out, where bridges were thrown across chasms and ravines, to maintain the general level. From Ty-Gwynn to Lake Ogwen, the road along the face of the rugged hill and across the river Ogwen was entirely new made, of a uniform width of 28 feet between the parapets, with an inclination of only 1 in 22 in the steepest place. A bridge was thrown over the deep chasm forming the channel of the Ogwen, the embankment being carried forward from the rock-cutting, protected by high breastworks. From Capel - Curig to near the great waterfall over the river Lugwy, about a mile of new road was cut; and a still greater length from Bettws across the river Conway and along the face of Dinas Hill to Rhyd-dlanfair, a distance of 3 miles, its steepest descent being 1 in 22, diminishing to 1 in 45. By

ROAD DESCENT TO THE VALLEY OF THE LUGWY.

[By Percival Skelton]

this improvement, the most difficult and dangerous pass along the route through North Wales was rendered safe and commodious. Another point of almost equal difficulty occurred near Ty-Nant, through the rocky pass of Glynn Duffrws, where the road was confined between

steep rocks and rugged precipices : there the way was
widened and flattened by blasting, and thus reduced to

ROAD ABOVE NANT FFRANCON, NORTH WALES

[By Percival Skelton, after his original Drawing.]

the general level; and so on eastward to Llangollen
and Chirk, where the main Shrewsbury road to London
was joined.[1]

---

[1] The Select Committee of the
House of Commons, in reporting as to
the manner in which these works
were carried out, stated as follows :—
" The professional execution of the
new works upon this road greatly sur-
passes anything of the same kind in
these countries.   The science which
has been displayed in giving the
general line of the road a proper inclination through a country whose
whole surface consists of a succession
of rocks, bogs, ravines, rivers, and
precipices, reflects the greatest credit
upon the engineer who has planned
them ; but perhaps a still greater degree of professional skill has been
shown in the construction, or rather
the building, of the road itself.   The
great attention which Mr. Telford
has bestowed to give to the surface of
the road one uniform and moderately
convex shape, free from the smallest
inequality   throughout   its   whole
breadth ; the numerous land drains,
and, when necessary, shores and tunnels of substantial masonry, with
which all the water arising from
springs or falling in rain is instantly
carried off; the great care with which
a sufficient foundation is established
for the road, and the quality, solidity,
and disposition of the materials that
are put upon it, are matters quite new
in the system of roadmaking in these
countries."—' Report from the Select
Committee on the Road from London
to Holyhead in the year 1819.'

By means of these admirable roads the traffic of North Wales continues to be mainly carried on to this day. Although railways have superseded coach-roads in the more level districts, the hilly nature of Wales precludes their formation in that quarter to any considerable extent; and even in the event of railways being constructed, a large part of the traffic of every country must necessarily continue to pass over the old high roads. Without them even railways would be of comparatively little value; for a railway station is of use chiefly because of its easy accessibility, and thus, both for passengers and merchandise, the common roads of the country are as useful as ever they were, though the main post-roads have in a great measure ceased to be employed for the purpose for which they were originally designed.

The excellence of the roads constructed by Mr. Telford through the formerly inaccessible counties of North Wales was the theme of general praise; and their superiority, compared with those of the richer and more level districts in the midland and western English counties, becoming the subject of public comment, he was called upon to execute like improvements upon that part of the post-road which extended between Shrewsbury and the metropolis. A careful survey was made of the several routes from London northward by Shrewsbury as far as Liverpool; and the short line by Coventry, being 153 miles from London to Shrewsbury, was selected as the one to be improved to the utmost. Down to 1819 the road between London and Coventry was in a very bad state, being so laid as to become a heavy slough in wet weather. There were also many steep hills to be cut down, in some parts of deep clay, in others deep sand. A mail-coach had been tried to Banbury; but the road below Aylesbury was so bad, that the Post-Office authorities were obliged to give it up. The twelve miles from Towcester to Daventry

were still worse. The line of way was covered with
banks of dirt; in winter it was a puddle of from four to
six inches deep—quite as bad as it had been in Arthur
Young's time; and when horses passed along the road,
they came out of it a mass of mud and dirt.[1]   There were
also several steep and dangerous hills to be crossed.
The loss of horses by fatigue in travelling by that
route was represented at the time to be very great.
Even the roads in the immediate neighbourhood of the
metropolis were little better, those under the Highgate
and Hampstead trust being pronounced in a wretched
state.   They were badly formed, on a clay bottom, and
being undrained, were almost always wet and sloppy.
The gravel was usually tumbled on and spread unbroken,
so that the materials, instead of becoming consolidated,
were only rolled about by the wheels of the carriages
passing over them.   Mr. Telford applied the same
methods in the reconstruction of these roads that he had
already adopted in Scotland and Wales, and the same
improvement was shortly experienced in the more easy
passage of vehicles of all sorts and the great acceleration
of the mail service.

In addition to the reconstruction of these roads, that
along the coast from Bangor, by Conway, Abergele,
St. Asaph, and Holywell, to Chester, was greatly improved.   It formed the mail road from Dublin to Liverpool, and it was considered of importance to render it as
safe and level as possible.   The principal new cuts on
this line were those along the rugged skirts of the huge
Penmaen-Mawr; around the base of Penmaen-Bach to
the town of Conway; and between St. Asaph and Holywell, to ease the ascent of Rhyall Hill.   But more
important than all, as a means of completing the main

---

[1] Evidence of William Waterhouse before the Select Committee, 10th March,
1819.

line of communication, there were the great bridges over the Conway and the Menai Straits to be constructed. These dangerous ferries had still to be crossed in open boats, sometimes in the night, when the luggage and mails were exposed to great risks, and passengers occasionally lost with them. It was therefore determined, after long consideration, to erect bridges over these formidable straits, and Mr. Telford was employed to execute the works,—in what manner we propose to describe in the succeeding chapter.

CHAPTER XII.

THE MENAI AND CONWAY BRIDGES.

THE erection of a bridge over the Straits of Menai had long been matter of speculation amongst engineers. As early as 1776 Mr. Golborne proposed his plan of an embankment, with a bridge in the middle of it ; and a few years later, in 1785, Mr. Nichols proposed a wooden viaduct, furnished with drawbridges, at Cadnant Island. Later still, Mr. Rennie proposed his design of a cast iron bridge. But none of these plans were carried out, and the whole subject remained in abeyance until the year 1810, when a commission was appointed to inquire and report as to the state of the roads between Shrewsbury, Chester, and Holyhead. The result was, that Mr. Telford was called upon to report as to the most effectual method of bridging the Menai Strait, and thus completing the communication with the port of embarkation for Ireland.

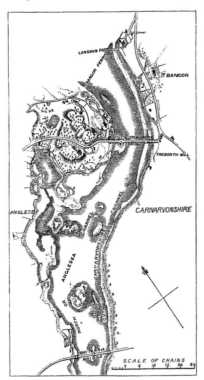

MAP OF MENAI STRAIT. [Ordnance Survey.]

Mr. Telford submitted alternative plans for a bridge

over the Strait: one at the Swilly Rock, consisting of
three cast iron arches of 260 feet span, with a stone
arch of 100 feet span between each two iron ones, to
resist their lateral thrust; and another at Ynys-y-moch,
to which he himself attached the preference, consisting
of a single cast iron arch of 500 feet span, the crown
of the arch to be 100 feet above high water of spring
tides, and the breadth of the roadway to be 40 feet.

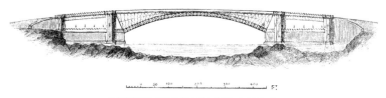

TELFORD'S PROPOSED CAST IRON BRIDGE.

The principal objection taken to this plan by engineers
generally, was the supposed difficulty of erecting a proper
centering to support the arch during construction; and
the mode by which Mr. Telford proposed to overcome
this may be cited in illustration of his ready ingenuity
under such circumstances.   He proposed to suspend the
centering from above instead of supporting it from below
in the usual manner—a contrivance afterwards revived
by another very skilful engineer, the late Mr. Brunel.
Frames, fifty feet high, were to be erected on the top of
the abutments, and on these strong blocks, or rollers and
chains, were to be fixed, by means of which, and by the
aid of windlasses and other mechanical powers, each
separate piece of centering was to be raised into, and
suspended in, its proper place.   Mr. Telford regarded

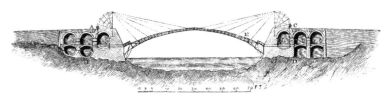

PROPOSED PLAN OF SUSPENDED CENTERING.

this method of constructing centres as applicable to stone as well as to iron arches; and indeed it is applicable, as Mr. Brunel held, to the building of the arch itself.[1] Mr. Telford anticipated that, if the method recommended by him were successfully adopted on the large scale proposed at Menai, all difficulties with regard to carrying bridges over deep ravines would be done away with, and a new era in bridge-building begun. For this and other reasons—but chiefly because of the much greater durability of a cast iron bridge compared with the suspension bridge afterwards adopted—it is matter of regret that he was not permitted to carry out this novel and grand design. It was, however, again objected by mariners that the bridge would seriously affect, if not destroy, the navigation of the Strait; and this plan, like Mr. Rennie's, was eventually rejected.

Several years passed, and during the interval Mr. Telford was consulted as to the construction of a bridge over Runcorn Gap on the Mersey, above Liverpool. As the river was there about twelve hundred feet wide, and much used for purposes of navigation, a bridge of the ordinary construction was found inapplicable. But as he was required to furnish a plan of the most suitable structure, he proceeded to consider how the difficulties

---

[1] In an article in the 'Edinburgh Review,' No. cxli., from the pen of Sir David Brewster, the writer observes:—"Mr. Telford's principle of suspending and laying down from above the centering of stone and iron bridges is, we think, a much more fertile one than even he himself supposed. With modifications, by no means considerable, and certainly practicable, it appears to us that the voussoirs or arch-stones might themselves be laid down from above, and suspended by an appropriate mechanism till the keystone was inserted. If we suppose the centering in Mr. Telford's plan to be of iron, this centering itself becomes an iron bridge, each rib of which is composed of ten pieces of fifty feet each; and by increasing the number of suspending chains, these separate pieces or voussoirs having been previously joined together, either temporarily or permanently, by cement or by clamps, might be laid into their place, and kept there by a single chain till the road was completed. The voussoirs, when united, might be suspended from a general chain across the archway, and a platform could be added to facilitate the operations." This is as nearly as possible the plan afterwards revived by Mr. Brunel, and for the originality of which, we believe, he has generally the credit, though it clearly belongs to Telford.

of the case were best to be met.   The only practicable
plan, he thought, was a bridge constructed on the principle of suspension.   Expedients of this kind had long
been employed in India and America, where wide rivers
were crossed by means of bridges formed of ropes and
chains; and even in this country a suspension bridge,
though of a very rude kind, had long been in use near
Middleton on the Tees, where, by means of two common
chains stretched across the river, upon which a footway
of boards was laid, the colliers were enabled to pass
from their cottages to the colliery on the opposite bank.
Captain (afterwards Sir Samuel Brown) took out a patent
for forming suspension bridges in 1817; but it appears
that Telford's attention had been directed to the subject
before this time, as he was first consulted respecting the
Runcorn Bridge in the year 1814, when he proceeded to

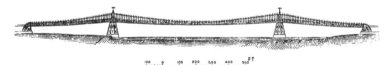

OUTLINE OF TELFORD'S PROPOSED BRIDGE AT RUNCORN

make an elaborate series of experiments on the tenacity
of wrought iron bars, with the object of employing this
material in his proposed structure.   After he had made
upwards of two hundred tests of malleable iron of various
qualities, he proceeded to prepare his design of a bridge,
which consisted of a central opening of 1000 feet span,
and two side openings of 500 feet each, supported by pyramids of masonry placed near the low water lines.   The
roadway was to be 30 feet wide, divided into one central
footway and two distinct carriageways of 12 feet each.
At the same time he prepared and submitted a model of
the central opening, which satisfactorily stood the various
strains which were applied to it.   This Runcorn design
of 1814 was of a very magnificent character, perhaps
superior even to that of the Menai Suspension Bridge,

afterwards erected; but unhappily the means were not forthcoming to carry it into effect. The publication of his plan and report had, however, the effect of directing public attention to the construction of bridges on the suspension principle; and many were shortly after designed and erected by Telford and other engineers in different parts of the kingdom.

Mr. Telford continued to be consulted by the Commissioners of the Holyhead Roads as to the completion of the last and most important link in the line of communication between London and Holyhead, by bridging the Straits of Menai; and at one of their meetings in 1815, shortly after the publication of his Runcorn design, the inquiry was made whether a bridge upon the same principle was not applicable in this particular case. The engineer was instructed again to examine the Straits and submit a suitable plan and estimate, which he proceeded to do in the early part of 1818. The site selected by him as the most favourable was that which had previously been fixed upon for the projected cast iron bridge, namely at Ynys-y-moch—the shores there being bold and rocky, affording easy access and excellent foundations, whilst by spanning the entire channel between the low water lines, and the roadway being kept uniformly 100 feet above the highest water at spring tide, the whole of the navigable waterway would be left entirely uninterrupted. The distance between the centres of the supporting pyramids was proposed to be of the then unprecedented width of 550 feet, and the height of the pyramids 53 feet above the level of the roadway. The main chains were to be sixteen in number, with a deflection of 37 feet, each composed of thirty-six bars of

MENAI BRIDGE.

half-inch-square iron, so placed as to give a square of
six on each side, making the whole chain about four inches
in diameter, welded together for their whole length,
secured by bucklings, and braced round with iron wire;
whilst the ends of these great chains were to be secured
by a mass of masonry, built over stone arches between
each end of the supporting piers and the adjoining shore.
Four of these arches were to be on the Anglesea, and
three on the Caernarvonshire side, each of them fifty-two
feet six inches span. The roadway was to be divided, as
in the Runcorn design—a carriageway twelve feet wide
on each side, and a footpath of four feet in the middle.
Mr. Telford's plan was supported by Mr. Rennie and
other engineers of eminence; and the Select Committee
of the House of Commons, being satisfied as to its practicability, recommended Parliament to pass a Bill and
make a grant of money to enable the work to be carried
into effect.

The necessary Act passed in the session of 1819,
and Mr. Telford immediately proceeded to Bangor to
make preparations for commencing the works. The
first proceeding was to blast off the inequalities from
the surface of the rock, called Ynys-y-moch, situated on
the western or Holyhead side of the Strait, at that time
only accessible at low water. The object was to form an
even surface upon it for the foundation of the west main
pier. It used to be at this point, where the Strait
was narrowest, that horned cattle were driven down,
preparatory to swimming them across the channel to
the Caernarvon side, when the tide was weak and at
its lowest ebb. Many cattle were, nevertheless, often
carried away when the current was too strong for the
animals to contend against.

At the same time a landing-quay was erected on
Ynys-y-moch, which was connected with the shore by
an embankment carrying lines of railway. Along this
horses drew the sledges laden with stone required for the

work; the material being brought in barges from the quarries opened at Penmon Point, on the north-eastern extremity of the Isle of Anglesea, a little to the westward of the northern opening of the Strait. When the surface of the rock had been levelled and the causeway completed, the first stone of the main pier was laid by Mr. W. A. Provis, the resident engineer, on the 10th of August, 1819; but not the slightest ceremony was observed on the occasion.

Later in the autumn preparations were made for proceeding with the foundations of the eastern main pier on the Bangor side of the Strait. After excavating the beach to a depth of seven feet, a solid mass of rock was reached, which served the purpose of an immoveable foundation for the pier. At the same time workshops were erected; builders, artisans, and labourers were brought together from distant quarters; vessels and barges were purchased or built for the special purpose of the work; a quay was constructed at Penmon Point for loading the stones for the piers; and all the requisite preliminary arrangements were made for proceeding with the building operations in the ensuing season.

A careful specification of the masonry work was drawn up, and the contract first let to Messrs. Stapleton and Hall; but as they did not proceed satisfactorily, and desired to be released from the contract, it was re-let on the same terms to Mr. John Wilson, one of Mr. Telford's principal contractors for mason work on the Caledonian Canal. The building operations were begun with great vigour early in 1820. The three arches on the Caernarvonshire side and the four on the Anglesea side were first proceeded with. They are of immense magnitude, and occupied four years in construction, having been finished late in the autumn of 1824. These piers are 65 feet in height from high water line to the springing of the arches, the span of each being 52 feet 6 inches. The work of the main piers also made satisfactory progress, and the masonry proceeded so rapidly that stones could scarcely be got from the quarries in sufficient quantity to keep the builders at work. By the end of June about three hundred men were employed.

The two main piers, each 153 feet in height, upon which the main chains of the bridge were to be suspended, were built with great care and under rigorous inspection. In these, as indeed in most of the masonry of the bridge, Mr. Telford adopted the same practice which he had employed in his previous bridge structures, that of leaving large void spaces, commencing above high water mark and continuing them up perpendicularly nearly to the level of the roadway. " I have elsewhere expressed my conviction," he says, when referring to the mode of constructing these piers, " that one of the most important improvements which I have been able to introduce into masonry consists in the preference of cross-walls to rubble, in the structure of a pier, or any

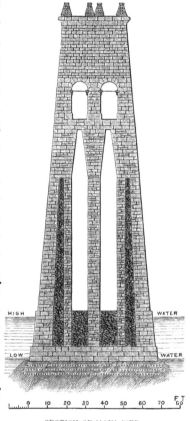

SECTION OF MAIN PIER

other edifice requiring strength. Every stone and joint in such walls is open to inspection in the progress of the work, and even afterwards, if necessary; but a solid filling of rubble conceals itself, and may be little better than a heap of rubbish confined by side walls."[1] The

---

[1] ' Life of Telford,' p. 221.

walls of these main piers were built from within as well
as from without all the way up, and the inside was as
carefully and closely cemented with mortar as the external face.   Thus the whole pier was bound firmly
together, and the utmost strength given, while the
weight of the superstructure upon the lower parts of
the work was reduced to its minimum.

Over the main piers the small arches intended for the
roadways were constructed, each being 15 feet to the
springing of the arch, and 9 feet wide.   Upon these
arches the masonry was carried upwards, in a tapering
form, to a height of 53 feet above the level of the road.
As these piers were to carry the immense weight of the
suspension chains, great pains were taken with their
construction,   and each stone, from top to bottom, was
firmly bound together with iron dowels to prevent the
possibility of their being separated or bulged by the
immense pressure they had to withstand.[1]

The most important point in the execution of the
details of the bridge, where the engineer had no past
experience to direct him, was in the designing and fixing
of the wrought iron work.   Mr. Telford had continued his
experiments as to the tenacity of bar iron, until he had
obtained several hundred distinct tests; and at length,
after the most mature deliberation, the patterns and
dimensions were finally arranged by him, and the contract for the manufacture of the whole was let to Mr.
Hazeldean, of Shrewsbury, in the year 1820.   The iron
was to be of the best Shropshire, drawn at Upton forge,

---

[1] To guard against the effects of
lateral pressure resulting from this
heavy mass of masonry bearing on
the arches forming the roadway, six
strong wrought iron ties, four inches
wide and two inches thick, firmly
bolted at each end, were introduced
horizontally at the springing of the
arches over the carriage ways, and
thus any deficiency of strength at that
point was effectually provided for.

Strong cast iron blocks or saddles
were placed upon the top of the
piers to bear the suspension chains,
and they were fitted with wrought
iron self-acting rollers and brass
bushes, for the purpose of regulating
the contraction and expansion of the
iron, by moving themselves either
way according to the temperature of
the atmosphere, without the slightest
derangement to any part of the work.

and finished and proved at his works, under the inspection of a person appointed by the engineer.[1]

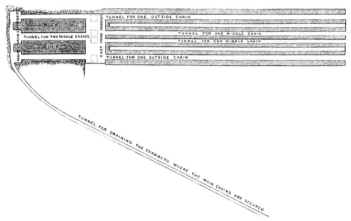

CUT SHOWING FIXING OF THE CHAINS IN THE ROCK.

The mode by which the land ends of those enormous suspension chains were rooted to the solid ground on either side of the Strait, was remarkably ingenious and effective. Three oblique tunnels were made by blasting the rock on the Anglesea side; they were each about six feet in diameter, the excavations being carried down an inclined plane to the depth of about twenty yards. A considerable width of rock lay be-

---

[1] The cross sections of the bars forming the main chains were 3¼ inches square; they were proved to be capable, according to Mr. Telford's experiments, of bearing a strain of not less than 87¾ tons before fracture; but in order not to strain the iron unduly, the proof was limited to 35 tons, or about 11 tons to every square inch of cross section. Every piece of iron introduced in the work was submitted to this test by means of a very accurate and powerful proving machine constructed for the purpose; while under the strain, it was frequently struck with a hammer, and after being proved, every separate piece was well cleansed, put into a stove, and when brought to a gentle heat was immersed in a trough con- taining linseed oil. It was then taken out after a short time, again put into the stove, and, when dried, appeared as if covered with varnish. A coat of linseed oil paint was finally put over all, and the iron bar was in this state sent to the bridge for use. "No precautions were spared," writes Mr. Telford, "to render every part perfectly true, and therefore secure; for as any variation in the length of the numerous bars would produce unequal bearings, each was subjected to a fresh adjustment by means of a steel model, upon which they were bored when cold, so that a cross-bolt passed through a certain number, in most cases through eight bars, so as to form four chains, thus accurately attached to each other."

tween each tunnel, but at the bottom they were all
united by a connecting horizontal avenue or cavern,
sufficiently capacious to enable the workmen to fix the
strong iron frames, composed principally of thick flat
cast iron plates, which were engrafted deeply into the
rock, and strongly bound together by the iron work
passing along the horizontal avenue; so that, if the
iron held, the chains could only yield by tearing up the
whole mass of solid rock under which they were thus
firmly bound.

A similar method of anchoring the main chains was
adopted on the Caernarvonshire side. A thick bank
of earth had there to be cut through, and a solid
mass of masonry built in its place, the rock being
situated at a greater distance from the main pier; thus
involving a greater length of suspending chain, and a
disproportion in the catenary or chord line on that side
of the bridge. The excavation and masonry thus rendered necessary proved a work of vast labour, and its
execution occupied a considerable time; but by the
beginning of the year 1825 the suspension pyramids,
the land piers and arches, and the rock tunnels, had all
been completed, and the main chains firmly secured in
them; the work being sufficiently advanced to enable
the suspending of the chains to be proceeded with. This
was by far the most difficult and anxious part of the
undertaking.

With the same careful forethought and provision for
every contingency which had distinguished the engineer's procedure in the course of the work, he had
made frequent experiments to ascertain the actual
power which would be required to raise the main chains
to their proper curvature. A valley lay convenient
for the purpose, a little to the west of the bridge on
the Anglesea side. Fifty-seven of the intended vertical suspending rods, each nearly ten feet long and
an inch square, having been fastened together, a piece

of chain was attached to one end to make the chord
line 570 feet in length; and experiments having been
made and comparisons drawn, Mr. Telford ascertained
that the absolute weight of one of the main chains of the
bridge between the points of suspension was 23½ tons,
requiring a strain of 39½ tons to raise it to its proper
curvature. On this calculation the necessary apparatus
required for the hoisting was prepared. The mode of
action finally determined on for lifting the main chains,
and fixing them into their places, was to build the central portion of each upon a raft 450 feet long and 6 feet
wide, then to float it to the site of the bridge, and lift it
into its place by capstans and proper tackle.

At length all was ready for hoisting the first great
chain, and about the middle of April, 1825, Mr. Telford
left London for Bangor to superintend the operations.
An immense assemblage collected to witness the sight;
greater in number than any that had been collected
in the same place since the men of Anglesea, in their
war-paint, rushing down to the beach, had shrieked
defiance across the Straits at their Roman invaders
on the Caernarvon shore. Numerous boats arrayed
in gay colours glided along the waters; the day—the
26th of April—being bright, calm, and in every way
propitious. At half-past two, about an hour before
high water, the raft bearing the main chain was cast
off from near Treborth Mill, on the Caernarvon side.
Towed by four boats, it began gradually to move from
the shore, and with the assistance of the tide, which
caught it at its further end, it swung slowly and majes-
tically round to its position between the main piers,
where it was moored. One end of the chain was then
bolted to that which hung down the face of the Caernarvon pier; whilst the other was attached to ropes
connected with strong capstans fixed upon the Anglesea
side, the ropes passing by means of blocks over the top
of the pyramid of the Anglesea pier. The capstans for

hauling in the ropes bearing the main chain, were two in number, manned by about 150 labourers.   When all

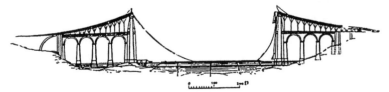

CUT OF BRIDGE, SHOWING STATE OF SUSPENSION CHAIN

was ready, the signal was given to "go along!"   A band of fifers struck up a lively tune; the capstans were instantly in motion, and the men stepped round in a steady trot. All went well. The ropes gradually coiled in. As the strain increased, the pace slackened a little; but "Heave away! now she comes!" was sung out. Round went the men, and steadily and safely rose the ponderous chain. The tide had by this time turned, and bearing upon the side of the raft, now getting freer of its load, the current floated it away from under the middle of the chain still resting on it, and it swung easily off into the water. Until this moment a breath- less silence pervaded the watching multitude; and nothing was heard amongst the working party on the Anglesea side but the steady tramp of the men at the capstans, the shrill music of the fife, and the occasional order to "Hold on!" or "Go along!"   But no sooner was the raft seen floating away, and the great chain safely swinging in the air, than a tremendous cheer burst forth along both sides of the Straits. The rest of the work was only a matter of time. The most anxious moment had passed. In an hour and thirty-five minutes after the commencement of the hoisting, the chain was raised to its proper curvature, and fastened to the land portion of it which had been previously placed over the top of the Anglesea pyramid. Mr. Telford ascended to the point of fastening, and satisfied himself that a continuous and safe connection had been formed

from the Caernarvon fastening on the rock to that on
Anglesea. The announcement of the fact was followed
by loud and prolonged cheering from the workmen,
echoed by the spectators, and extending along the Straits
on both sides, until it seemed to die away along the
shores in the distance. Three foolhardy workmen,
excited by the day's proceedings, had the temerity to
scramble along the upper surface of the chain—which
was only nine inches wide and formed a curvature of
590 feet—from one side of the Strait to the other!

Far different were the feelings of the engineer who
had planned this magnificent work. Its failure had been
predicted; and, like Brindley's Barton Viaduct, it had
been freely spoken of as a " castle in the air." Telford
had, it is true, most carefully tested every point by
repeated experiment, and so conclusively proved the
sufficiency of the iron chains to bear the immense weight
they would have to support, that he was thoroughly convinced as to the soundness of his principles of construction; and satisfied that, if rightly manufactured and properly put together, the chains would hold together and
the piers would sustain them. Still there was necessarily
an element of uncertainty in the undertaking. It was
the largest structure of the kind that had ever been
attempted. There was the contingency of a flaw in the
iron; some possible scamping in its manufacture; some
little point which, in the multiplicity of details to be
attended to, he might have overlooked, or which his
subordinates might have neglected. It was, indeed,
impossible but that he should feel intensely anxious as
to the result of the day's operations. Mr. Telford afterwards stated to a friend, only a few months before his
death, that for some time previous to the opening of the
bridge, his anxiety was so extreme that he could scarcely
sleep; and that a continuance of that condition must
have very soon completely undermined his health. We
are not, therefore, surprised to learn that when his

friends rushed to congratulate him on the result of the first day's experiment, which decisively proved the strength and solidity of the bridge, they should have found the engineer upon his knees engaged in prayer. A vast load had been taken off his mind; the perilous enterprise of the day had been accomplished without loss of life; and his spontaneous act was thankfulness and gratitude.

MENAI BRIDGE.

[By Percival Skelton, after his original Drawing.]

The suspension of the remaining fifteen chains was accomplished without difficulty. The last was raised and fixed on the 9th of July, 1825, when the entire line was completed. On fixing the final bolt, a band of music descended from the top of the suspension pier on the Anglesea side to a scaffolding erected over the centre of the curved part of the chains, and played the National Anthem amidst the cheering of many thousand persons assembled along the shores of the Strait; whilst the workmen marched in procession along the bridge, upon

which a temporary platform had been placed, and the
St. David steam-packet of Chester passed under the chains
towards the Smithy Rocks and back again, thus reopen-
ing the navigation of the Strait. In August the road
platform was commenced, and in September the trussed
bearing bars were all suspended. The road was constructed of timber in a substantial manner, the planking
being spiked together, with layers of patent felt between
the planks, and the carriage-way being protected by oak
guards placed seven feet and a half apart. Side railings
were added and toll-houses and approach-roads completed
by the end of the year; and the bridge was opened for
public traffic on Monday, the 30th of January, 1826,
when the London and Holyhead mail-coach passed over
it for the first time, followed by the Commissioners of
the Holyhead roads, the engineer, several stage-coaches,
and a multitude of private persons too numerous to
mention.

We may briefly add a few facts as to the quantities of
materials used, and the dimensions of this remarkable
structure. The total weight of iron was 2187 tons, in
33,265 pieces. The total length of the bridge is 1710 feet,
or nearly a third of a mile; the distance between the
points of suspension of the main bridge being 579 feet.
The total sum expended by Government in its erection,
including the embankment and about half a mile of new
line of road on the Caernarvon side, together with the
toll-houses, was 120,000*l.*

Shortly after the Menai Bridge was commenced, it was
determined by the Commissioners of the Holyhead road
that a bridge of similar design should be built over the
estuary of the Conway, immediately opposite the old
castle at that place, and which had formerly been crossed
by an open ferry boat. The first stone was laid on the
3rd of April, 1822, and, the works having proceeded
satisfactorily, the bridge and embankment approaching
it were completed by the summer of 1826. But the

operations being of the same kind as those connected with the larger structure above described, though of a much less difficult character, it is unnecessary to enter into any details as to the several stages of its construction. In this bridge the width between the centres of the supporting towers is 327 feet, and the height of the under side of the roadway above high water of spring tides only 15 feet. The heaviest work was an embankment at its eastern approach, 2015 feet in length and about 300 feet in width at its highest part.

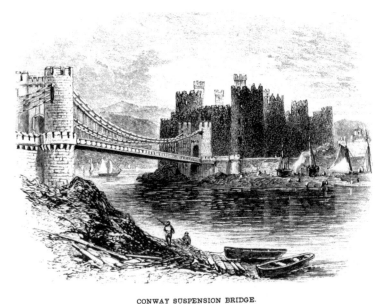

CONWAY SUSPENSION BRIDGE.

[By Percival Skelton, after his original Drawing.]

CHAPTER XIII.

Docks, Drainage, and Bridges.

It will have been observed, from the Lives of those Engineers which we have thus far been enabled to lay before the reader, how much has been done by skill and industry to open up and develope the material resources of the kingdom.   The stages of improvement which we have recorded exhibit a measure of the vital energy which has from time to time existed in the nation.   In the earlier periods the war was with nature; the sea was held back by embankments; the Thames, instead of being allowed to overspread the wide marshes on either bank, was confined within limited bounds, by which the navigable depth of its channel was increased at the same time that a wide extent of land was rendered available for agriculture.

In those early days the great object was to render the land more habitable, comfortable, and productive. Marshes were reclaimed, and wastes subdued. But so long as the country remained comparatively closed, and intercourse was restricted by the want of bridges and roads, improvement was extremely slow. Whilst roads are the consequence of civilization, they are also among its most influential causes. We have seen even the blind Metcalf acting as an effective instrument of progress in the northern counties by the formation of long lines of road. Brindley and the Duke of Bridgewater carried on the work in the same districts, and conferred upon the north and north-west of England the blessings of cheap and effective water communication. Smeaton followed and carried out similar undertakings in still remoter places, joining the east and west coasts of Scotland by the Forth and Clyde Canal, and building bridges in the far north. Rennie made harbours, built bridges, and hewed out docks for shipping, the increase in which had kept pace with the growth of our home and foreign trade. He was followed by Telford, whose long and busy life, as we have seen, was occupied in building bridges and making roads in all directions, in districts of the country formerly inaccessible, and therefore comparatively barbarous. At length the wildest districts of the Highlands and the most rugged mountain valleys of North Wales were rendered as easy of access as the comparatively level counties in the immediate neighbourhood of the metropolis.

During all this while the wealth and industry of the country had been advancing with rapid strides. London had grown in population and importance. Many improvements had been effected in the river, but the dock accommodation was still found insufficient; and, as the recognised head of his profession, Mr. Telford, though now grown old and fast becoming infirm, was called upon to supply the requisite plans. He had been engaged upon great works for upwards of thirty years, previous to which he had led the life of a working mason. But he had been a steady, temperate man all his life; and though nearly seventy, when consulted as to the proposed new docks, his mind was as able to deal with the subject in all its bearings as it had ever been; and he undertook the work.

In 1824 a new Company was formed to provide a dock nearer to the heart of the City than any of the existing ones. The site selected was the space between the Tower and the London Docks, which included the property of St. Katherine's Hospital. The whole extent of land available was only twenty-seven acres of a very irregular figure, so that when the quays and warehouses were laid out, it was found that only about ten acres remained for the docks; but these, from the nature of the ground, presented an unusual amount of quay room. The necessary Act was obtained in 1825; the works were begun in the following year; and on the 25th of October, 1828, the new docks were completed and opened for business.

The St. Katherine Docks communicate with the river by means of an entrance tide-lock, 180 feet long and 45 feet wide, with three pair of gates, admitting either one very large or two smaller vessels at a time. The lock-entrance and the sills under the two middle lock-gates are fixed at the depth of ten feet under the level of low water of ordinary spring tides. The formation of these dock-entrances was a work of much difficulty, demanding great skill on the part of the engineer. It was necessary to excavate the ground to a great depth below low water for the purpose of getting in the foundations, and the cofferdams were therefore of great strength, to enable them, when pumped out by the steam-engine, to resist the lateral pressure of forty feet of water at high tide. The difficulty was, however, effectually overcome, and the wharf walls, locks, sills, and bridges of the St. Katherine Docks are generally regarded as a masterpiece of harbour construction.[1]

---

[1] In the arrangement of the wharves and warehouses every improvement was introduced and adopted which was calculated to facilitate the transaction of business at the least labour and expense. Alluding to the rapidity with which these works were constructed, Mr. Telford says:—" Seldom, indeed never within my knowledge, has there been an instance of an undertaking of this magnitude, in a very confined situation, having been perfected in so short a time; nor could it have been accomplished in any other place than London, where materials and labour to any extent are always to be procured, as also the command of capital in the power of intelligent directors accustomed to transactions on a large scale. The extreme rapidity with which every operation was forced on was doubtless defensible, as useful and desirable in a mercantile speculation, with the view of having a speedy return for the advance of large sums, and for encouraging future advances, also for meeting the urgent demands of increasing commerce; but, as a practical engineer, responsible for the success of difficult operations, I must be allowed to protest against such haste, pregnant as it was, and ever will be, with risks, which, in more instances than one, severely taxed all my experience and skill, and dangerously involved the reputation of the directors as well as their engineer."—'Telford's Life,' p. 155.

Amongst the remaining bridges executed by Mr. Telford, towards the close of his professional career, may be mentioned those of Tewkesbury and Gloucester. The former town is situated on the Severn, at its confluence with the river Avon, about eleven miles above Gloucester. The surrounding district was rich and populous; but being intersected by a large river, without a bridge, the inhabitants applied to Parliament for powers to provide so necessary a convenience. The design first proposed by a local architect was a bridge of three arches; but Mr. Telford, when called upon to advise the trustees, recommended that, in order to interrupt the navigation as little as possible, the river should be spanned by a single arch; and he submitted a design of such a character, which was approved and subsequently erected. It was finished and opened in April, 1826.

This is one of the most extensive as well as graceful of Mr. Telford's numerous cast iron bridges. It is of a single span of 170 feet, with a rise of only 17 feet. It consists of six ribs of about three feet three inches deep, the spandrels being filled in with light diagonal work. The narrow Gothic arches in the masonry of the abutments give the bridge a very light and graceful appearance, at the same time that they afford an enlarged passage for the high river floods.

The bridge at Gloucester consists of one large stone arch of 150 feet span. It replaced a structure of great antiquity, of eight arches, which had stood for about 600 years. The roadway over it was very narrow, and the number of piers in the river and the small dimensions of the arches offered considerable obstruction to the navigation. To give the largest amount of waterway, and at the same time reduce the gradient of the road over the bridge to the greatest extent, Mr. Telford adopted the following expedient. He made the general body of the arch an ellipse, 150 feet on the chord-line and 35 feet rise, while the voussoirs, or external archstones, being in the form of a segment, have the same chord, with only 13 feet rise. " This complex form," says Mr. Telford, " converts each side of the vault of the arch into the shape of the entrance of a pipe, to suit the contracted passage of a fluid, thus lessening the flat surface opposed to the current of the river whenever the tide or upland flood rises above the springing of the middle of the ellipse, that being at four feet above low water; whereas the flood of 1770 rose twenty feet above low water of an ordinary spring-tide, which, when there is no upland flood, rises only eight or nine feet." [1]   The bridge was finished and opened in 1828.

The last structures erected after our engineer's designs were at Edinburgh and Glasgow ; his Dean Bridge at the former place, and his Jamaica Street Bridge at the latter, being regarded as among his most successful works.   Since his employment as a journeyman mason at the building of the houses in Princes Street, Edinburgh, the New Town had spread in all directions.   At each visit to it on his way to or from the Caledonian Canal or the northern harbours, he had been no less surprised than delighted at the architectural improvements which he found going forward.   A new quarter had risen up during his lifetime, and had extended northward and westward in long lines of magnificent buildings of freestone, until in 1829 its further progress was checked by the deep ravine running along the back of the New Town, in the bottom of which runs the little Water of Leith.   It was determined to throw a stone bridge across this stream, and Telford was called upon to supply the design.   The point of crossing the valley was immediately behind Moray Place, which stands almost upon its verge, the sides being bold, rocky, and finely wooded. The situation was well adapted for a picturesque structure, such as Telford was well able to supply.   The depth of

---

[1] 'Telford's Life,' p. 261.

the ravine to be spanned involved great height in the piers, the roadway being 106 feet above the level of the river. The bridge was of four arches of 90 feet span each, and its total length 447 feet; the breadth between the parapets for the purposes of the roadway and foot-

DEAN BRIDGE, EDINBURGH.

[By E. M. Wimperis, after a Sketch by J S. Smiles.]

paths being 39 feet.[1] The bridge was completed and opened in December, 1831.

But the most important, as it was the last, of Mr. Telford's stone bridges was that erected across the Clyde at the Broomielaw, Glasgow. Little more than fifty years since, the banks of the river at that place were literally covered with broom—and hence its name—whilst the

---

[1] The piers are built internally with hollow compartments, as at the Menai Bridge, the side walls being 3 feet thick and the cross walls 2 feet. Projecting from the piers and abutments are pilasters of solid masonry. The main arches have their springing 70 feet from the foundations, and rise 30 feet; and at 20 feet higher, other arches, of 96 feet span and 10 feet rise, are constructed; the face of these, projecting before the main arches and spandrels, producing a distinct external soffit of 5 feet in breadth. This, with the peculiar piers, constitutes the principal distinctive feature in the bridge.

stream was scarcely deep enough to float a herring-buss. Now, the Broomielaw is a quay frequented by ships of the largest burden, and bustling with trade and commerce. Skill and enterprise have deepened the Clyde, dredged away its shoals, built quays and wharves along its banks, and rendered it one of the busiest streams in the world. It has become a great river thoroughfare, worked by steam. On its waters the first steamboat ever constructed for purposes of traffic in Europe was launched by Henry Bell in 1812; and the Clyde boats to this day retain their original prestige.

The deepening of the river at the Broomielaw had led to a gradual undermining of the foundations of the old bridge, which was situated close to the principal landing-place. A little above it was an ancient overfall weir, which had also contributed to scour away the foundations of the piers. Besides, the bridge was felt to be narrow, inconvenient, and ill-adapted for accommodating the immense traffic passing across the Clyde at that point. It was, therefore, determined to take down the old structure, and build a new one; and Mr. Telford was called upon to supply the design. The foundation was laid with great ceremony on the 18th of March, 1833, and the new bridge was completed and opened on the 1st of January, 1836, rather more than a year after the engineer's death. It is a very fine work,[1] consisting of seven arches, segments of circles, the central arch being 58 feet 6 inches; the span of the adjoining arches diminishing to 57 feet 9 inches, 55 feet 6 inches, and 52 feet respectively. It is 560 feet in length, with an open waterway of 389 feet, and its total width of carriageway and footpath is 60 feet, or wider than any river bridge in the kingdom.

Like most previous engineers of eminence—like Perry, Brindley, Smeaton, and Rennie—Mr. Telford was in the course of his life extensively employed upon the drainage

---

[1] See illustration at p 286.

2 H 2

of the Fen districts.   He had been jointly concerned with
Mr. Rennie in carrying out the important works of the
Eau Brink Cut,[1] and at Mr. Rennie's death he succeeded
to much of his practice as consulting engineer.

It was principally in designing and carrying out the
drainage of the North Level that Mr. Telford distinguished himself in Fen drainage.   The North Level
includes all that part of the Great Bedford Level situated between Morton's Leam and the river Welland,
comprising about 48,000 acres of land.   The river Nene,
which brings down from the interior the rainfall of
almost the entire county of Northampton, flows through
nearly the centre of the district.   In some places the
stream is confined by embankments, in others it flows
along artificial cuts, until it enters the great estuary of
the Wash, about five miles below Wisbeach.   This town
is situated on another river which flows through the
Level, called the Old Nene.   Below the point of junction of these rivers with the Wash, and still more to
seaward, was South Holland Sluice, through which the
waters of the South Holland Drain entered the estuary.[2]
At that point a great mass of silt had accumulated,
which tended to choke up the mouths of the rivers
further inland, rendering their navigation difficult and
precarious, and seriously interrupting the drainage of
the whole lowland district traversed by both the Old
and New Nene.   Indeed the sands were accumulating
at such a rate, that the outfall of the Wisbeach river
threatened to become completely destroyed.

Such being the state of things, it was determined to
take the opinion of some eminent engineer, and Mr.
Rennie was employed to survey the district and recommend a measure for the remedy of these great evils.
He performed this service in his usually careful and
masterly manner ; but as the method which he proposed,
complete though it was, would have seriously inter-

---

[1] See 'Life of Rennie,' p. 168.   [2] See map at p. 50, vol. i.

fered with the trade of Wisbeach, by leaving it out of
the line of navigation and drainage which he proposed
to open up, the corporation of that town determined to
employ another engineer; and Mr. Telford was selected
to examine and report upon the whole subject, keeping
in view the improvement of the river immediately adjacent to the town of Wisbeach. Mr. Telford confirmed
Mr. Rennie's views to a large extent, more especially
with reference to the construction of an entirely
new outfall, by making an artificial channel from Kin-
dersley's Cut to Crab-Hole Eye anchorage, by which
a level lower by nearly twelve feet would be secured
for the outfall waters; but he preferred leaving the
river open to the tide as high as Wisbeach, rather than
place a lock with draw-doors at Lutton Leam Sluice, as
had been proposed by Mr. Rennie. He also suggested
that the acute angle at the Horseshoe be cut off and
the river deepened up to the bridge at Wisbeach,
making a new cut along the bank on the south side of
the town, which should join the river again immediately
above it, thereby converting the intermediate space, by
draw-doors and the usual contrivances, into a floating
dock. Though this plan was approved by the parties interested in the drainage, to Telford's great mortification
it was opposed by the corporation of Wisbeach, and like
so many other excellent schemes for the improvement
of the Fen districts, it eventually fell to the ground.

The cutting of a new outfall for the river Nene, however, could not much longer be delayed except with
great danger to the reclaimed lands of the North Level,
which, but for some relief of the kind, must shortly have
become submerged and reduced to their original waste
condition. The subject was revived in 1822, and Mr.
Telford was again called upon, in conjunction with Sir
John Rennie, whose father had died in the preceding
year, to submit a plan of a new Nene Outfall; but it
was not until the year 1827 that the necessary Act

was obtained, and then only with great difficulty and
cost, in consequence of the opposition of the town of
Wisbeach. The works consisted principally of a deep
cut or canal, about six miles in length, penetrating far
through the sandbanks into the deep waters of the
Wash. They were commenced in 1828, and brought
to completion in 1830, with the most satisfactory
results. A greatly improved outfall was secured by thus
carrying the mouths of the rivers out to sea, and the
drainage of the important agricultural districts through
which the Nene flows was greatly benefited; whilst
at the same time nearly 6000 acres of valuable corn-
growing land were added to the county of Lincoln.

But the opening of the Nene Outfall was only the first
of a series of improvements which eventually included
the whole of the valuable lands of the North Level, in
the district situated between the Nene and the Welland.
The opening at Gunthorpe Sluice, which was the outfall
for the waters of the Holland Drain, was not less than
eleven feet three inches above low water at Crab-Hole;
and it was therefore obvious that by lowering this opening, a vastly improved drainage of the whole of the level
district, extending from twenty to thirty miles inland,
for which that sluice was the artificial outlet, would
immediately be secured. Urged by Mr. Telford, an Act
for the purpose of carrying out the requisite improvement was obtained in 1830, and the excavations having
been commenced shortly after, were completed in 1834.
A new cut was made from Clow's Cross to Gunthorpe
Sluice, in place of the winding course of the old Shire
Drain; besides which, a bridge was erected at Cross Keys,
or Sutton Wash, and an embankment made across the Salt
Marshes, forming a high road, which, with the bridges
previously erected at Fossdyke and Lynn, effectually
connected the counties of Norfolk and Lincoln. The
result of the improved outfall was what the engineer
had predicted. A thorough natural drainage was

secured for an extensive district, embracing nearly a hundred thousand acres of fertile land, which had before been very ineffectually though expensively cleared of the surplus water by means of windmills and steam-engines. The productiveness of the soil was greatly increased, and the health and comfort of the inhabitants promoted to an extent that surpassed all previous expectation. The whole of the new cuts were easily navigable, being from 140 to 200 feet wide at bottom, whilst the old outlets had been variable and often choked with shifting sand. The district was thus effectually opened up for navigation, and a ready transit afforded for coals and other articles of consumption. Wisbeach became readily accessible to vessels of much larger burden, and in a few years after the construction of the Nene Outfall the trade of that port had more than doubled. Mr. Telford himself, towards the close of his life, spoke with natural pride of the improvements which he had thus been in so great a measure instrumental in carrying out, and which had so materially promoted the comfort, prosperity, and welfare of a very extensive district.[1]

We may mention, as a remarkable effect of the opening of the new outfall, that in a few hours the lowering of the waters was felt throughout the whole of the Fen level. The sluggish and stagnant drains, cuts, and leams in far distant places, began actually to *flow*; and the sensation created was such, that at Thorney,

---

[1] "The Nene Outfall channel," says Mr. Tycho Wing, "was projected by the late Mr. Rennie in 1814, and executed jointly by Mr. Telford and the present Sir John Rennie. But the scheme of the North Level drainage was eminently the work of Mr. Telford, and was undertaken upon his advice and responsibility, when only a few persons engaged in the Nene Outfall believed that the latter could be made, or if made, that it could be maintained. Mr. Telford distin-guished himself by his foresight and judicious counsels at the most critical periods of that great measure, by his unfailing confidence in its success, and by the boldness and sagacity which prompted him to advise the making of the North Level drainage, in full expectation of the results for the sake of which the Nene Outfall was undertaken, and which are now realised to the extent of the most sanguine hopes."

near Peterborough, some fifteen miles from the sea, the
intelligence penetrated even to the congregation at
church—for it was Sunday morning—that " the waters
were running;" when immediately the whole flocked
out, parson and all, to see the great sight, and acknowledge the blessings of science.[1] A humble Fen poet of
the last century thus quaintly predicted the moral results
likely to arise from the improved drainage of his native
district :—

> " With a change of elements, suddenly
> There shall a change of men and manners be;
> Hearts thick and tough as hides shall feel remorse,
> And souls of sedge shall understand discourse;
> New hands shall learn to work, forget to steal,
> New legs shall go to church, new knees to kneel."

The prophecy has indeed been amply fulfilled. The
barbarous race of Fen-men has disappeared before the
skill of the engineer. As the land has been drained,
the half-starved fowlers and fen-roamers have subsided
into the ranks of steady industry; become farmers,
traders, and labourers. The plough has passed over the
bed of Holland Fen, and the agriculturist reaps his
increase more than a hundred fold. Wide watery wastes,
formerly abounding in fish, are now covered with waving
crops of corn every summer. Sheep graze on the dry
bottom of Whittlesea Mere, and kine low where not
many years since the silence of the waste was only disturbed by the croaking of frogs and the screaming of
wild fowl. All this has been the result of the science of
the engineer, the enterprise of the landowner, and the
industry of our peaceful army of skilled labourers.[2]

---

[1] 'Land we Live In,' vol. i., p. 371.

[2] Now that the land actually won
has been made so richly productive,
the engineer is at work with magnificent schemes of reclamation of lands
at present submerged by the sea.
The Norfolk Estuary Company have
a scheme for reclaiming 50,000 acres;
the Lincolnshire Estuary Company, 30,000 acres; and the Victoria Level
Company, 150,000 acres—all from the
estuary of the Wash. By the process called warping, the land is
steadily advancing upon the ocean,
and before many years have passed,
thousands of acres of the Victoria
Level will have been reclaimed for
purposes of agriculture.

CHAPTER XIV.

Mr. Telford's later Years—his Death and Character.

When Mr. Telford had occasion to visit London on business during the early period of his career, his quarters were at the Salopian Coffee House, now the Ship Hotel, at Charing Cross. It is probable that his Shropshire connections led him in the first instance to the 'Salopian;' but the situation being near to the Houses of Parliament, and in many respects convenient for the purposes of his business, he continued to live there for no less a period than twenty-one years. During that time the Salopian became a favourite resort for engineers; and not only Telford's provincial associates, but numerous visitors from abroad (where his works attracted even more attention than they did in England) took up their quarters there. Several apartments were specially reserved for Telford's exclusive use, and he could always readily command any additional accommodation for purposes of business or hospitality. The successive landlords of the Salopian at length came to regard the engineer as a fixture, and even bought and sold him from time to time with the goodwill of the business. When he at length resolved, on the persuasion of his friends, to take a house of his own, and gave notice of his intention of leaving, the landlord, who had but recently entered into possession, almost stood aghast. " What! leave the house!" said he; " Why, Sir, I have just paid 750l. for you!" On explanation it appeared that this price had actually been paid by him to the outgoing landlord, on the assumption that Mr. Telford was a fixture of the hotel; the previous tenant having paid

450*l.* for him; the increase in the price marking very
significantly the growing importance of the engineer's
position.   There was, however, no help for the discon-
solate landlord, and Telford left the Salopian to take
possession of his new house at 24, Abingdon Street.
Labelye, the engineer of Westminster Bridge, had formerly occupied the dwelling; and, at a subsequent period,
Sir William Chambers, the architect of Somerset House.
Telford used to take much pleasure in pointing out to
his visitors the painting of Westminster Bridge, impa-
nelled in the wall over the parlour mantelpiece, made
for Labelye by an Italian artist whilst the bridge works
were in progress.   In that house Telford continued to
live until the close of his life.

One of the subjects in which he took much interest
during his later years was the establishment of the Institute of Civil Engineers.   In 1818 a society had been
formed, consisting principally of young men educated
to civil and mechanical engineering, who occasionally
met to discuss matters of interest relating to their
profession.   As early as the time of Smeaton, a social
meeting of engineers was occasionally held at an inn
in Holborn, which was discontinued, in 1792, in consequence of some personal differences amongst the
members.   It was revived in the following year, under
the auspices of Mr. Jessop, Mr. Naylor, Mr. Rennie,
and Mr. Whitworth, and joined by other gentlemen of
scientific distinction.   They were accustomed to dine
together every fortnight at the Crown and Anchor in
the Strand, spending the evening in conversation on
engineering subjects.   But as the numbers and importance of the profession increased, the desire began to be
felt, especially amongst the junior members, for an institution of a more enlarged character.   Hence the movement among the younger men to which we have alluded,
and which led to an invitation to Mr. Telford to accept
the office of President of their proposed Engineers' In-

stitute. To this he consented, and entered upon the duties of the office on the 21st of March, 1820.

During the remainder of his life, Mr. Telford continued to watch over the progress of the society, which gradually grew in importance and usefulness. He supplied it with the nucleus of a reference library, now become of great value to its members. He established the practice of recording the proceedings,[1] minutes of discussions, and substance of the papers read, which has led to the accumulation, in the printed records of the Institute, of a vast body of information as to engineering practice. In 1828 he exerted himself strenuously and successfully in obtaining a Charter of Incorporation for the society; and finally, at his death, he left the Institute their first bequest of 2000*l*., together with many valuable books, and a large collection of documents which had been subservient to his own professional labours.

In the distinguished position which he occupied, it was natural that Mr. Telford should be called upon, as he often was, towards the close of his life, to give his opinion and advice as to projects of public importance. Where strongly conflicting opinions were entertained on any subject, his help was occasionally found most valuable; for he possessed great tact and suavity of manner, which often enabled him to reconcile opposing interests when they stood in the way of important enterprises.

[1] We are informed by Joseph Mitchell, Esq., C.E., of the origin of this practice. Mr. Mitchell was a pupil of Mr. Telford's, living with him in his house at 24, Abingdon Street. It was the engineer's custom to have a dinner-party every Tuesday, after which his engineering friends were invited to accompany him to the Institution, the meetings of which were then held on Tuesday evenings in a house in Buckingham Street, Strand. The meetings did not usually consist of more than from twenty to thirty persons. Mr. Mitchell took notes of the conversations which followed the reading of the papers. Mr. Telford afterwards found his pupil extending the notes, on which he asked permission to read them, and was so much pleased that he took them to the next meeting, and read them to the members. Mr. Mitchell was then formally appointed reporter of conversations to the Institute; and the custom having been continued, a large mass of valuable practical information has thus been placed on record.

In 1828 he was appointed one of the commissioners to investigate the subject of the supply of water to the metropolis, in conjunction with Dr. Roget and Professor Brande, and the result was the very able report published in that year. Only a few months before his death, in 1834, he prepared and sent in an elaborate separate report, containing many excellent practical suggestions, which had the effect of strongly stimulating the water companies, and eventually led to great improvements.

On the subject of roads he was the very highest authority, his friend Southey jocularly styling him the "Colossus of Roads" as well as "Pontifex Maximus." The Russian Government frequently consulted him with reference to the new roads with which that great empire was being opened up. The Polish road from Warsaw to Briesc on the Russian frontier, 120 miles in length, was constructed after his plans, and it remains, we believe, the finest road in the Russian dominions to this day.

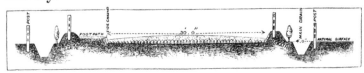

SECTION OF POLISH ROAD.

He was consulted by the Austrian Government on the subject of bridges as well as roads. Count Széchenyi recounts the very agreeable and instructive interview which he had with Telford when he called to consult him as to the bridge proposed to be erected across the Danube, between the towns of Buda and Pesth. On a suspension bridge being suggested by the English engineer, the Count, with surprise, asked if such an erection was *possible* under the circumstances he had described? "We do not consider anything to be impossible," replied Telford; "impossibilities exist chiefly in the prejudices of mankind, to which some are slaves, and from which few are able to emancipate themselves and enter on the path of truth." But supposing a suspension bridge were not deemed advisable under the circumstances, and it were considered necessary altogether to avoid motion, " then," said he, " I should recommend you to erect a cast iron bridge of three spans, each 400 feet; such a bridge will have no motion, and though half the world lay a wreck, it would still stand." [1] The suspension bridge was eventually resolved upon; it was constructed by one of Mr. Telford's ablest pupils, Mr. Tierney Clark, between the years 1839 and 1850, and is one of the greatest triumphs of English engineering to be found in Europe, the Buda-Pesth people proudly declaring it to be " the eighth wonder of the world."

At a time when speculation was very rife—in the year 1825—Mr. Telford was consulted respecting a grand scheme for cutting a canal across the Isthmus of Darien; and about the same time he was employed to resurvey the line for a ship canal—which had before occupied the attention of Whitworth and Rennie—between Bristol and the English Channel. But although he gave great attention to this latter project, and prepared numerous plans and reports upon it, and although an Act was actually passed enabling it to be carried out, the scheme was eventually abandoned, like the preceding ones with the same object, for want of the requisite funds.

Our engineer had a perfect detestation of speculative jobbing in all its forms, though on one occasion he could not help being used as an instrument by schemers. A public company was got up at Liverpool, in 1827, to form a broad and deep ship canal, of about seven miles in length, from opposite Liverpool to near Helbre Isle, in the estuary of the Dee: its object was to enable shipping to avoid the variable shoals and sand-banks which obstruct the entrance to the Mersey. Mr. Telford entered on the project with great zeal, and his name was widely quoted

---

[1] Supplement to Weale's ' Bridges,' Count Széchenyi's Report, p. 18.

in its support.　It appeared, however, that one of its principal promoters, who had secured the right of preemption of the land on which the only possible entrance to the canal could be formed on the northern side, suddenly closed with the corporation of Liverpool, who were opposed to the plan, and "sold" his partners as well as the engineer for a large sum of money.　Telford, disgusted at being made the instrument of an apparent fraud upon the public, destroyed all the documents relating to the scheme, and never afterwards spoke of it except in terms of extreme indignation.

About the same time the formation of locomotive railways was extensively discussed, and schemes were set on foot to construct them between several of the larger towns.　But Mr. Telford was now about seventy years old ; and, desirous of limiting the range of his business rather than extending it, he declined to enter upon this new branch of engineering.　Yet, in his younger days, he had surveyed numerous lines of railway—amongst others, one as early as the year 1805, from Glasgow to Berwick, down the vale of the Tweed.　A line from Newcastle-on-Tyne to Carlisle was also surveyed and reported on by him some years later ; and the Stratford and Moreton Railway was actually constructed under his direction.　He made use of railways in all his large works of masonry, for the purpose of facilitating the haulage of materials to the points at which they were required to be deposited or used.　There is a paper of his on the Inland Navigation of the County of Salop, contained in 'The Agricultural Survey of Shropshire,' in which he speaks of the judicious use of railways, and recommends that in all future surveys " it be an instruction to the engineers that they do examine the county with a view of introducing iron railways wherever difficulties may occur with regard to the making of navigable canals."　When the project of the Liverpool and Manchester Railway was started, we are informed that he was offered the appointment of engineer ; but he declined, partly because of his advanced age, but also out of a feeling of duty to his employers, the Canal Companies, stating that he could not lend his name to a scheme which, if carried out, must so materially affect their interests.

Towards the close of his life he was afflicted by deafness, which made him feel exceedingly uncomfortable in mixed society. Thanks to a healthy constitution, unimpaired by excess and invigorated by active occupation, his working powers had lasted longer than those of most men. He was still cheerful, clear-headed, and skilful in the arts of his profession, and felt the same pleasure in useful work that he had ever done. It was, therefore, with difficulty that he could reconcile himself to the idea of retiring from the field of honourable labour, which he had so long occupied, into a state of comparative inactivity. But he was not a man who could be idle, and he determined, like his great predecessor Smeaton, to occupy the remaining years of his life in arranging his engineering papers for publication. Vigorous though he had been, he felt that the time was shortly approaching when the wheels of life must stand still altogether. Writing to a friend at Langholm, he said, "Having now been occupied for about seventy-five years in incessant exertion, I have for some time past arranged to decline the contest; but the numerous works in which I am engaged have hitherto prevented my succeeding. In the mean time I occasionally amuse myself with setting down in what manner a long life has been laboriously, and I hope usefully, employed." And again, a little later, he writes : " During the last twelve months I have had several rubs ; at seventy-seven they tell more seriously than formerly, and call for less exertion and require greater precautions. I fancy that few of my age belonging to the valley of the Esk remain in the land of the living." [1]

One of the last works on which Mr. Telford was pro-

---

[1] Letter to Mrs. Little, Post Office, Langholm, 28th August, 1833.

fessionally consulted was at the instance of the Duke
of Wellington—not many years younger than himself,
but of equally vigorous intellectual powers—as to the
improvement of Dover Harbour, then falling rapidly to
decay. The long-continued south-westerly gales of
1833-4 had the effect of rolling an immense quantity
of shingle up Channel towards that port, at the entrance
to which it became deposited in unusual quantities, so as
to render it at times altogether inaccessible. The Duke,
as a military man, took a more than ordinary interest
in the improvement of Dover, as the military and naval
station nearest to the French coast; and it fell to him
as Lord Warden of the Cinque Ports to watch over
the preservation of the harbour, situated at a point in
the English Channel which he regarded as of great stra-
tegic importance in the event of a continental war.
He therefore desired Mr. Telford to visit the place and
give his opinion as to the most advisable mode of procedure with a view to improving the harbour. The
result was a report, in which the engineer recommended
a plan of sluicing, similar to that adopted by Mr. Smeaton
at Ramsgate, and which was afterwards carried out with
considerable success by Mr. James Walker, C.E.

This was his last piece of professional work. A few
months later he was laid up by bilious derangement of
a serious character, which recurred with increased violence towards the close of the year; and on the 2nd of
September, 1834, Thomas Telford closed his useful and
honoured career, at the advanced age of seventy-seven.
With that absence of ostentation which characterised him
through life, he directed that his remains should be laid,
without ceremony, in the burialground of the parish
church of St. Margaret's, Westminster. But the members
of the Institute of Civil Engineers, who justly deemed
him their benefactor and chief ornament, urged upon his
executors the propriety of interring him in Westminster
Abbey. He was buried there accordingly, near the
middle of the nave; where the letters, "Thomas Tel-

ford, 1834," mark the place beneath which he lies.[1]
The adjoining stone bears the inscription, " Robert
Stephenson, 1859," that engineer having during his life
expressed the wish that his body should be laid near
that of Telford; and the son of the Killingworth en-
gineman thus sleeps by the side of the son of the
Eskdale shepherd.

TELFORD'S BURIAL PLACE IN WESTMINSTER ABBEY.

[By Percival Skelton.]

It was a long, a successful, and a useful life which
thus ended. Every step in his upward career, from the

---

[1] A statue of him, by Bailey, has
since been placed in the east aisle of
the north transept, known as the Islip
Chapel. It is considered a fine work,
but its effect is quite lost in consequence of the crowded state of the
aisle, which has very much the look
of a sculptor's workshop. The subscription raised for the purpose of
erecting the statue was 1000l., of
which 200l. was paid to the Dean
for permission to place it within the
Abbey.

poor peasant's hut in Eskdale to Westminster Abbey, was nobly and valorously won. The man was diligent and conscientious; whether as a working mason hewing stone blocks at Somerset House, as a foreman of builders at Portsmouth, as a road surveyor at Shrewsbury, or as an engineer of bridges, canals, docks, and harbours. The success which followed his efforts was thoroughly well deserved. He was laborious, pains-taking, and skilful; but, what was better, he was honest and upright. He was a most reliable man; and hence he came to be extensively trusted. Whatever he undertook, he endeavoured to excel in. He would be a first-rate hewer, and he became so. He was himself accustomed to attribute much of his success to the thorough way in which he had mastered the humble beginnings of this trade. He was even of opinion that the course of manual training he had undergone, and the drudgery, as some would call it, of daily labour—first as an apprentice, and afterwards as a journeyman mason—had been of greater service to him than if he had passed through the curriculum of a University. Writing to his friend, Miss Malcolm, respecting a young man who desired to enter the engineering profession, he in the first place endeavoured to dissuade the lady from encouraging the ambition of her *protégé*, the profession being overstocked, and offering very few prizes in proportion to the large number of blanks. "But," he added, "if civil engineering, notwithstanding these discouragements, is still preferred, I may point out that the way in which both Mr. Rennie and myself proceeded, was to serve a regular apprenticeship to some practical employment—he to a millwright, and I to a general housebuilder. In this way we secured the means, by hard labour, of earning a subsistence; and, in time, we obtained by good conduct the confidence of our employers and the public; eventually rising into the rank of what is called Civil Engineering. This is the true way of

acquiring practical skill, a thorough knowledge of the
materials employed in construction, and last, but not
least, a perfect knowledge of the habits and dispositions
of the workmen who carry out our designs. This course,
although forbidding to many a young person, who be-
lieves it possible to find a short and rapid path to distinction, is proved to be otherwise by the two examples
I have cited. For my own part, I may truly aver that
' steep is the ascent, and slippery is the way.' " [1]

That Mr. Telford was enabled to continue to so advanced an age employed on laborious and anxious work,
was no doubt attributable in a great measure to the
cheerfulness of his nature. He was, indeed, a most
happy-minded man. It will be remembered that, when
a boy, he had been known in his valley as " Laughing
Tam." The same disposition continued to characterize
him even in his old age. He was playful and jocular,
and rejoiced in the society of children and young people,
especially when well-informed and modest. But when
they pretended to acquirements they did not possess, he
was quick to detect and see through them. One day a
youth expatiated to him in very large terms about a
friend of his, who had done this and that, and made so
and so, and could do all manner of wonderful things.
Telford listened with great attention, and when the
youth had done, he quietly asked, with a twinkle in his
eye, " Pray, can your friend lay eggs ? "

When in society he gave himself up to it, and thoroughly enjoyed it. He did not sit apart, a moody and
abstracted " lion ; " nor desire to be regarded as " the
great engineer," pondering new Menai Bridges ; but he
appeared in his natural character of a simple, intelligent,
cheerful companion ; as ready to laugh at his own jokes
as at other people's ; and he was as communicative to a
child as to any philosopher of the party.

---

[1] Letter to Miss Malcolm, Burnfoot, Langholm, dated 7th Oct., 1830.

Robert Southey, than whom there was no better judge of a loveable man, said of him, "I would go a long way for the sake of seeing Telford and spending a few days in his company." Southey had the best opportunity of knowing him well; for he performed a long tour with him through Scotland, in 1819. And a journey in company, extending over many weeks, is, probably, better than anything else, calculated to bring out the weak as well as the strong points of a friend: indeed, many friendships have completely broken down under the severe test of a single week's tour. But Southey on that occasion firmly cemented a friendship which lasted until Telford's death. On one occasion the latter called at the poet's house, in company with Sir Henry Parnell, when engaged upon the survey of one of his northern roads. Unhappily Southey was absent at the time; and, writing about the circumstance to a correspondent, he said, "This was a great mortification to me, inasmuch as I owe Telford every kind of friendly attention, and like him heartily." [1]

Campbell, the poet, was another early friend of our engineer; and the attachment seems to have been mutual. Writing to Dr. Currie, of Liverpool, in 1802, Campbell says: "I have become acquainted with Telford the engineer, 'a fellow of infinite humour,' and of strong enterprising mind. He has almost made me a bridge-builder already; at least he has inspired me with new sensations of interest in the improvement and ornament of our country. Have you seen his plan of London Bridge? or his scheme for a new canal in the North Highlands, which will unite, if put in effect, our Eastern and Atlantic commerce, and render Scotland the very emporium of navigation? Telford is a most useful cicerone in London. He is so universally acquainted, and so popular in his manners, that he can

---

[1] 'Selections from the Letters of Robert Southey.' Edited by J. W. Warter, B.D. Vol. iii., p. 326.

introduce one to all kinds of novelty, and all descriptions of interesting society." [1] Shortly after, Campbell named his first son after Telford, who stood godfather for the boy. Indeed, for many years, Telford played the part of Mentor to the young and impulsive poet, advising him about his course in life, trying to keep him steady, and holding him aloof as much as possible from the seductive allurements of the capital. But it was a difficult task, and Telford's numerous engagements necessarily left the poet at many seasons very much to himself. It appears that they were living together at the Salopian when Campbell composed the first draft of his poem of Hohenlinden; and several important emendations made in it by Telford were adopted by Campbell. Although the two friends pursued different roads in life, and for many years saw little of each other, they often met again, especially after Telford took up his abode at his house in Abingdon Street, where Campbell was a frequent and always a welcome guest.

When engaged upon his surveys, our engineer was the same simple, cheerful, laborious man. While at work, he gave his whole mind to the subject in hand, thinking of nothing else for the time; dismissing it at the close of each day's work, but ready to take it up afresh with the next day's duties. This was a great advantage to him as respected the prolongation of his working faculty. He did not take his anxieties to bed with him, as many do, and rise up with them in the morning; but he laid down the load at the end of each day, and resumed it all the more cheerfully when refreshed and invigorated by natural rest. It was only while the engrossing anxieties connected with the suspension of the Menai Bridge were weighing heavily upon his mind, that he could not sleep; and then, age

---

[1] Beattie's 'Life and Letters of Thomas Campbell,' vol i., p. 451.

having stolen upon him, he felt the strain almost more
than he could bear.   But that great anxiety over, his
spirits speedily resumed their wonted elasticity.

When engaged upon the construction of the Carlisle
and Glasgow road, he was very fond of getting a few of
the "navvy men," as he called them, to join him at an
ordinary at the Hamilton Arms Hotel, Lanarkshire, each
paying his own expenses; and though Telford told them
he could not drink, yet he would carve and draw corks
for them.   One of the rules he laid down was, that no
business was to be introduced from the moment they sat
down to dinner.   All at once, from being the plodding,
hard-working engineer, with responsibility and thought
in every feature, Telford unbended and relaxed, and
became the merriest and drollest of the party.   He possessed a great fund of anecdote available for such occasions, had an extraordinary memory for facts relating to
persons and families, and the wonder to many of his
auditors was, how in all the world a man living in London
should know so much better about their locality and
many of its oddities than they did themselves.

In his leisure hours at home, which were but few, he
occupied himself a good deal in the perusal of miscellaneous literature, never losing his taste for poetry.
He continued to indulge in the occasional composition
of verses until a comparatively late period of his life;
one of his most successful efforts being a translation of
the 'Ode to May,' from Buchanan's Latin poems, executed in a very tender and graceful manner.   That he
might be enabled to peruse engineering works in French
and German, he prosecuted the study of those languages,
and with such success that he was shortly able to read
them with comparative ease.   He occasionally occupied
himself in literary composition on subjects connected
with his profession.   Thus he wrote for the Edinburgh
Encyclopedia, conducted by his friend Sir David (then
Dr.) Brewster, the elaborate and able articles on Archi-

tecture, Bridge-building, and Canal making. Besides his contributions to that work, he advanced a considerable sum of money to aid in its publication, which remained a debt due to his estate at the period of his death.

Although occupied as a leading engineer for nearly forty years—having certified contractors' bills during that time amounting to many millions sterling—he died in comparatively moderate circumstances. Eminent constructive ability was not very highly remunerated in Telford's time, and his average income did not amount to more than is paid to the resident engineer of any modern railway. But Telford's charges were perhaps unusually low—so much so that a deputation of members of the profession on one occasion formally expostulated with him on the subject.

Although he could not be said to have an indifference for money, he yet estimated it as a thing worth infinitely less than character. His wants were few, and his household expenses small; and though he entertained many visitors and friends, it was in a quiet way and on a moderate scale. The small regard he had for personal dignity may be inferred from the fact, that to the last he continued the practice, which he had learnt when a working mason, of darning his own stockings.[1] But he had nevertheless the highest idea of the dignity of his profession; not, however, because of the money it would

---

[1] Mr. Mitchell says : " He lived at the rate of about 1200l. a year. He kept a carriage, but no horses, and used his carriage principally for making his journeys through the country on business. I once accompanied him to Bath and Cornwall, when he made me keep an accurate journal of all I saw. He used to lecture us on being independent, even in little matters, and not ask servants to do for us what we might easily do for ourselves. He carried in his pocket a small book containing needles, thread, and buttons, and on an emergency was always ready to put in a stitch. A curious habit he had of mending his stockings, which I suppose he acquired when a working mason. He would not permit his housekeeper to touch them, but after his work at night, about nine or half-past, he would go upstairs, and take down a lot, and sit mending them with great apparent delight in his own room till bed-time. I have frequently gone in to him with some message, and found him occupied with this work."

produce, but of the great things it was calculated to accomplish. In his most confidential letters we find him often expatiating on the noble works he was engaged in designing or constructing, and the national good likely to flow from them, but never on the pecuniary advantages he himself was to reap. He doubtless prized, and prized highly, the reputation they would bring him; and, above all, there seemed to be uppermost in his mind, especially in the earlier part of his career, whilst many of his schoolfellows were still alive, the thought of "What will they say of this in Eskdale?" but as for the money results to himself, Telford seemed, to the close of his life, to regard them as of comparatively small moment.

During the twenty-one years that he acted as principal engineer for the Caledonian Canal, we find from the Parliamentary returns that the amount paid to him for his reports, detailed plans, and superintendence, was exactly 237*l.* a-year. When he conceived the works to be of great public importance, and promoted by public-spirited persons at their own expense, he refused to receive any payment for his labour, or even repayment of the expenses incurred by him. Thus, while employed by the Government in the improvement of the Highland roads, he persuaded himself that he ought at the same time to promote the similar patriotic objects of the British Fisheries Society, which were carried out by voluntary subscription; and for many years he acted as their engineer, refusing to accept any remuneration whatever for his trouble.[1]

Mr. Telford held the sordid money-grubber in perfect detestation. He was of opinion that the adulation paid

---

[1] "The British Fisheries Society," adds Mr. Rickman, "did not suffer themselves to be entirely outdone in liberality, and shortly before his death they pressed upon Mr. Telford a very handsome gift of plate, which, being inscribed with expressions of their thankfulness and gratitude towards him, he could not possibly refuse to accept."—'Life of Telford,' p. 283.

to mere money was one of the greatest dangers with which modern society was threatened. "I admire commercial enterprise," he would say; "it is the vigorous outgrowth of our industrial life: I admire everything that gives it free scope, as wherever it goes, activity, energy, intelligence—all that we call civilization—accompany it; but I hold that the aim and end of all ought not to be a mere bag of money, but something far higher and far better."

Writing once to his Langholm correspondent about an old schoolfellow, who had grown rich by scraping, Telford said: "Poor Bob L——! His industry and sagacity were more than counterbalanced by his childish vanity and silly avarice, which rendered his friendship dangerous and his conversation tiresome. He was like a man in London, whose lips, while walking by himself along the streets, were constantly ejaculating ' Money! Money!' But peace to Bob's memory: I need scarcely add, confusion to his thousands!" He himself was most careful in resisting the temptations to which men in his position are frequently exposed; but he was preserved by his honest pride, not less than by the purity of his character. He would not receive anything in the shape of presents or testimonials from persons employed under him as contractors; for he would not have even the shadow of an obligation stand in the way of performing his rigid duty to those who employed him to watch over and protect their interests.

Yet Telford was not without a proper regard for money, as a means of conferring benefits on others, and especially as a means of being independent. By the close of his life he had accumulated as much as, invested at interest, would bring him in about 800*l.* a-year, and enable him to occupy the house in Abingdon Street until he died. This was amply sufficient for his wants, and more than enough for perfect independence. It enabled him also to continue those secret acts of bene-

volence which constituted perhaps the most genuine
pleasure of his life.   It is one of the most delightful
traits in this excellent man's career to find him so constantly occupied in works of spontaneous charity, in
places so remote that it is impossible the slightest feeling
of ostentation could ever have sullied the purity of his
acts.   Among the large mass of Telford's private letters
which have been submitted to us, we find constant reference to sums of money transmitted for the support of
poor people in his native valley.   At new year's time he
regularly sent a remittance of from 30*l.* to 50*l.*, to be
distributed by the kind Miss Malcolm of Burnfoot, and,
after her death, by Mr. Little, the postmaster at Langholm; and the contributions thus so kindly made did
much to fend off the winter's cold, and surround with
many small comforts those who most needed help, but
were perhaps too modest to ask it.

Many of those in the valley of the Esk had known of
Telford in his younger years as a poor barefooted boy;
yet though become a man of distinction, he had too
much good sense to be ashamed of his humble origin;
perhaps he even felt proud that, by dint of his own
valorous and persevering efforts, he had been able to rise
so much above it.   Throughout his long life his heart
always warmed at the thought of Eskdale.   He rejoiced
at the honourable rise of Eskdale men as reflecting
credit upon his "beloved valley."   Thus, writing to his
Langholm correspondent, with reference to the honours
conferred on the different members of the family of
Malcolm, he said: "The distinctions so deservedly bestowed upon the Burnfoot family, establish a splendid
era in Eskdale; and almost tempt your correspondent to
sport his Swedish honours, which that grateful country
has repeatedly, in spite of refusal, transmitted."[1]   It
might be said that there was narrowness and provin-

---

[1] Letter to Mr. William Little, Langholm, 24th January, 1815.

cialism in this; but when young men are thrown into
the world, with all its temptations and snares, it is well
that the recollections of home and kindred should survive
to hold them in the path of rectitude, and cheer them in
their onward and upright course in life. And there is
no doubt that Telford was borne up on many occasions
by the thought of what the folks in the valley would say
about him and his progress in life, when they met together at market, or at the Westerkirk church porch on
Sabbath mornings. In this light, provincialism or
local patriotism is a prolific source of good; and may
be regarded as among the most valuable and beautiful
emanations of the parish life of our country. Although
Telford was honoured with the titles and orders of
merit conferred upon him by foreign monarchs, what he
esteemed beyond them all was the respect and gratitude
of his own countrymen; and, not least, the honour
which his really noble and beneficent career was calculated to reflect upon "the folks of the nook," the remote
inhabitants of his "beloved Eskdale."

When the engineer proceeded to dispose of his savings
by will, which he did a few months before his death, the
distribution was a comparatively easy matter. The total
amount of his bequeathments was 16,600*l*. About one-
fourth of the whole he set apart for educational purposes,—2,000*l*. to the Civil Engineers' Institute, and
1,000*l*. each to the ministers of Langholm and Westerkirk, in trust for the parish libraries. The rest was
bequeathed, in sums of from 200*l*. to 500*l*., to different
persons who had acted as clerks, assistants, and sur-
veyors, in his various public works; and to his intimate
personal friends. Amongst these latter were Colonel
Pasley, the nephew of his early benefactor; Mr. Rickman, Mr. Milne, and Mr. Hope, his three executors; and
Robert Southey and Thomas Campbell, the poets. To
both of these last the gift was most welcome. Southey
said of his: "Mr. Telford has most kindly and unex-

pectedly left me 500*l.*, with a share of his residuary property, which I am told will make it amount in all to 850*l.* This is truly a godsend, and I am most grateful for it. It gives me the comfortable knowledge that, if it should please God soon to take me from this world, my family would have resources fully sufficient for their support till such time as their affairs could be put in order, and the proceeds of my books, remains, &c., be rendered available. I have never been anxious overmuch, nor ever taken more thought for the morrow than it is the duty of every one to take who has to earn his livelihood; but to be thus provided for at this time I feel to be an especial blessing." [1]

Among the most valuable results of Telford's bequests in his own district, was the establishment of the libraries at Langholm and Westerkirk on a firmer foundation. That at Westerkirk had been originally instituted in the year 1792, by the miners employed to work an antimony mine (since abandoned) on the farm of Glendinning, within sight of the place where Telford was born. On the dissolution of the mining company, in 1800, the little collection of books was removed to Kirkton Hill; but on receipt of Telford's bequest, a special building was erected for their reception at New Burtpath, near the village of Westerkirk. The annual income derived from the Telford fund enabled additions of new volumes to be made to it from time to time; and its uses as a public institution were thus greatly increased. The books are exchanged once a month, on the day of the full moon; on which occasion readers of all ages and conditions,—farmers, shepherds, ploughmen, labourers, and their children,—resort to it from far and near, taking away with them as many volumes as they desire for the month's reading.

---

[1] 'Selections from the Letters of Robert Southey,' vol. iv., p. 391. We may here mention that the last article which Southey wrote for the 'Quarterly' was his review of the 'Life of Telford.'

Thus there is scarcely a cottage in the valley in which good books are not to be found under perusal; and we are told that it is a common thing for the Eskdale shepherd to take a book in his plaid to the hill-side— a volume of Shakespeare, Prescott, or Macaulay—and read it there, under the blue sky, with his sheep and the green hills before him. And thus, as long as the bequest lasts, the good, great engineer will not cease to be remembered with gratitude in his beloved Eskdale.

TELFORD'S NATIVE VALLEY ESKDALE.

[By Percival Skelton.]

INDEX TO VOL. II.

ABERDEEN.

A.

ABERDEEN,—early history of, 400; ancient bridges at, 103; Smeaton's harbour works at, 70; Rennie's report on the harbour at, 205; Telford's harbour works at, 399.

ADAM, ROBERT, architect, 309, 312.

ADMIRALTY, Mr. Rennie and the, 238, 266, 268.

AGRICULTURE,—early Scotch, 94, 104, 107; Lincolnshire, 153, 157, 164; Highland, 375, 388.

AIR-PUMP, Smeaton's improvements in, 12.

ALBION MILLS, 129, 134.

ANCHOR FORGE, Rennie's, at Woolwich, 265.

ANGLESEA, roads through, 434.

ANTON'S GOWT, Boston, 161.

AQUEDUCTS,—at Limpley Stoke, 145; Lune, 148; Chirk, 342; Pont-Cysylltau, 344.

ARBROATH, monks of, and Inchcape Rock, 223.

ARCH,—theory of equilibrium of, 171; arching of Waterloo Bridge, 182; Southwark, 191.

ARCHITECTURE, Telford's study of, 305, 332.

ARDROSSAN CANAL, 418.

ARMSTRONG, JOHNNIE, Eskdale, 289.

ARTISANS OF BIRMINGHAM, 132.

AUSTHORPE LODGE, Leeds, Smeaton's house at, 4, 74.

AXHOLME, Smeaton's drainage of Isle of, 52.

AYRSHIRE, state of, last century, 97.

B.

BALGOWNIE BRIG, Aberdeen, 103.

BANFF,—bridge, 2, 61; harbour, 398.

BANK OF ENGLAND note machinery, 265.

BANKS, Sir JOSEPH, and drainage works in Lincolnshire, 155.

BARLEY MILLS introduced in Scotland, 105.

BATH, Telford's visit to, 333.

BEAULEY BRIDGE, 384.

BEDFORD LEVEL, drainage works, 166, 467.

BELL, Diving,—Smeaton's use of, 69; Rennie's improvement of, 219.

BELL ROCK LIGHTHOUSE,—dangers of the Inchcape Rock, 223; Captain Brodie's de-

BRINDLEY.

sign of a cast-iron lighthouse, 225; Mr. Rennie consulted, ib.; his design adopted, 227; the works begun, 228; Mr. Rennie's visits of inspection, 230; the lighthouse completed, 232; Mr. Stevenson's publication, 234.

BENTHAM, Sir SAMUEL, 208, 256.

BETTWS-Y-COED, iron bridge at, 438.

BEWDLEY BRIDGE, 366.

BIRMINGHAM, — its industry and artisans, 121; Telford's notice of, 334; canal, 423; and Liverpool Junction Canal, 425.

BLACK, Dr., Edinburgh, 126.

BLACKMAIL,—Highland, 98; border, 289.

BLACKSTONE EDGE, Yorkshire, 147.

BLANKNEY FEN, Lincolnshire, 153.

BONAR BRIDGE, 387.

BORDER RAIDS, 288; improvements, 290.

BOSTON,—Rennie's plan for improvement of Witham at, 164; bridge at, 176.

BOULTON and WATT, Soho, 129, 131.

BRADFORD TRADE, 1724, 5.

BREAKWATER, Plymouth, 92, 252.

BRECHIN, ancient bridge at, 103.

BRIDGENORTH, church built by Telford at, 331, 339.

BRIDGES,—Banff, 2, 61; Bristol, 53; London (old), 53; Blackfriars, 54; Perth, 56; Coldstream, 59; Hexham, 62; early Scotch, 102; Rennie's first bridge at Stevenhouse Mill, 131; Kelso, 172; Musselburgh, 174; Boston, 176; Waterloo, 178; Southwark, 187; New London, 268; Langholm, 302; Montford, 331; Coalbrookdale, 356; Wear Bridge, Sunderland, 358; Buildwas, 359; Bewdley, 366; Tongueland, 368; Dunkeld, 380, 384; Beauley and Conan, ib.; Craigellachie, 385; Bonar, 387; Wick, 391; Galton Bridge, Birmingham, 424; Cartland Crags, Lanarkshire, 431; Pathhead, 433; Morpeth, ib.; Bettws-y-Coed, 438; Menai (designs), 444; Runcorn (design), 447; Telford's Menai Bridge, 449; Conway, 459; Tewkesbury, 464; Gloucester, ib.; Dean Bridge, Edinburgh, 465; Glasgow Bridge, 286, 466.

BRIDGE-BUILDING, theory of, 170.

BRINDLEY and SMEATON, 3.

BRISTOL.

BRISTOL,—old bridge, 53, 64 ; Telford's notice of, 369.
BRITISH CHANNEL, lighting of, 46.
BRITISH FISHERIES SOCIETY, 204, 377, 392, 488.
BRODIE, Captain, proposed iron lighthouse on the Bell Rock, 229.
BROTHERS OF THE BRIDGE, Scotland, 102.
BROOMIELAW, Glasgow,—Rennie's proposed docks at, 206 ; Telford's bridge at, 466.
BROSELEY, iron bridge at, 356.
BROWNRIGG, General, and Mr. Rennie, 283.
BUCCLEUGH, Duke of, landlord in Eskdale, 289, 290, 299.
BUDA-PESTH suspension bridge, 476.
BUILDWAS BRIDGE, 359.
BURGH SCHOOLS, Scotland, 122.
BURNS, ROBERT, Telford's verses on, 329, 353, 371.

C.

CAITHNESS, county of, in eighteenth century, 379.
CALDER, Smeaton's improvement of its navigation, 51.
CALEDONIAN CANAL, 198, 409.
CAMPBELL, T., poet, Telford's intimacy with, 372, 484, 491.
CANALS,—Forth and Clyde, 57 ; Birmingham and Worcester, 87 ; Kennet and Avon, 144 ; Rochdale, 145 ; Lancaster, 148 ; Grand Trunk, 149, 420 ; Royal Canal of Ireland, 150 ; Ellesmere, 340 ; Shrewsbury, 346 ; Caledonian, 409 ; Glasgow and Ardrossan, 418 ; Weaver, ib. ; Gotha, 419 ; Gloucester and Berkeley, 421 ; Birmingham, 423 ; Macclesfield, 424 ; Birmingham and Liverpool Junction, 425.
CANAL SHARES, rage for, 341.
CARLISLE and Glasgow Road, 427.
CARRON Iron Works, 61.
CARTLAND CRAGS BRIDGE, 431.
CARTMEL, Cumberland, 337.
CAS-CHROM, ancient agricultural implement, 375.
CATCHWATER drains, Rennie's system of, 159.
CENTERING, — Waterloo Bridge, 182 ; Telford's plan of suspended, 445.
CHAD'S, ST., Shrewsbury, fall of, 320.
CHAMBERS, Sir WILLIAM, architect, 309, 474.
CHARLESTON HARBOUR, Rennie's report on, 216.
CHATHAM DOCKYARD, 241 ; Rennie's plan of improvement of Medway at, 250.
CHIRK AQUEDUCT, 342.
CIVIL ENGINEERS' INSTITUTE, 474, 480, 491.
CLACHNAGARRY, canal basin at, 413.
CLARK, Mr. TIERNEY, engineer, 477.
CLOTHING TRADE of Leeds, 5.

DRAINAGE.

CLYDE NAVIGATION,—Rennie's report on, 205 ; Telford on, 418 ; bridge at Broomielaw and, 467.
COAST DEFENCES, Rennie's reports on, 235.
COALBROOKDALE, first iron bridge at, 356.
COCKBURN, Lord,—on Professor Robison, 127 ; on Highland roads, 378.
COCKBURN, Mr., East Lothian agriculturist, 107.
COFFER-DAMS,—Waterloo Bridge, 181 ; Southwark Bridge, 188 ; Sheerness Docks, 249.
COLDSTREAM BRIDGE, 59.
COLQUHOUN, on the traffic of the Thames, 196.
CONAN BRIDGE, 384.
CONCENTRATION of dockyard work, Rennie's report on, 239, 246.
CONWAY BRIDGE, 175, 459.
COPE, Sir JOHN, march to Edinburgh, 101.
CORPACH, canal basin at, 411.
COUNTRY-BRED BOYS, engineers, 292.
COVENANTERS, in Eskdale, 290.
CRAIG, Mr., Arbigland, 96.
CRAIGELLACHIE BRIDGE, 385.
CRAMOND BRIDGE, 177.
CROOKS, the, Eskdale, 295, 306.
CURRIE, Dr., Liverpool, 371, 484.

D.

DARLASTON BRIDGE, 177.
DARWIN, Dr., Liverpool, 371.
DAVIDSON, MATTHEW, Telford's foreman, 347, 352, 414.
DEAN BRIDGE, Edinburgh, 465.
DEE, Pont-Cysylltau aqueduct over the, 344.
DEPTFORD DOCKYARD, 241, 250.
DERWENTWATER ESTATES, Smeaton receiver for, 49.
DIVING BELL, Smeaton's employment of it at Ramsgate Harbour, 69 ; its improvement by Rennie, 218.
DIXON, Colonel, 377.
DOCKS,—on the Thames, 136, 197 ; West India, 197 ; London, 198 ; East India, 202 ; St. Katherine's, 462 ; Grimsby, 207 ; Hull, 213 ; Greenock, 214 ; Leith, 215 ; Aberdeen, 403 ; Dundee, 408.
DOCKYARDS, Royal,—their machinery, 238 ; Rennie's report on, 239 ; Plymouth, 240 ; Portsmouth, ib. ; Deptford, 241 ; Woolwich, ib. ; Sheerness, 242 ; Chatham, 250.
DODD's design of Strand Bridge, 178.
DOUGLAS, JAMES, inventor and projector, 362.
DOWNS, proposed harbour of refuge in, 65.
DOVER HARBOUR, 64, 480.
DRAGON-SLAYER, Rennie the greatest, 169.
DRAINAGE, works of, Smeaton's, 50 ; Meikle's, Kincardine Moss, 115 ; Rennie's, in Lincoln Fens, 152, 166 ; Telford's, in North Level, 467.

DREDGING.

DREDGING-MACHINE, improvement by Rennie, 214.
DROWNED lands in the Fens, 153, 167, 468.
DUMFRIES, in eighteenth century, 98.
DUNBAR Grammar School, Rennie at, 122.
DUNDEE HARBOUR, 407.
DUNKELD BRIDGE, 380, 384.

E.

EAST FEN, Lincoln, drainage by Rennie, 153.
EAST INDIA DOCKS, 202.
EAST LOTHIAN, notices of, 94, 97, 101, 108.
EAU BRINK CUT, 166.
EDINBURGH,—Water supply, 57; consumption of butcher's meat, eighteenth century, 97; Edinburgh and Glasgow caravan, 101; University professors, 121; building of New Town, 305; Dean Bridge, 465.
EDUCATION of early engineers, 141.
EDUCATION, Scotch, 121.
EDDYSTONE,—dangers of the rock, 15; Winstanley's lighthouse, 18; Rudyerd's, 21; Smeaton's, 26; the works begun, 34; lighthouse completed, 44; Smeaton's narrative, 82.
EIL, Loch, Caledonian Canal, 411.
ELLESMERE CANAL,—Telford appointed engineer, 334; description of the route, 340; construction of the works, 342.
ELLIOTS, Borderers, 289.
EMIGRATION, Highland, 376.
ENCYCLOPEDIA BRITANNICA, Robison's articles in, 170.
ENGINEERING works of Holland, Smeaton's inspection of, 14.
ENGINEERS, first club of, 79.
ENGINEERS, Institute of Civil, 474, 480, 491.
ESK, river, Dumfriesshire, 287.
ESKDALE,—description of, 287; early inhabitants, 288; modern, 290; farmers' houses in, 299; Telford's poem on, 304, 329; Telford revisits, 351; Libraries in, 492.
EYEMOUTH HARBOUR, 71.

F.

FANNERS introduced into Scotland, 105.
FELONS at Shrewsbury managed by Telford, 323.
FEN scenery, characteristics of, 169.
FENS,—Smeaton's drainage of, 51; Rennie's, 152; Telford's, 468.
FERRIES, Highland, 378, 385.
FIDLER'S BURN BRIDGE, 431.
FINANCIAL issues of engineering undertakings, 417.
FINDHORN RIVER, its sudden floods, 386.
FISHERIES SOCIETY, British, 204, 377, 392.

HARRISON.

FLETCHER of Saltoun, 95; introduces barley mills in Scotland, 105.
FOCHABERS, ferry at, 385.
FOLKESTONE HARBOUR, 391.
FORTH AND CLYDE CANAL, 57.
FOSSDYKE, Smeaton's drainage, 52; Wash Bridge, 177.
FOUNDATIONS of Bridges,—Smeaton's, 55, 57, 63; Rennie's, 181, 188.
FRAZERBURGH HARBOUR, 397.
FRENCH REVOLUTION, Telford influenced by, 326.
FULTON,—his Torpedo, 237; Rennie's opinion of, ib.

G.

GALTON BRIDGE, Birmingham Canal, 424.
GAOL at Shrewsbury, Telford rebuilds, 316.
GERMAN'S BRIDGE, 167.
GIBB, Mr. J., Engineer, 404.
GILNOCKIE TOWER, 289, 351.
GLASGOW,—early communication with Edinburgh by coach, 100; old bridge at, 103; Rennie's proposed docks at, 205; Telford's Bridge at, 466.
GLASGOW AND ARDROSSAN CANAL, 418.
GLASGOW and Carlisle road, 427.
GLENDINNING sheep farm, Eskdale, 292.
GLOUCESTER,—canal, 420; bridge, 464.
GOTHA CANAL, Sweden, 419.
GRAEMES, the border, 289.
GRAMMAR SCHOOLS, Scotland, 122.
GRAND CANAL OF IRELAND, 150.
GRAND TRUNK CANAL, 149, 420.
GREAT Glen of Scotland, 375, 409.
GREAT Level of the Fens,—Rennie's drainage of, 166; Telford's drainage of North-Level, 467.
GREENLAND FISHERY, Peterhead, 393.
GREENOCK quays and docks, 214.
GRIMSBY Harbour and docks, 207.

H.

HADDINGTON, 93, 100.
HALL, Captain BASIL, on the dangers of the Bell Rock, 224.
HARBOUR construction, Rennie's principles of, 216.
HARBOURS,—St. Ives, 63; Ramsgate, 65; Aberdeen, 70, 205; Eyemouth, 71; Wick, 204, 390; Holyhead, 208; Howth and Kingstown, 213; Charleston, 216; Folkestone, 391; Peterhead, 393; Frazerburgh, 397; Banff, 398; Aberdeen, 399; Dundee, 407.
HARECASTLE TUNNEL, 420.
HARRISON, JOHN, Leeds, 7.

HIGHLANDS.

HIGHLANDS and Lowlands, feuds between, 98.
HIGHLAND roads and bridges, 375; moral results of Telford's Highland contracts, 389; Highland harbours, 391.
HOBHOLE DRAIN, Lincolnshire, 162.
HOE at Plymouth, Smeaton on the, 42.
HOLLOW walls, Rennie's invention of, for sea works, 207.
HOLMES, Mr., Smeaton's partner, 7.
HOLYHEAD,—harbour, 208; roads to, 434.
HORSE-LOAD, Smeaton's account of the, 5.
HOUSTON MILL, East Lothian, 107.
HOWARD, JOHN, philanthropist, visit to Shrewsbury, 317.
HOWTH HARBOUR, 213.
HUDDART, Captain, Memoir of, 265.
HULL DOCKS, 213.
HUTCHESON BRIDGE, Glasgow, 208.
HYTHE MILITARY CANAL, 236.

I.

IMPLEMENTS of husbandry, early Scotch, 104, 375.
INDOLENCE of Scotch in eighteenth century, 95.
INDIA DOCKS,—West, 197; East, 202.
INLAND NAVIGATION. (See *Canals.*)
INVERNESS, road communications, 379, 388.
IRELAND,—transit to, by Holyhead, 208, 434; canals in, 150, 370.
IRON BRIDGES,—Rennie's, 175, 176, 187; Telford's, 346, 359, 361, 387, 445, 452.
IRON MANUFACTURE, 337, 355.
IRON-WORK,—Smeaton and Murdock's use of iron in mill work, 138; Rennie's, 139; in roofing, 203; Telford's, in canal aqueducts and locks, 346, 359, 361.
IRON VESSEL, the first constructed, 337.
ISLEWORTH BRIDGE, 177.
IVES, ST., Harbour, 63.

J.

JACKSON family,—Janet, Telford's mother, 294, 315, 325, 350; Thomas, Telford's cousin, 298, 307.
JAMES V. and Johnnie Armstrong, 289.
JESSOP, JOSIAS, engineer, 30, 34.
JESSOP, WILLIAM, engineer,—notices of, 30, 65, 178, 342, 347; brief memoir of, 197.
JOHNSTONE family, of Westerhall, 289, 307, 312, 320.
JORDAN, Mrs., Telford's notice of, 324.

K.

KAIMES, Lord, 95, 104, 115.
KATHERINE'S DOCKS, ST., 462.

LOSTWITHIEL.

KELSO BRIDGE, 172.
KEN BRIDGE, New Galloway, 177.
KENNET AND AVON CANAL, 144.
KINCARDINE MOSS, drained by Meikle's machinery, 115.
KINCARDINE O'NEILL, ancient bridge at, 103.
KING'S LYNN, 167.
KINGSTOWN HARBOUR, 213.
KINLOCH, Mr., his improvements of the thrashing machine, 110.
KINNEIL HOUSE, Boroughsto'ness, 61.
KIRKCUDBRIGHT, in eighteenth century, 97.

L.

LANG, OLIVER, first steam boat for royal navy constructed by, 267.
LANCASTER CANAL, 148.
LANGHOLM, Eskdale, 287, 299, 301, 328, 351, 491.
LATHE, Smeaton's Turning, 77.
LEA, river, Rennie's plans for defence of London, 235.
LEEDS,—in 1724, 4; grammar school, 7; Wellington Bridge, 177.
LEGGING Harecastle Tunnel, 421.
LEITH Harbour and docks, 215.
LEVEL, Bedford,—Rennie's reports on, 166; Telford's drainage of North, 467.
LEWIS, invention of the, 22.
LIGHTHOUSES,—Winstanley's on the Eddystone, 16; Rudyerd's, 21; Smeaton's, 26; on Spurn Point, 72; on the Scotch coasts, 224; on the Bell Rock, 225.
LIGHTING of the English Channel, 46.
LIGHTNING, Smeaton's protection of the Eddystone Lighthouse against, 37.
LIMPLEY STOKE AQUEDUCT, 145.
LINCOLN Fens drained by Rennie, 152.
LINTON, East, 93, 119.
LITTLE FAMILY, Eskdale, 297, 328, 490.
LITTLEBOROUGH TUNNEL, 146.
LIVERPOOL,—Rennie's docks at, 204, 215; Telford's notice of, 369; Mersey job at, 477.
LLANGOLLEN, canal works in vale of, 342.
LOCH, Mr., Inspector of Fisheries, his notice of Rennie as a scholar, 123.
LOCHS, Caledonian Canal, 411.
LOCKS,—Rochdale Canal, 146; London Docks, 200; Ellesmere Canal, 361; Caledonian Canal, 414.
LONDON,—Old Bridge, Smeaton's improvements on, 53; London in 1785, 134; docks, 136; commerce of the port of, 195; East and West India Docks, 197; Waterloo Bridge, 178; Southwark Bridge, 187; New London Bridge, 268; Telford's proposed one-arched bridge at, 361.
LOSTWITHIEL CHURCH spire struck by lightning, Smeaton's paper on, 37.

LOTHIANS.

LOTHIANS, farming in, 94, 107.
LOUIS XIV.—anecdote of, and the Eddystone Lighthouse, 23.
LOVET, Captain, lessee of Eddystone Lighthouse, 21, 26.
LUCKNOW BRIDGE, 177.
LUGWY, North Wales, road near, 439.
LUNE AQUEDUCT, 148.
LYONS, first iron bridge projected at, 355.

M.

MACADAM; Mr., Roadmaker, 430.
MACCLESFIELD, Earl of, 26.
MACCLESFIELD CANAL, 420.
MACHINERY,—of Albion Mills, 137; of royal dockyards, 238.
MALCOLM FAMILY, Eskdale, 287, 296, 482, 490.
MAPS,—Smeaton's native district, 4; coast of Devon and Cornwall, 15; Rennie's native district, 93; Lincolnshire Fen district, 155; Eau Brink Cut, 168; London and St. Katherine Docks, 199; East and West India Docks, 203; Holyhead Harbour, 210; proposed Royal Docks at Northfleet, 244; Sheerness Docks, 247; proposed Medway improvement, 250; Plymouth Sound, 253; Telford's native district, 288; Ellesmere Canal, 340; Telford's Highland roads, 382; Peterhead Harbour, 395; Banff Harbour, 398; Aberdeen Harbour, 404; Dundee Harbour, 406; Caledonian Canal, 412; Menai Straits, 444.
MARINER'S COMPASS, Smeaton's improvements of, 12.
MARYKIRK, ancient bridge at, 103.
MAUD FOSTER DRAIN, Boston, 161.
MECHANICAL work in Scotland, early state of, 104, 128.
MECHANICAL EDUCATION, advantages of, 122, 281, 330, 483.
MEDWAY improvement, Rennie's plan of, 250.
MEGGAT, valley of the, Eskdale, 292, 306.
MEIKLE, JAMES, millwright, 105.
MEIKLE, ANDREW,—miller at Houston Mill, 106; skill as a millwright, 109; takes out patent for dressing grain, 109; invents the thrashing machine, 111; improvements in the sails of windmills, 114; his machinery employed to drain Kincardine Moss, 115; his odd mechanical contrivances, 116; his death, 117.
MELVILLE, Lord, 257, 267.
MENAI STRAITS,—Rennie's design of a cast-iron bridge across, 175; Telford's design, 445; construction of the Suspension Bridge, 448.
MICKLE, translator of " Lusiad," 304.
MILITARY CANAL, Hythe, 236.

PATHHEAD.

MILLS,—erected by Smeaton, 73; Houston Mill, 107; Invergowrie, 125; Bonnington, ib.; Albion, 129, 134; mills erected and repaired by Rennie, 139; Rotherhithe, 235; powder mills, Waltham, 237.
MINT MACHINERY, Rennie's, 139.
MONEY,—scarcity of, prevents improvements, 50, 59, 96, 172; Smeaton's estimate of, 80; Rennie's, 283; Telford's, 487.
MOORINGS for Royal navy, Rennie's, 238.
MORPETH BRIDGE, 433.
MUSSELBURGH BRIDGE, 174.

N.

NANT FFRANCON, North Wales, road near, 440.
NAPOLEON's opinion of Plymouth Breakwater, 258.
NAVY, Royal,—dockyards, 238; steam power introduced in, 267.
NENE RIVER,—Smeaton's report on improvement of, 52; new outfall constructed, 468.
NEPTUNE'S STAIRCASE, Caledonian Canal, 412.
NEUILLY, Peyronnet's bridge of, 178.
NEW GALLOWAY BRIDGE, 177.
NEW LONDON BRIDGE, 268.
NEWTON STEWART BRIDGE, 177.
NICHOLSON, PETER, pupil of Andrew Meikle, 117.
NITH, river, Dumfries, Smeaton's report, 50.
NORE, light at the, 48.
NORTHFLEET DOCKS, Rennie's plan of, 243.
NORTH LEVEL,—Smeaton's report on drainage of, 52; Rennie's report, 165; Telford's drainage of, 467.

O.

ORKHILL on the Spey, ancient bridge at, 103.
OUSE, river, Rennie's improvement of outfall, 167.
OUTFALLS,—Rennie's drainage by improved, 159; Eau Brink, 167; Nene, 468.
OWEN, ROBERT, on state of the Glasgow roads, 428.
OXFORD, Telford's visit to, 333.

P.

PACK-HORSES, carriage by, 5, 99.
PAINE, THOMAS,—" Rights of Man," 326; his iron bridge, 357.
PAISLEY CANAL, 418.
PALMER's mail-coaches, 135.
PARNELL, Sir H., and Welsh roads, 437.
PASLEY, family of, 297, 300, 308.
PATHHEAD BRIDGE, 433.

PEDEN.

PEDEN, "prophet," in Eskdale, 290.
PERRY'S DOCK, Blackwall, 136.
PERTH,—new bridge, 56; ancient bridge at, 103.
PESTH SUSPENSION BRIDGE, 476.
PETERHEAD HARBOUR, 393.
PHANTASSIE, Rennie's birthplace at, 93, 101.
PICKERNELL, Mr., resident engineer, Hexham Bridge, 62.
PLANS. (See *Maps*.)
PLAYFAIR, Professor, his views on arches of iron, 365.
PLYMOUTH, 15, 31, 36; Hoe at, 41; dockyard, 240; breakwater, 92, 252.
POLISH roads, Telford's, 476.
PONT-CYSYLLTAU AQUEDUCT, 344.
PONT-Y-PRYDD, Rennie's notice of, 178.
PORTSMOUTH,—Rennie's report on dockyard, 240; Telford employed at, 312.
POST-OFFICE, 135, 432, 435.
PRESTONKIRK, East Lothian, 93, 117, 121.
PRESTONPANS, field of, 108.
PRITCHARD, Mr., architect of first iron bridge, 358.
PROVIS, Mr., resident engineer of Menai Bridge, 234, 450.
PULTENEY, Sir WILLIAM, 312, 316, 323, 328, 335, 377.
PUMPING machinery, Smeaton's, 56, 72.

Q.

QUARANTINE establishment in the Medway, 233.
QUEENSBERRY, Duchess of, and Smeaton, 83.

R.

RAILROADS,—used by Rennie to carry on his works, 150, 203; Croydon and Merstham, 198; Telford and railroads, 425, 478.
RALPH THE ROVER, Southey's ballad, 223.
RAMSGATE HARBOUR,—constructed by Smeaton, 65; repaired by Rennie, 218.
RENNIE, family of, 93, 112, 118.
RENNIE, JOHN,—birth and parentage, 93; education, 119; apprenticeship, 122; begins business as millwright, 125; student at Edinburgh University, 126; tour in England, 128; his first bridge, 131; erects machinery of Albion Mills, 136; mechanical engineer in London, 141; constructs Kennet and Avon Canal, 144; Rochdale Canal, 145; Lancaster and other canals, 148; drainage of Lincoln Fens, 152; Kelso Bridge, 172; Musselburgh Bridge, 174; Boston Bridge, 176; bridges at Newton Stewart, &c., 177; Waterloo Bridge, 178; Southwark Bridge, 187; London Docks, 197;

SCOTTS.

East India Docks, 202; report on improvement of Clyde, 206; hollow walls, Grimsby Docks, 207; Holyhead Harbour, 208; Howth and Kingstown Harbours, 213; Hull Docks, and improvement of dredging machine, 213; Greenock Quays and Docks, 214; report on Southampton, 215; Leith Docks, 215; Charleston Harbour, and principles of harbour construction, 216; Ramsgate Harbour, and improvement of divingbell, 219; Bell Rock Lighthouse, 223; reports on coast defences, and works, 235; report on Royal dockyards, 238; grand plan of Northfleet Docks, 243; Sheerness Docks, 246; Plymouth Breakwater, 253; design of New London Bridge, 268; foreign tour, 274; death and character, 277.
RENNIE, Sir JOHN, — drainage works by, 165; finishes Plymouth Breakwater, 262; constructs New London Bridge, 272.
REVESBY, Sir J. Banks's mansion at, 155.
REYNOLDS, Mr., Coalbrookdale, 356.
ROADS,—in England, 49; in Scotland, 99, 101; general construction of turnpike roads, 135; Mr. Rennie's system of constructing, 185; Highland, 375, 378; Glasgow and Carlisle, 427; Telford's system of constructing, 429; Macadam's, 430; London and Edinburgh, 432; Welsh, 434; Shrewsbury and Holyhead, 436; London and Shrewsbury, 441; Telford's Polish roads, 476.
ROBISON, Professor, Edinburgh, 126, 137, 170.
ROCHDALE CANAL, 145.
ROEBUCK, Dr., Carron Works, 60.
ROMILLY, Sir Samuel, on Scotch progress, 374.
ROYAL DOCKYARDS. (See *Dockyards*.)
ROYAL CANAL of Ireland, 150.
ROYAL NAVY, introduction of steam power in, 267.
ROYAL SOCIETY, Smeaton and, 11, 20, 37, 75.
RUDYERD, JOHN, — his lighthouse on the Eddystone, 21.
RUNCORN, Telford's proposed suspension bridge at, 446.

S.

ST. KATHERINE'S DOCKS, 462.
SALOPIAN COFFEE-HOUSE, 473.
SCHOOLS,—Scotch parochial, 121; burgh or grammar, 122.
SCOTLAND,—state of, in eighteenth century, 94; ancient bridges in, 102; parish schools of, 121; grammar schools, 122; progress of, 374.
SCOTTISH BORDERERS, 288.
SCOTTS, the Border, 289.

SELF.

SELF-EDUCATION of early engineers—Smeaton's, 8, 13; Rennie's, 126, 141; Telford's, 314, 486.
SELKIRK common carrier, 100.
SHEERNESS DOCKYARD, 242; its re-construction by Rennie, 246.
SHEET-PILING, description of, 60.
SHIPPING of the Thames, 136, 195.
SHREWSBURY,—castle, 316, 329; Shrewsbury and Holyhead road, 436; and London, 441.
SKIRVING, ARCHIBALD, portrait painter, memoir of, 278.
SLAVE-TRADE of Aberdeen, eighteenth century, 401.
SLUICES, in drainage of Lincoln Fens, 159.
SMEATON, JOHN,—birth and education, 3; placed in a solicitor's office, 10; learns the trade of mathematical instrument maker, 11; makes improvements in mechanics, 12; receives the Royal Society's Gold Medal, ib.; tour in Holland, 13; employed to rebuild the Eddystone Lighthouse, 26; his design, 29; visits the rock, 31; his model, 33; the works begun, 34; is nearly shipwrecked, 35; his courage, 39; completion of the building, 43; extensively employed as a civil engineer, 49; in Fen drainage, ·51; repairs London Bridge, 53; his pumping engines, 56; builds a bridge at Perth, 56; constructs Forth and Clyde Canal, 57; bridge at Coldstream, 59; improvements at Carron works, 61; bridge at Banff, ib.; Hexham Bridge, its failure, 62; St. Ives Harbour, 63; Ramsgate Harbour, 65; Eyemouth Harbour, 71; his machinery, 72; improves the steam-engine, 73; private life, habits, and character, 74.
SMOLLETT'S journey to London, 100.
SOMERSET HOUSE,—its river front, 179; Telford works as a mason at, 309.
SOUTHAMPTON HARBOUR, Rennie's report on, 215.
SOUTHEY and Telford, 427, 484, 491.
SOUTHWARK IRON BRIDGE, 187.
SPEY, river,—ancient bridge over, 103; floods, 386; Craigellachie bridge over, 387.
STANGATE CREEK, quarantine establishment at, 238.
STANHOPE, Earl, his mechanical ingenuity, 142.
STEAM-ENGINE,—Smeaton's improvements of, 73; Watt's engine applied to driving machinery, 129, 137; used in navigation, 143; introduced in Royal navy, 267.
STEIN, Mr., Kilbeggie, orders the first thrashing machine, 112.
STEPHENSON, ROBERT, his estimate of Smeaton, 86.
STEVENHOUSE MILL BRIDGE, 131.
STEVENSON, Mr., resident engineer of Bell Rock Lighthouse, 225, 229, 232.
STIRLING, ancient bridge at, 103.

TRADE.

SUNDERLAND, iron bridge over the Wear at, 358.
SUSPENDED CENTERING, Telford's plan of, 445.
SUSPENSION BRIDGES, 446.
SUTHERLAND, county of, in eighteenth century, 379.
SWEDEN, King of, confers knighthood on Telford, 419.
SZÉCHENYI, Count, his interview with Telford, 476.

T.

TELFORD, THOMAS,—his native district, 287; birthplace in Eskdale, 292; his boyhood, 295; school education, 296; apprenticed to a stone-mason, 298; journeyman, 301; writes poetry, 304; mason at Edinburgh, 305; rides to London, 307; mason at Somerset House, 308; foreman of masonry at Portsmouth, 312; his studies, 314; prints letters to his mother, 315; surveyor to county of Salop, 316; habits and studies, 318; his dutifulness as a son, 325; his politics, 326; appointed engineer of Ellesmere Canal, 334; construction of the works, 340; re-visits Eskdale, 350; tour in Wales, 352; his poetry, 353; builds an iron bridge at Buildwas, 359; proposes a one-arch iron bridge over the Thames, 361; erects Bewdley Bridge, 366; Tongueland Bridge, 368; Highland roads and bridges, 374; Scotch harbours, 390; Caledonian Canal, 409; Gotha Canal, 419; Harecastle Tunnel, 420; constructs roads in England, Scotland, and Wales, 427; bridge over Menai Strait and river Conway, 444; St. Katherine's Docks, 462; Tewkesbury and Gloucester bridges, 464; Dean Bridge, Edinburgh, 465; Glasgow Bridge, 466; drainage of the North Level of the Fens, 467; Telford's residence in London, 473; President of Institute of Civil Engineers, 474; decline of health, 479; death, 480; character, 481.
TELFORD, WILLIAM, Surgeon, 296, 320.
TEMPLE NEWSAM, Leeds, 6; Smeaton's pumping engine at, 72; gates of, hung by Smeaton, 77.
TEWKESBURY BRIDGE, 464.
THAMES SHIPPING, 136, 195.
THEORY, applied to bridge building, 170.
THOMSON, ANDREW (Telford's master), mason, Langholm, 298, 302.
THRASHING MACHINE, Meikle's invention of, 109.
TODMORDEN, canal works near, 146.
TONGUELAND BRIDGE, 368.
TORBAY HARBOUR, 207.
TRADE,—of London, 136, 195; Glasgow, 205, 467.

TRAVELLING.

TRAVELLING,—in Scotland, 100; in England, 135, 427; in Wales, 434.
TURNPIKE ROADS, 101, 135, 427.
TYNE, RIVER, East Lothian, 93, 120.

U.

UNION of Scotland with England, 95.
URICONIUM, Roman city of, 322.

V.

VIRGINIA WATER, Rennie's bridge over, 177.

W.

WAGES of workmen, 57, 96, 327.
WALES, roads in, 434.
WAINFLEET drainage, Lincolnshire, 154, 163, 166.
WALDERSEA DRAINAGE, Lincolnshire, 154.
WALLS, hollow,—employed by Rennie in sea works, 208; by Telford in bridge piers, 344, 347, 369, 451.
WALTHAM POWDER MILLS, 237.
WASH, drainage of lands adjoining the, 165, 468.
WATER AND WIND, Smeaton's paper on power of, to turn mills, 12.
WATER supply of towns,—Rennie's reports, 264; Telford's, 369, 476.
WATERLOO BRIDGE, 178.
WATT, JAMES,—survey of canal between Forth and Clyde, 59; intimacy with Smeaton, 79; Rennie's visit to, at Soho, 129; report on improvement of Clyde, 206; Telford's notices of, 371, 410.
WEAR IRON BRIDGE, Sunderland, 358.
WEAVER NAVIGATION, 418.
WELLINGTON BRIDGE, Leeds, 177.

YOUNG.

WEST FEN,—Smeaton employed to drain, 51; its drainage by Rennie, 152.
WEST INDIA DOCKS, 197.
WESTERKIRK, Eskdale, 291; school, 296; manse, 303; library, 492.
WESTON, Mr., lessee of Eddystone Lighthouse, 26.
WHARVES, dock, 197, 463.
WHIDBEY, Mr., 254, 257, 276.
WHITKIRK, Leeds, Smeaton's parish church, 4, 9; his burial place at, 90.
WICK HARBOUR,—Rennie's report on, 204; Telford's improvements of, 390.
WIGTON, Earl of, his fear of the effects of the Union, 95.
WILDMORE FEN,—Smeaton employed to drain, 51; drainage by Rennie, 152.
WILKINSON, JOHN, ironmaster, 337, 356.
WILLIAMSON, PETER, the Aberdeen slave, 401.
WINNOWING MACHINE introduced into Scotland, 105.
WINSTANLEY, HENRY,—his eccentricities, 16; his waterworks, 17; his lighthouse on the Eddystone, 18; his death, 19.
WISBEACH, drainage works near, 468.
WITHAM, RIVER,—Smeaton employed to improve, 51; Rennie's drainage of lands adjoining the, 153.
WOOLWICH DOCKYARD,—Rennie's improvements of, 241; report on, ib.; anchor forge at, 265.
WORKMEN, Telford's views as to, 310.
WROXETER, discovery of Roman city near, 322.

Y.

YNYS-Y-MOCH, bridge at, Menai Straits, 449.
YORK BUILDINGS WATERWORKS, Rennie's report on, 264.
YOUNG, ARTHUR,—his description of drowned lands in the Fens, 154, 156.

END OF VOL. II.

LONDON: PRINTED BY WILLIAM CLOWES AND SONS, STAMFORD STREET, AND CHARING CROSS.

Milton Keynes UK
Ingram Content Group UK Ltd.
UKHW032320161024
449665UK00001B/26

9 781108 052931